W9-BAA-660

ALSO BY KEN CROSWELL

*The Alchemy of the Heavens (1995)*

*Planet Quest (1997)*

*Magnificent Universe (1999)*

*See the Stars (2000)*

# THE UNIVERSE AT MIDNIGHT

## OBSERVATIONS ILLUMINATING THE COSMOS

KEN CROSWELL

THE FREE PRESS
NEW YORK LONDON TORONTO SYDNEY SINGAPORE

fP

THE FREE PRESS
A Division of Simon & Schuster, Inc.
1230 Avenue of the Americas
New York, NY 10020



For information about discounts for bulk purchases,
please contact Simon & Schuster Special Sales:
1-800-456-6798 or business@simonandschuster.com

Designed by Karolina Harris

Manufactured in the United States of America

10 9 8 7 6 5 4 3 2 1

Library of Congress Cataloging-in-Publication Data

Croswell, Ken.
The universe at midnight : observations illuminating the cosmos /
Ken Croswell.
p. cm.
Includes bibliographical references and index.
1. Cosmology. I. Title.

QB981 .C884 2001
523.1—dc21 2001040232

ISBN 0-684-85931-9 (alk. paper)

*Cosmologists are often in error,*
*but never in doubt.*

—LEV LANDAU, RUSSIAN PHYSICIST

# CONTENTS

# INTRODUCTION:
# A GREAT ATTRACTION

*Will you sleep at night?*
*With the Plough and the stars alight?*

—MARILLION ("EASTER")

SOMETIMES, your own Galaxy gets in the way. Between the Milky Way's glowing stars weave clouds of gas sprinkled with the dark ash of ancient suns. Although these interstellar clouds can assume fanciful shapes—eagles, horseheads, seagulls, swans—they obscure the universe beyond, cloaking a fifth of the sky in perpetual fog.

"We knew that finding galaxies behind the Milky Way would be very difficult," said Renée Kraan-Korteweg, a Dutch-born astronomer at the University of Guanajuato in Mexico. "At optical wavelengths, it was not believed possible to uncover these galaxies, because they would be so hard to see." Indeed, this daunting region which girdled the sky had long been labeled the "zone of avoidance," since galaxies seemed to shun it.

So did galaxy-seeking astronomers. In 1994, however, Kraan-Korteweg and her colleagues trespassed into the forbidden zone and took aim at the constellation Cassiopeia with a radio telescope. The radio waves zipped past the gas and dust, betraying a large galaxy, never before seen, only 10 million light-years from Earth—in astronomical terms, just around the block. It looked like the letter S, as if to say it had surrendered. Without the Milky Way's gas and dust, the galaxy would appear as one of the ten brightest in Earth's sky. Meanwhile, other astronomers dug up the wreckage of a small galaxy in Sagittarius, on the far side of the Milky Way. It proved to be the closest galaxy ever seen, a mere 80,000 light-years from Earth. It got torn apart when it strayed too close to the Milky Way's shores.

Even with these discoveries, a far mightier entity lurked behind the Milky Way. Astronomers had earlier discovered that the gravitational pull of something they christened the Great Attractor was trying to suck in all galaxies, including our own, for hundreds of millions of light-years around. The Great Attractor had accelerated our Galaxy to a speed of 1 million miles per hour, but its exact center eluded detection by concealing itself behind the southern Milky Way.

Kraan-Korteweg and her colleagues got to work, projecting magnified images of optical photographic plates onto a small screen in a darkened room. "It was a very slow process going through these plates—finding galaxies, identifying them, classifying them," she said. "They're smudges that you can't see without enlarging them." Not only did the Milky Way's gas and dust dim the galaxies, but also hordes of foreground stars speckled the photographs, raindrops on a celestial windshield, further obstructing the view.

Nevertheless, she and her colleagues persevered. "We were charting thousands and thousands of new galaxies," said Kraan-Korteweg, "and it became very clear that there was an enormous concentration of galaxies very close to the predicted center of the Great Attractor." Deploying telescopes in Chile, South Africa, and Australia, the astronomers then measured the new galaxies' distances, finding them all 250 million light-years from Earth, nestled in a dense cluster like candles lighting an immense chandelier.

The cluster's galaxies dwelt in and around Norma, a forgettable southern constellation invented by an eighteenth-century astronomer best known for inventing forgettable southern constellations. Norma lies southwest of Scorpius, the striking zodiacal constellation that bears the brilliant red star Antares.

"The Norma cluster is like Manhattan in New York," said Kraan-Korteweg: "the business district with all the tall buildings." The cluster abounds with giant elliptical galaxies, the celestial equivalent of skyscrapers, which throng the richest galaxy clusters. The Norma cluster isn't the Great Attractor any more than Manhattan is New York City; it accounts for only a tenth of the Great Attractor's mass, but it does mark the monster's heart—supergalactic downtown.

Strangely, astronomers had catalogued the Norma cluster a few years earlier, but because of the Milky Way's dusty veil, they thought it only a minor affair. In actuality, the Norma cluster rivals the greatest clusters known, such as the famous Coma cluster, yet until recently no one knew it existed. "In astronomy," said Kraan-Korteweg, "there's still so much to explore."

TIME after time, the universe has astonished those who explore it—offering hidden galaxy clusters, mysterious halos of dark matter, even a bizarre "antigravity" force that seems to pervade empty space. For millennia, of course, people have gazed heavenward and contemplated great cosmological questions—How did the universe begin? What is it made of? What will be its ultimate fate?—but only in the last hundred years have astronomers begun to acquire the data about the stars and galaxies that may answer these provocative questions.

*The Universe at Midnight* aims to tell this story of cosmological inquiry and discovery. During the day, scientists and philosophers can construct elegant theories of how they think the universe should operate; but at night, at midnight, when powerful telescopes swing toward distant galaxies, the universe delivers its verdict. Sometimes it validates existing observations and theories. Often, though, it repudiates them, forcing scientists to devise new conceptions of the cosmos. Drawing upon extensive interviews with the scientists who made the key discoveries, *The Universe at Midnight* tells the twisted, tangled, riveting story as it happened. It is part mystery novel, part detective story, part human drama. It is also, I hope, an up-to-date portrait of the state of cosmology today and how observers and theorists have arrived there. The book therefore proceeds chronologically, as one surprising discovery led to another. It starts with cosmology's oldest observation—of the darkness that falls every evening—and ends with the recent discovery that the universe's expansion may be speeding up.

Nearly half of *The Universe at Midnight* examines work from the past decade, illustrating how rapidly cosmology has progressed. But the book also sounds a note of caution, for this progress has often come at the expense of overturning previous truths. A hundred years ago, for example, astronomers thought that the universe was static, when today we know that it is expanding; forty years ago, they thought that the glittering stars within galaxies constituted the bulk of the universe, when today we think they are mere gems floating on a black velvet sea of dark matter; and ten years ago, they thought that the universe's expansion must be slowing, as the gravitational attraction of the galaxies braked its speed. Furthermore, some features of modern cosmology, such as the mysterious dark matter which sheathes the galaxies, and the repulsive force which seems to drive them apart, are so peculiar that they suggest crucial elements in cosmology remain missing. Although revolutionary discoveries add drama to the story, they also make one wonder which truths presented herein may themselves be overturned, the next time astronomers atop tall mountains point their telescopes at the heavens, at midnight.

I THANK the scientists I interviewed, who gave me fascinating insights behind their work: Charles Alcock, Ralph Alpher, Neta Bahcall, Charles Bennett, Michael Bolte, Hermann Bondi, David Branch, Ruth Daly, Alan Dressler, Alex Filippenko, Wendy Freedman, Kenneth Freeman, Margaret Geller, Thomas Gold, Steve Gregory, Edward Guinan, Alan Guth, Edward Harrison, Fred Hoyle, George Jacoby, Christopher Kochanek, Renée Kraan-Korteweg, Andrei Linde, Brian Mason, John Mather, Edward Olszewski, Donald Osterbrock, Jeremiah Ostriker, Bohdan Paczyński, Saul Perlmutter, Morton Roberts, Vera Rubin, Allan Sandage, Wallace Sargent, Rudolph Schild, Brian Schmidt, Joseph Silk, George Smoot, François Spite, Krzysztof Stanek, Gary Steigman, Gustav Tammann, Laird Thompson, John Tonry, Brent Tully, Edwin Turner, Henry Tye, Anthony Tyson, Don VandenBerg, Paul Wesson, Ned Wright, and Donald York.

I thank those who read the manuscript and offered their comments: David Hudgins, Charles Liu, and Richard Pogge.

Finally, I thank my editor of many years, Stephen Morrow, for his passion, and my production editor of many years, Loretta Denner, for her precision.

# ONE

# HIGH MIDNIGHT

*Stars in my pocket like grains of sand*
*Radiant pinpoints, they sting my hand*
*I grab a fistful, a gift for the breeze*
*Who will carry them far and away with ease*

—WHITE WILLOW ("THE BOOK OF LOVE")

IT IS just before midnight, and stars spangle the sky: newborn stars emerging from magenta gas clouds, middle-aged suns dutifully towing planets through space, elderly red giants about to puff their atmospheres into the void. Every star the naked eye can see races around a giant black hole buried behind the dust clouds of the constellation Sagittarius. In the next hour, we too will dash half a million miles through space, as the Sun pursues its orbit around the Sagittarian black hole. The Sun, the Earth, and the canopy of stars all belong to the same celestial kingdom, the Milky Way, which boasts more stars than the Earth does people.

Yet most of the universe lies beyond the Milky Way. In the constellation Andromeda, a faint wisp betrays another Milky Way, another titanic empire of stars, planets, and possibly people, all anchored to one another by gravity. Call it a *galaxy*, after the Greek for "Milky Way." So distant is the Andromeda Galaxy that the light we see tonight set out from its spiral shores 2.4 million years ago, before the human race arose. Both Andromeda and the Milky Way belong to a flock of galaxies called the Local Group. Among Local Group galaxies, Andromeda ranks number one in size and brilliance, our own Milky Way number two—not bad, since altogether the Local Group houses three dozen galaxies.

Yet most of the universe lies beyond the Local Group. Long ago, stargazers pictured a star pattern south of the Big Dipper's handle as a woman holding a spike of grain, and they named her Virgo. Here cluster thousands of galaxies that rule the center of a huge galactic metropolis which a maverick astronomer named the Local Supercluster. At the time, other astronomers didn't believe him, but in fact galaxies speckle the Local Supercluster the way stars do a galaxy, with the Milky Way residing near its edge. The Local Supercluster's most beautiful galaxies spin so fast that they whip up exquisite spirals and resemble celestial whirlpools. The Milky Way is a spiral, as is its partner in Andromeda. Other galaxies, less fortunate, look round or oval; they are called elliptical galaxies. And still others, less orderly, less organized, are dubbed irregular. When the light we see from the Local Supercluster's most remote galaxies set out into space, dinosaurs ruled the Earth.

Yet most of the universe lies beyond the Local Supercluster. An even larger supercluster drapes through the constellations Pisces and Perseus, and in the opposite direction a cluster of superclusters—a superdupercluster?—tries to pull us its way. Its name is the Great Attractor. Superclusters string the cosmos like glowing cobwebs, with vast voids of mostly empty space between. If galaxies were trees, the superclusters and the voids would be forests and meadows.

Far beyond the superclusters and voids, like the backdrop to a mighty play, shines the afterglow of the universe's creation, the redshifted remains of light that tore free of matter 300,000 years after the universe's birth. Tiny fluctuations in this light reveal that the early universe was not perfectly smooth. Instead, some regions were denser than others. The denser regions, through their gravitational force, attracted additional material, eventually conglomerating into the galaxies, galaxy clusters, and galaxy superclusters that adorn the cosmos today. The sparser regions lost material to the denser ones and became the voids.

## COSMOLOGY: THE UNIVERSE AT LARGE

Beautiful though each star, galaxy, or supercluster is, cosmologists aim to understand the entire universe—its origin, composition, evolution, and ultimate fate. This is no small task, but one that has tripped up astronomers before and will likely do so again. As one senior cosmologist said, "Cosmology and comedy, cosmic and comic, are not far apart."

Most of what cosmologists know about the universe they learned only in the last hundred years. Their observations indicate that the whole material universe—what is now a myriad of stars, galaxies, and planets—was once compressed into a single point that burst forth 12 to 16 billion years ago in the big bang. Relics of this primordial fireball lie close to home: the hydrogen in a glass of water—the H in that $H_2O$—came from the big bang, the oxygen from brilliant stars that exploded billions of years afterward.

Ever since the big bang, the universe has been expanding. Each hour, 3 million miles of new space opens up between us and the Virgo cluster, 12 million miles of new space between us and the Pisces-Perseus supercluster. Not everything moves away, however. The Milky Way's gravity binds its stars to the homeland and even forces several smaller galaxies to revolve around it, colonies paying tribute to the Galaxy's empire. The Milky Way derives its gravitational muster from every star, every planet, every person within it: your weight contributes approximately 0.0000000000000000000000000000000000003 percent of the Milky Way's total, so if you left the Milky Way, its stars would revolve slightly more slowly, its satellite galaxies would swing around the Galaxy slightly more sluggishly, the Galaxy's grip on its farthest-flung outposts would be slightly less secure. Thank you for choosing to live in the Milky Way.

Gravity herds the entire Local Group, preventing its galaxies from fleeing the flock. Gravity might do the same to the universe itself. If the universe is dense enough, the gravity of all its galaxies will eventually halt its expansion. Then the universe will start to collapse, ending its days in a fiery inverse big bang—a "big crunch." For decades astronomers have known that the universe harbors far more matter than meets the eye, dark material whose gravity tugs on the stars and galaxies. However, despite all this dark matter, the universe is not dense enough to collapse. Moreover, empty space seems to exert a repulsive force that speeds up the universe's expansion. If so, not only will the universe expand forever, it will expand forever faster. In only about 150 billion years, all galaxies beyond the Local Supercluster will vanish from the sky, because the space between them and us will be expanding so fast that their light can no longer reach the Earth.

These remarkable deductions come from large telescopes, some perched atop mountains, others lofted into space, that have helped decipher the tangled clues left behind in the wake of the big bang. The first cosmological observation, however, required no telescope at all. Look at the night sky: it is dark. Why?

# TWO

# IN THE DARK

*Here I opened wide the door;—*
*Darkness there and nothing more.*
—EDGAR ALLAN POE ("THE RAVEN")

SUNSET again. Vermilion and gold veil the western horizon, rose clouds drift overhead, turquoise and lavender blanket the east. Lamps light in homes, candles at dinner tables, and streetlights along roads, pearls on strings threading the twilight. As dusk deepens, stars emerge in the east, then the west. Why is the sky dark at night?

It's a simple question, the sort a child might ask and a parent dismiss; yet cosmologist Edward Harrison has spent decades studying the problem. "My interest in the tantalizing riddle of the dark night sky began many years ago," he wrote. "I puzzled over the problem of why the universe is not filled with light, and too soon thought I had it solved. But I was not to escape from the coils of this very old riddle so easily. Sometimes for hours, sometimes for days, I returned, lured by its power and subtlety."

Harrison grew up in London during the 1920s and 1930s, interested in a variety of subjects. "One didn't make distinctions in those days," he said. "Writing poetry, drawing pictures, wondering about steam engines: they all sort of fused into one. There were no boundaries—not very sharp, anyway." His background influenced his views on schooling. "My ideas of education are different from current ideas. I think one should foster the creative urge by pushing kids into the arts, not into the sciences; and then later on, if they want to, they can move into the sciences. But if you're going to go right into the sciences from the beginning, you've got such a long way to go before you can start being creative that the creative impulse can die."

Harrison became a physicist. "In those days, doing physics was a pleasure. It was not the rat race it became later: a scramble for funding, very competitive—which has got its advantages—but it also tends to take the joy out of doing science. The modern-day physicist is trapped, locked into a special field; he or she has got to stay in that field in order to build up a reputation, to get the funding and the promotions, and has got to spend an inordinate amount of time on doing this. And the family suffers. That is what I call a rat race. There's no longer joy in that kind of career."

During a more relaxed era, Harrison began contemplating the dark night sky. "In England I had lots of leisure time with the family, going on long vacations, and I was free to pursue my own interests, to study different aspects of astronomy. It was in that sort of climate that Olbers' paradox was one of the things one played with."

Olbers' paradox is the modern term for the puzzle of the dark night sky. At first it might seem obvious why the night is dark: the Sun has set. However, other suns shine elsewhere. If the universe is infinite, filled with infinitely many stars, then stars should cover the sky just as trees line a forest dweller's horizon, and the night should blaze with light. This contradiction between theory and observation bears the name of a nineteenth-century doctor and astronomer who swept the sky for solar system debris.

## FROM COMETS TO COSMOLOGY

Heinrich Wilhelm Olbers slept only four hours a night. By day he tended the sick in the northern German town of Bremen, winning accolades for battling cholera epidemics. By night he scanned the sky from his home, discovering comets and two of the first four asteroids.

In 1823, Olbers wrote a paper whose scope spanned the entire cosmos. Imagine, he said, that space is uniformly sprinkled with stars. Stars closest to Earth look largest, but farther stars outnumber them, because more space exists at greater distances. Olbers showed that the larger number of distant stars exactly compensates their lesser size, so stars in each spherical shell of space around the Earth cover an equal area of the sky. If the universe is infinite, then so is the number of these spherical shells, which means that stars blanket the sky, making every point glisten like the Sun.

Wrote Olbers, "How fortunate for us that nature has arranged matters differently! How fortunate that the Earth does not receive starlight from every point of the celestial vault! Yet, with such unimaginable brightness and heat, amounting to 90,000 times more than what we now experience, the Almighty could easily have designed organisms capable of adapting to such extreme conditions. But astronomy for the inhabitants of the Earth would remain forever in a primitive state; nothing would be known about the fixed stars; only with difficulty would the Sun be detected by virtue of its spots; and the Moon and planets would be distinguished merely as darker disks against a background as brilliant as the Sun's disk."

In his time Olbers' work attracted little attention, nor was he the first to puzzle over the darkness of night. In 1543, Polish astronomer Nicolaus Copernicus had ushered in a revolution by publishing his bold claim that the Sun, not the Earth, occupied the universe's center, contradicting a literal interpretation of several biblical passages. Copernicus never considered the implications for the night sky, but other astronomers could now regard the Sun as a mere star among stars that punctuated space out to infinity.

The first astronomer to do so was Englishman Thomas Digges. In 1576 he attached an appendix to a book his father had written. The father had adopted an Earth-centered cosmos, so in the appendix the son sought to set things right by advocating Copernicus's Sun-centered system. Digges went further than Copernicus, though, by scattering the stars throughout space. He then attempted to explain why these stars did not light up the night: "Especially of that fixed Orbe garnished with lightes innumerable and reachinge vp in *Sphaericall altitude* without ende. Of whiche lightes Celestiall it is to bee thoughte that we onely behoulde sutch as are in the inferioure partes of the same Orbe, and as they are hygher, so seeme they of lesse and lesser quantity, euen tyll our sighte beinge not able farder to reache or conceyue, the greatest part rest by reason of their wonderfull distance inuisible vnto vs." Thus, said Digges, the night was dark because distant stars were too faint to see. Although his explanation seemed reasonable, it was actually wrong. The combined light of invisible stars can itself be visible, just as this paper is visible even though each of its atoms is not. In fact, the unaided eye can see the Andromeda Galaxy, even though none of its stars is bright enough for the unaided eye to see.

Shortly after Digges, Italian priest and philosopher Giordano Bruno advocated an infinite universe with infinitely many stars, planets, and living beings: "Thus is the excellence of God magnified and the greatness of his kingdom made manifest; he is glorified not in one, but in countless suns; not in a single earth, a single world, but in a thousand thousand, I say in an infinity of worlds." Bruno never pondered the consequences of an infinite universe for the night's darkness. Nor did he fully realize the consequences for his own life: in 1600 the Catholic Church found him guilty of heresy and burned him alive.

Although Digges and Bruno embraced infinity, the concept horrified Johannes Kepler, the great German astronomer who worked out the laws of planetary motion and recognized that the planets traveled around the Sun on elliptical rather than circular orbits. Kepler complained that in an infinite universe, the Sun would be lost amid a wilderness of stars. The night was dark, he said, because a dark wall surrounded the starry space we inhabit.

Despite Kepler's antipathy toward infinity, the same physicist whose law of universal gravitation explained Kepler's laws would resurrect the infinite universe. Isaac Newton was perhaps the greatest scientist who ever lived. He owed his brilliance to several factors. For one thing, he loved knowledge so intensely that while working he sometimes forgot to eat or sleep. For another, he roamed over so many fields of inquiry that some, like mathematics and physics, were bound to prove fruitful, even if others, like alchemy, did not. Furthermore, Newton was probably gay, offering him from childhood a different perspective on life, perhaps inducing him to ask the sort of questions no one else did, such as those Thomas Mann's Tonio Kröger put to himself: "Why is it I am different, why do I fight everything, why am I at odds with the masters and like a stranger among the other boys?" And, for the import of science: why did the apple fall *down*?

Whether that apple ever existed—it's been called the second most famous apple in history—no one knows, let alone whether it triggered Newton's concept of universal gravitation. According to this idea, every mass attracted every other. The apple fell because the Earth's mass attracted the apple's mass. Newton extended this idea into the heavens. Just as the Earth pulled the apple down, so the Earth attracted the Moon and kept it in orbit; so the Sun attracted the Earth and kept *it* in orbit; and indeed so every star attracted every other.

Gravity's full cosmic consequences, however, went unmentioned in Newton's 1687 masterpiece, the *Principia*. Newton only began to contemplate the trouble five years later, when a preacher named Richard

Bentley wrote to him for help. Bentley was sermonizing against atheism and sought to use the existence of the Sun, planets, and stars to prove God's power. Newton sympathized with this goal, for he also believed in God. But Bentley had a problem that arose from Newton's own law of gravity. Suppose, said Bentley, that the universe was finite. Wouldn't gravity pull every star toward the universe's center, causing the cosmos to collapse? No, not at all, replied Newton, for the universe was infinite and thus stable—it had no center to collapse into. Long before Newton, Roman philosopher Lucretius had argued the same way for an infinite universe. However, Bentley questioned Newton's answer. Stars were not distributed evenly; instead, some regions harbored more than others, so the star-rich regions should collapse. This problem Newton never solved, forcing him to appeal to God: "a continual miracle is needed to prevent the Sun and the fixt stars from rushing together through gravity." Centuries later, the same problem would bedevil another great physicist, Albert Einstein.

Newton never addressed the dark night sky, but his astronomer friend Edmond Halley recognized the problem posed by an infinite, star-filled universe. Halley had edited Newton's *Principia* and paid for its publication, but he is best known for predicting the return of the bright comet that now bears his name. Halley had hoped merely that people would remember the prediction was made by an Englishman.

In 1721, with Newton presiding, Halley spoke before the Royal Society and offered two ways to reconcile Newton's infinite, star-filled universe with a dark night sky. In the first, he made a mathematical error and incorrectly said that the combined light from distant stars did not equal that from nearby ones. In the second, he repeated Thomas Digges' erroneous claim that invisible stars contributed no light to the night: "their Beams . . . are not sufficient to move our Sense." Thus neither of Halley's solutions explained why the night was dark.

The next to ponder the dark night sky, and the last before Olbers, was Swiss astronomer Jean-Philippe Loys de Chéseaux. In 1744, Chéseaux published his discussion in an appendix to a book about a six-tailed comet that had passed Earth that year. Unlike Digges and Halley, Chéseaux recognized that stars too distant to be seen did contribute light to the sky. The night was dark, he said, because space was not transparent. Instead, it contained material that absorbed enough light to darken the night. In his 1823 paper, Olbers himself offered the same solution. Strangely, Olbers owned Chéseaux's book but failed to cite it, even though he did mention Halley. Perhaps Olbers had read Chéseaux's discussion decades earlier, and it survived in his mind as a subconscious memory.

In any event, both Chéseaux and Olbers were wrong: absorbing material in space cannot keep the night dark. As astronomers later recognized, the very light this material blocked would heat the material and make it glow, so eventually it would radiate as profusely as the stars. It's like the ground beneath a tree in the rain. At first the leaves shelter the ground, but soon the leaves themselves start to drip.

## Thus Quoth the Raven

Surprisingly, the first correct resolution to Olbers' paradox came not from an astronomer at a well-equipped European observatory but from a poet and writer in America. During his short life, Edgar Allan Poe achieved fame for imbuing his stories and poems with the macabre and supernatural. Darkness permeated Poe's life: his mother died when he was two, his gambling and drinking ended a brief stay in college, and afterward he often lived in poverty. His wife died in her twenties, and Poe himself died after a drinking bout, only forty years old.

Darkness also permeated Poe's work. Witness these words from his horror story "The Fall of the House of Usher": *dull, dark, oppressively, dreary, melancholy, insufferable, gloom, sternest, desolate, terrible, bleak, vacant, decayed, depression, bitter, hideous, iciness, sinking, sickening, dreariness, torture, shadowy, sorrowful, precipitous, black, ghastly*—all in just the opening paragraph! Likewise, his famous poem "The Raven" said:

> *Deep into that darkness peering, long I stood there wondering, fearing,*
> *Doubting, dreaming dreams no mortal ever dared to dream before;*
> *But the silence was unbroken, and the darkness gave no token.*

It was cosmologist Edward Harrison who unearthed the Poe reference that correctly explained what is now called Olbers' paradox. "Americans are so fond of Edgar Allan Poe," said Harrison, "so we British ought to find out why. I started reading his works, and I was very struck by *Eureka*, an essay he wrote only a year before he died. A remarkable essay—everyone ought to read it." In one of his books, Harrison recounted the experience. "When I first read Poe's words I was astounded: How could a poet, at best an amateur scientist, have perceived the right explanation 140 years ago when in our colleges the wrong explanation . . . is still being taught?"

Poe published *Eureka* in 1848. Said Poe, "I have no desire to live since I have done 'Eureka.' I could accomplish nothing more." Poe wanted his publisher to print 50,000 copies, but the actual print run was only 500. Reviews were mixed: some called it new and startling; others dismissed it as nonsense. In more recent times, British astronomer Arthur Eddington derided it as a crank theory, while Harrison hailed it as an imaginative masterpiece.

To Poe, God was a poet, and the universe, as he wrote in *Eureka*, "the most sublime of poems." He also stated Olbers' paradox: "Were the succession of stars endless, then the background of the sky would present us an uniform luminosity, like that displayed by the Galaxy—*since there could be absolutely no point, in all that background, at which would not exist a star.* The only mode, therefore, in which, under such a state of affairs, we could comprehend the *voids* which our telescopes find in innumerable directions, would be by supposing the distance of the invisible background so immense that no ray from it has yet been able to reach us at all." In short, Poe proposed, light from distant stars failed to brighten the night because it had not had time to reach us; we cannot see farther than the universe is old. To use modern numbers: if the universe is 15 billion years old, then astronomers can see no farther than 15 billion light-years. The dark night sky therefore testified to the universe's creation.

However, Poe immediately questioned his own solution. "That this *may* be so, who shall venture to deny?" he wrote. "I maintain, simply, that we have not even the shadow of a reason for believing that it *is* so."

In *Eureka*, Poe several times mentioned German astronomer Johann Mädler, who owed his interest in astronomy to a bright comet that had passed Earth when he was in his teens. Mädler mapped the Moon's surface and published a popular book on astronomy that went through six editions. Early editions of this book, *Populäre Astronomie*, explained the dark night as Olbers had, from interstellar material that absorbed starlight. But in another book, in 1858, ten years after Poe's *Eureka*, and more vigorously in an 1861 edition of *Populäre Astronomie*, Mädler offered an explanation similar to Poe's: "The velocity of light is *finite*; a *finite* amount of time has passed from the beginning of Creation until our day, and we, therefore, can only perceive the heavenly bodies out to the distance that light has traveled during that finite amount of time. As the dark background of the heavens is sufficiently explained in this manner, indeed presents itself as necessary, the compulsion to assume light absorption is eliminated. Instead of saying that the light from those distances does *not* reach us, one must say: it has *not yet* reached us."

In 1901, Scottish mathematician and physicist Lord Kelvin published a quantified version of this argument. In order for the night to be bright, Kelvin calculated, we would need to see out to a distance of hundreds of trillions of light-years. Because the universe is much younger than a trillion years, the night is dark.

Poe, Mädler, and Kelvin all recognized that astronomers see the universe as it *was*, not as it *is*. The farther they look, the further they peer into the past—a concept carrying religious implications that Edward Harrison said slowed the resolution of Olbers' paradox. "We can look back and wonder why it is people were being so obtuse about the whole thing, but it was a problem that was looked at within a cultural context," said Harrison. "In any other context, they would have recognized the importance of the speed of light, but it was running in conflict with deep-rooted beliefs concerning the age of the universe. If you realized that there was a conflict with the biblical testimony, you just kept quiet about it." Literally interpreted, the Bible suggested that the universe was only 6,000 years old, which by Poe's argument meant astronomers should not be able to see farther than 6,000 light-years.

## The Rediscovery of Olbers' Paradox

Even though Olbers' paradox had a centuries-long history, it did not become famous until the 1950s. Indeed, had you mentioned Olbers' paradox to astronomers in Olbers' day, they wouldn't have known what you were talking about.

The term first appeared in a 1952 cosmology book by Hermann Bondi. Bondi was one of the original proponents of the steady state cosmology. Unlike the big bang cosmology, which held that the universe burst into being some 15 billion years ago, the steady state theory postulated that the universe had existed forever. In such an eternal universe, Edgar Allan Poe's solution to Olbers' paradox—that the night was dark because distant starlight had not had time to reach us—failed: the universe was infinitely old and astronomers should see infinitely far.

For the steady state theory, salvation lay in the universe's expansion.

Expanding space stretches light waves to longer, or redder, wavelengths, so the more a light wave has traveled through space—the farther the galaxy is from Earth—the greater will be the light's redshift. Red light carries less energy than yellow or blue light, so the redshift weakens light from distant galaxies and darkens the night. As steady state cosmologist Fred Hoyle put it in 1955, "*The sky is dark at night because the Universe expands.* This is the unexpected resolution of the puzzle—so unexpected that it never occurred to the scientists of the nineteenth century."

Edward Harrison believes that the apparent connection between the night's darkness and the universe's expansion—the first so trivial, the second so profound—contributed to cosmology's popularity during the 1950s. "Go out at night, astronomers urged members of their audience, and look up at the dark starlit sky," he wrote. "The darkness of the night sky proves that the universe is expanding. Here was a theme, prefaced with a few words about the Doppler effect, that captured the imagination of a wide audience."

Enchanting though it is, the redshift fails to solve Olbers' paradox. It applies *only* in a steady state universe, which cosmologists later disproved. In a big bang universe, the expansion of space barely affects the night's darkness: if the universe's expansion stopped, the night would stay dark. Yet popular books and textbooks often get it wrong. A 1987 survey found that only three out of ten astronomy textbooks correctly explained why the night is dark.

Incredibly, an entire book on Olbers' paradox also got it wrong. *The Paradox of Olbers' Paradox* was written by Stanley Jaki, a Benedictine monk and historian of science. Jaki's book lambasted scientists for failing to research the subject's history and for missing the message of the dark night sky, which he said made the study of Olbers' paradox as paradoxical as the paradox itself. But as Harrison wryly noted, "While on this theme, one might also mention the paradox of historians of natural science knowing little or no natural science." Jaki never mentioned Edgar Allan Poe, belittled Johann Mädler, and incorrectly stated that the resolution to Olbers' paradox was a universe of finite size. Jaki found an infinite universe as distasteful as had his Catholic predecessors who burned Giordano Bruno at the stake. "Intoxication with cosmic infinity," Jaki wrote, "not only produced a crass myopia for its obvious contradictions but also contributed heavily to the loss of confidence in ethical values and standards, and in the validity of purposes and goals."

Jaki might have done well to follow Olbers' advice: "It is always dubious and dangerous to specify shortsightedly what God should do."

## THE UNIVERSE'S ENERGY CRISIS

Starting with Edgar Allan Poe, astronomers had one correct explanation for the dark night sky, that the universe was too young. In 1964, Edward Harrison discovered a second reason, that it has too little energy.

Harrison first read about Olbers' paradox in Hermann Bondi's cosmology book. "As a parting gesture before turning to other interests," Harrison wrote, "I estimated the amount of energy that would be needed to create a bright sky of the kind imagined by Wilhelm Olbers. The result at first seemed unbelievable. Then it dawned on me that something was seriously wrong with the way we were looking at the whole problem."

Harrison focused on the amount of energy generated by all the stars in the observable universe. It was minuscule. For the night to be bright, he calculated, the observable universe would need 10 trillion times more energy—every star would have to shine 10 trillion times more brightly than it does, or the universe would have to have 10 trillion times more stars. In addition, stars don't live forever. Even if the universe were infinitely old, the night would still be dark, said Harrison, because the stars would have burned out.

Stars like the Sun generate their light via nuclear reactions, converting a small mass $m$ into an energy $E$ equal to $mc^2$, where $c$ is the speed of light. Harrison showed that even if all mass in the universe were converted into energy, the nocturnal sky would be no brighter than a moonlit night. Thus, in the end, the night is doubly protected against brightness—it is too young, and it lacks the fuel—just as a vegetarian runner is doubly protected against obesity. Lighting the cosmos with starlight is like trying to heat a cold house with a single candle and waiting an hour: an hour isn't long enough, and even if you waited longer, the candle would fizzle out before it did the job.

Nevertheless, one can imagine an alternate universe that would have a bright night. Suppose the speed of light, $c$, were greater. Then astronomers would see more of the cosmos, and stars—which generate their energy via $E = mc^2$—would live longer. In fact, if the speed of light were infinite, as some scientists and philosophers once believed, astronomers would see the entire universe, stars would live forever, and if the universe were sufficiently large the night would be bright.

Harrison, who entered science as a physicist, never regretted switching to astronomy. "Astronomy has given me a wide-open opportunity to pursue all sorts of ideas," he said. "I've lived in a period when astronomy was a wonderful field of inquiry. It has not always been and may not always be; but just in my lifetime, it's emerged as a remarkable field of study."

During the first decades of the twentieth century, several remarkable discoveries shook cosmology. The first: in the depths of space, far beyond the shores of the Milky Way, other titanic galaxies shone, each an island of starlight in an immense cosmic sea.

# THREE

# GALAXIES BEYOND

LIKE streetlights dotting a country town, galaxies take a valiant stand against the dark. Galaxies house most of the stars in the universe and generate nearly all of the little light that illuminates space. The most spectacular galaxies sport breathtaking spirals that look like glowing hurricanes. To cosmologists, however, galaxies are also practical, helping them probe the universe's dynamic nature. Just as lilies on a lake rise and fall with the waves, so galaxies ride the cosmic sea and reveal the expansion of space. Furthermore, their gravitational pull tries to slow the universe's expansion and dictate its fate, whether it will expand forever or eventually collapse. Moreover, distant galaxies preserve images from ancient epochs, allowing astronomers to investigate the universe's past. In short, galaxies are so useful to cosmology that if they did not exist, they would have to be invented.

The Sun, the Earth, and every individual star the unaided eye can see all belong to one galaxy, the Milky Way, so named because its brightest stars inhabit a thin disk that paints the night with a white band of light. When observers look above or below the disk, though, they see fewer stars and the sky is darker. In the fifth century B.C., Greek philosopher Democritus had first proposed that the band of milky sky glow consisted of stars too faint to see. In 1609, Italian astronomer Galileo Galilei proved him right when he turned his telescope on the Milky Way.

## BEYOND THE MILKY WAY

Just as other stars spangle the sky beyond the Sun, so other galaxies throng the cosmos beyond the Milky Way. The first to suggest this idea appears to have been English architect Christopher Wren, who in 1657 contemplated the design of the entire universe. While addressing Gresham College in London, Wren predicted that powerful telescopes would find "every nebulous Star appearing as if it were the Firmament of some other World, at an incomprehensible Distance, bury'd in the vast Abyss of intermundious vacuum." A nebulous star, or nebula—Latin for *cloud*—is a patch of misty light. It's just what a galaxy of stars would look like from a distance of millions of light-years. In 1734, Swedish philosopher Emanuel Swedenborg likewise speculated about "innumerable spheres . . . or starry heavens in the finite universe."

Stronger advocacy of this concept came from Thomas Wright, an Englishman who pondered the structure of the Milky Way. He had met with initial resistance: when Wright was an adolescent, his father so objected to his interest in astronomy that he burned his son's books. Nevertheless, in 1750 Wright published an elaborate book attempting to explain how the Milky Way's stars were arranged. In this he was nearly alone, for most astronomers of his time focused on the planets rather than the stars beyond.

In Wright's favored model, the Milky Way was spherical, its stars distributed like specks on a balloon's surface. When we look tangent to this balloon, said Wright, we see the many stars that make up the band of glowing light called the Milky Way; when we look away from the balloon's surface, we see fewer stars. Wright went beyond science, though, in connecting his theory to God. Heaven, Wright said, resided at the Milky Way's center—the center of the balloon—and hell lay in the dark, at great distances from the balloon. The Sun and the Earth therefore lay between heaven and hell, experiencing the influence of both.

In the book's last two pages, almost as an afterthought, Wright proposed the existence of other "Creations," or Milky Ways, choosing words that echoed Giordano Bruno's bold declaration of infinitely many worlds scattered throughout an infinite universe: "And farther since without any Impiety; since as the Creation [the Milky Way] is, so is the Creator also magnified, we may conclude in Consequence of an Infinity, and an infinite all-active Power; that as the visible Creation [the Milky Way] is supposed to be full of siderial Systems and planetary

Worlds, so on, in like similar Manner, the endless Immensity is an unlimited Plenum of Creations [galaxies] not unlike the known Universe."

Furthermore, Wright identified those other galaxies as the nebulae: "That this in all Probability may be the real Case, is in some Degree made evident by the many cloudy Spots, just perceivable by us, as far without our starry Regions, in which tho' visibly luminous Spaces, no one Star or particular constituent Body can possibly be distinguished; those in all likelyhood may be external Creation, bordering upon the known one, too remote for even our Telescopes to reach."

Wright's work triggered further speculations. Prussian philosopher Immanuel Kant read a review of Wright's book that prompted him to propose, in 1755, what is now known to be the correct shape for the luminous Milky Way, a disk. Like Wright, Kant suggested that the nebulae were other Milky Ways. At the time, however, the world heard little of Kant's ideas, because his publisher went bankrupt, forcing the confiscation of most copies of Kant's book.

Even though Wren, Wright, and Kant had all correctly suggested that other galaxies would appear as nebulae, few nebulae were then known—and most of those, like the prominent one in Orion, have turned out to lie within the Milky Way. In fact, when Wright's book appeared, astronomers knew only four objects that have proved to be other galaxies: in the northern hemisphere, the Andromeda Galaxy and one of its satellites, M32; and in the southern hemisphere, the Large and Small Magellanic Clouds, two satellites of the Milky Way named for the Portuguese voyager who died while circumnavigating the globe. But a French comet hunter was about to catalogue dozens more.

## A COMET HUNTER MAPS THE HEAVENS

Charles Messier came to Paris at age twenty-one, fatherless and nearly penniless. Messier loved comets—a brilliant one had blazed across the heavens when he was a teen—but he had little more to offer than legible handwriting and an ability to draw. Nevertheless, he was given a job. In 1757 he began searching the sky, hoping to be the first to glimpse the great comet whose return Edmond Halley had slated for 1758 or 1759. Unfortunately for Messier, misguided by his boss's predictions, he lost out to a prosperous German farmer. Still, Messier went on to discover so many other comets that French king Louis XV called him the "ferret of comets."

An oft-told story, false in fact but true in spirit, testified to Messier's cometary zeal. Messier supposedly missed discovering his thirteenth comet to a rival comet hunter, Montaigne, because Messier had been tending his dying wife. When a visitor offered condolences for his wife's death, Messier's eyes filled with tears—not for his wife, but for the comet: "I had discovered twelve; alas, to be robbed of the thirteenth by that Montaigne!" When he realized he should be mourning for his wife, he rushed to add, "Ah! That poor woman!" Cute though the story is, it cannot be true, for Montaigne discovered no comets the year Messier's wife died.

In 1758, before Halley's Comet returned, Messier thought he had discovered his first comet. It lay in the constellation Taurus and looked fuzzy and nebulous, as comets do, rather than sharp and starlike. But the "comet" never moved. Instead, it became the first entry on a list of nebulae and star clusters Messier drew up to prevent comet hunters like himself from making the same mistake. As did most astronomers of his time, Messier found comets far more interesting than nebulae. After all, the former sashayed through the heavens flaunting beautiful tails, while the latter merely formed part of a stationary backdrop against which astronomers viewed the planets and comets.

Today, however, astronomers remember Messier not for his comets but for his list of 103 nebulae and star clusters. "A more modest publication of greater ultimate significance is difficult to imagine," wrote Harvard astronomer Charles Whitney. Messier left his name all over the sky, for many an astronomical landmark bears an M: the Orion Nebula (M42 and M43), the Pleiades star cluster (M45), the Ring Nebula (M57), the Crab Nebula (M1, the object Messier stumbled across in Taurus), and the Andromeda Galaxy (M31), the nearest giant galaxy to our own. Although Messier never knew it, about a third of his objects would prove to be galaxies beyond our own.

Messier completed the list in 1781, a year that marked another astronomical watershed. German-born English musician William Herschel discovered the planet Uranus, doubling the size of the known solar system. Ironically, just as Messier had little interest in the nebulae he catalogued, Herschel had little interest in the planets; he had found Uranus while studying the celestial vault beyond—the stars and nebulae of the Milky Way. Messier's list inspired Herschel to catalogue some 2,500 nebulae and star clusters. Later his son John doubled that number, publishing the *General Catalogue* in 1864. In 1888, Danish-born astronomer

John Dreyer augmented this work and created the *New General Catalogue* of 7,840 objects. In 1895 and 1908, Dreyer published two supplements, the *Index Catalogues*, upping the total to 13,226. As a result, most well-known galaxies have an M, NGC, or IC in their names.

Dreyer did some of his work at an Irish castle that held the world's largest telescope. Equipped with a mirror measuring 72 inches across, the "Leviathan of Parsonstown" had been built by William Parsons, the third Earl of Rosse, decades earlier. With this telescope, Lord Rosse scrutinized nebulae, especially those in Messier's catalogue. He scored his greatest triumph in 1845, just after the telescope's construction, when he saw that M51 sported a dramatic spiral—so dramatic that today it is called the Whirlpool Galaxy. Shortly after the discovery, however, the potato famine hit Ireland, killing a million people, and Rosse directed his attention to matters terrestrial. When the famine subsided, Rosse and his assistants discovered spiral shapes in dozens of other nebulae, including M33, a delicate face-on spiral near Andromeda, and M101, a spectacular face-on spiral near the Big Dipper. Rosse's discoveries renewed speculation that some of the nebulae were external galaxies. They also gave rise to the first suggestion, in 1852 by American astronomer Stephen Alexander, that the Milky Way itself was a spiral—a speculation that took ninety-nine years to confirm.

During the latter half of the 1800s, however, in one of those curious backward steps that astronomy occasionally takes, several events conspired to undermine belief in external galaxies. As a result, most astronomers swung to the opposite point of view—that the Milky Way was the only galaxy in the universe, and the spiral nebulae Rosse had observed were small systems located within the Galaxy.

## GALACTIC RETROGRESSION

One of the first blows against the existence of other galaxies came in 1864, the same year that the *General Catalogue* saw publication. English astronomer William Huggins turned his telescope on several nebulae, including the famous Orion Nebula, a patch of misty light in the Hunter's sword. Huggins had armed his telescope with a spectroscope, a device that split light into a rainbow of colors. The resulting spectra revealed that many of the nebulae, including the one in Orion, were made of gas, not stars. Extrapolating from this correct observation, many astronomers concluded that *all* nebulae, even the spirals, were gaseous, too. That meant they couldn't be star-filled galaxies like the Milky Way.

Five years later, in 1869, British astronomer and author Richard Proctor produced additional evidence suggesting that the spiral nebulae lay within the Milky Way. Following up on earlier work by others, Proctor plotted the positions of thousands of nebulae from the *General Catalogue*. His plot showed how they shunned the Milky Way's disk and instead clustered around what astronomers call the Galactic poles, the regions of space perpendicular to the Milky Way's disk. Proctor concluded that the spiral nebulae couldn't be other galaxies. Instead, they must belong to the Milky Way, for otherwise their distribution would take no notice of its disk.

A further sign that the spirals lay within the Milky Way came in 1885, when the light of an exploding star—now known to have been a supernova—reached the Earth. The explosion erupted inside the closest spiral, Andromeda, and reached the threshold of naked-eye brilliance. In the peculiar lexicon of astronomy, the first variable star found within a constellation is designated R. The 1885 explosion was the second, so it received the designation S Andromedae. Astronomers already knew of exploding stars called novae, which popped off now and then in the Milky Way. If the Andromeda nebula were a small system within the Milky Way, then S Andromedae could be an ordinary nova—no problem. But if Andromeda were another galaxy, millions of light-years distant, then in order to be seen from so far S Andromedae had to shine with the fury of a billion Suns. To astronomers of 1885, who had not even coined the word *supernova*, so much light coming from a single star seemed impossible. As if to confirm the point, ten years later a similar explosion was sighted in another nebula, NGC 5253 in the constellation Centaurus. The fireworks of S Andromedae and the new one, Z Centauri, further convinced astronomers that the host spirals lay close by, within the Milky Way, and so were not galaxies.

The flare-up in Andromeda intensified interest in the nebula. In 1887, British astronomer Isaac Roberts photographed Andromeda and discovered that, like M51, it was a spiral. Its spiral had previously eluded detection because its disk was nearly edge-on to our line of sight, hiding the spiral. When Roberts displayed his photographs in London, they created a sensation. "Here we (apparently) see a new solar system in process of condensation from a nebula," he wrote—"the central sun is now seen in the midst of nebulous matter which in time will be either absorbed or further separated into rings." Furthermore, Roberts said, two nebulae near Andromeda might be newborn planets, orbiting Andromeda's central sun.

These observations convinced astronomers that the spiral nebulae lay within the Milky Way. "The question whether nebulae are external galaxies hardly any longer needs discussion," wrote British author Agnes Clerke in 1890. "It has been answered by the progress of discovery. No competent thinker, with the whole of the available evidence before him, can now, it is safe to say, maintain any single nebula to be a star system of coordinate rank with the Milky Way." In short, there was only one galaxy in the universe: our own.

## ANDROMEDA'S "PARALLAX"

If astronomers had measured the spirals' actual *distances*, they would have realized their error: a determination that the Andromeda nebula was 2.4 million light-years distant would have shown not only that it lay outside the Milky Way but also that it must rival the Milky Way in order to be seen from so far.

However, measuring accurate distances had thwarted astronomers for millennia. Only in the 1760s, for example, did they ascertain even the distance to the Sun. This they did by exploiting the nearest planet, Venus. In 1761 and again in 1769, Venus passed between the Earth and the Sun, events so rare they never occurred during the twentieth century. Edmond Halley, of comet fame, had earlier recognized that observers at different locations on Earth would see Venus cross different parts of the Sun: to observers in the southern hemisphere, Venus would appear farther north on the Sun's disk than it would to northern observers, by an amount that depended on the Earth's distance from the Sun. Astronomers implemented Halley's plan and found that the Sun was eight light-minutes away. Halley, though, didn't live to see the success, just as he never saw the return of his comet.

Even after ascertaining the distance to the Sun, astronomers failed to fathom the stars beyond, for most of the bright stars were millions of times farther. Still, because the Earth goes around the Sun, astronomers view a star from one side of the Sun in January and from the other side in July. This induces a small shift, or parallax, in the star's apparent position. The closer the star is to the Earth, the larger is this parallax, so parallax provided a way to measure a star's distance. It also proved that the Earth went around the Sun. Indeed, those who thought otherwise, such as Aristotle, cited the *lack* of observed parallax to argue that the Earth did *not* circle the Sun.

In fact, the stars are so distant that their parallaxes are too tiny for ancient astronomers to have seen. In 1838, after repeated failures by others, Prussian astronomer Friedrich Wilhelm Bessel scored the first success. Bessel's skill in mathematics had been obvious to his parents, who wanted him to become a merchant. At the age of twenty, though, he met Heinrich Wilhelm Olbers, the German physician who contemplated the dark night sky. Olbers encouraged Bessel to pursue astronomy and later called him his greatest astronomical discovery.

Bessel's discovery of parallax came as Olbers turned eighty. "How can, how should I thank you enough for making my 80th birthday so pleasant?" Olbers wrote to Bessel. "Accept my congratulations for that great discovery, which now gives for the first time a solid and secure foundation to our ideas about the universe." Bessel succeeded where others had failed in part because he pursued a contrarian strategy. Most other parallax seekers had focused on the brightest stars, reasoning that these should be the closest to Earth and thus possess the largest and easiest-to-measure parallaxes. In contrast, Bessel chose a star so faint he could barely see it with his unaided eye. This star, named 61 Cygni, possessed a large movement, or proper motion, across the sky. The high proper motion suggested that the star was close to the Sun. In the same way, a speeding driver sees nearby road signs whiz past him, whereas distant mountains appear to hold stationary. Shortly after Bessel succeeded in detecting 61 Cygni's parallax—the star is just 11 light-years from Earth—Wilhelm Struve in Estonia and Thomas Henderson in South Africa measured the parallaxes of Vega and Alpha Centauri, respectively.

Parallax might solve the riddle of the spiral nebulae. If they lay within the Milky Way, as was thought during the late 1800s, they might be close enough to Earth to have detectable parallaxes. Sure enough, in 1907 Swedish astronomer Karl Bohlin, making one of the wrongest astronomical measurements ever, reported that he had detected the parallax of the Andromeda nebula. It was, he said, 19 light-years from Earth.

## The Racing Spiral Nebulae

The first strong evidence that the spiral nebulae were distant galaxies resembling the Milky Way came the following decade. Ironically, the work was motivated by the view that the spirals were instead newborn solar systems within the Milky Way. Furthermore, the discovery was made at an outcast Arizona observatory, what one astronomer called "the traditional 'bad boy' in astronomy."

Lowell Observatory had a bad name because it had been founded by the flamboyant Percival Lowell. The wealthy Bostonian believed that the planet Mars had intelligent beings who constructed canals to carry water from the Martian polar caps to the equator: "That Mars is inhabited by beings of some sort or other we may consider as certain as it is uncertain what those beings may be." Lowell had built the observatory primarily to study Mars, and in books, articles, and lectures he pressed his case for Martian intelligence—to the bewilderment and frustration of other astronomers. Because of his fondness for planets, Lowell wanted to know more about the spiral nebulae, which he and most other astronomers thought were newborn solar systems. He therefore asked his assistant, Vesto Slipher, to observe them.

Slipher complemented Lowell. Whereas Lowell was a Boston patrician, Slipher had been born on an Indiana farm; whereas Lowell was daring and provocative, Slipher was calm, cautious, and patient. The last trait would prove crucial in achieving Lowell's goal. Lowell wanted Slipher to see whether the spiral nebulae spun, as they should if they were newborn solar systems.

Slipher took aim at the spirals with a spectroscope, the same weapon deployed by William Huggins to discern the gaseous nature of the Orion Nebula. Slipher knew that spreading a star's light into a spectrum, or rainbow of starlight, did more than just reveal the star's composition; it also measured the star's speed. Light waves from a star that moves away from Earth get stretched to longer, or redder, wavelengths, producing a redshift. Conversely, light waves from a star that moves toward Earth get scrunched together into shorter, bluer wavelengths, producing a blueshift. Furthermore, the greater the object's speed along our line of sight, the greater is this Doppler shift.

By the time Slipher began trying to record the spectra of the spirals, astronomers had measured the Doppler shifts, and therefore the line-of-sight velocities, of over a thousand stars. Doing the same for the spirals was far more difficult, because they were faint and their light spread out. Slipher chose to start with the brightest spiral. In 1912 he captured Andromeda's spectrum, but it was so startling that he repeated the measurement three times before announcing the result.

Astronomers express speeds in kilometers per second. Even 1 kilometer per second is swift: a plane this fast could travel from Boston to San Francisco in just over an hour. Relative to the Sun, most stars in the Milky Way move at under 50 kilometers per second. Slipher found, however, that Andromeda was racing toward Earth at the astonishing speed of 300 kilometers per second. "It looks as if you had made a great discovery," wrote Lowell. "Try some more spiral nebulae for confirmation." Slipher did. His next target—what is now called the Sombrero Galaxy—lay on the opposite side of the sky from Andromeda, in the constellation Virgo. This time Slipher found an even more startling result: the Sombrero had such a large redshift that it must be darting away from Earth at 1,000 kilometers per second. A Boston-to-San Francisco flight this fast would take just five *seconds*.

Over the following months, Slipher observed other spiral nebulae. Not only did most have high velocities, but redshifts began to dominate blueshifts—most of the spirals were moving away from Earth. In 1914, Slipher presented his results at a meeting of the American Astronomical Society, where he received a standing ovation. All but three of his fifteen spirals showed redshifts.

Slipher never interpreted these redshifts as a sign of the universe's expansion. Instead, he believed the Milky Way was drifting southward through space, because several nebulae south of the Galactic plane, such as Andromeda, M32, and M33, showed blueshifts, whereas most nebulae north of the Galactic plane, such as the Sombrero, showed redshifts. He clung to this belief even after finding that many southern nebulae had redshifts.

Slipher's discovery convinced most astronomers that the spirals lay outside the Milky Way. After all, how could the Milky Way hold on to such speedy objects? The spirals could not be newborn solar systems, Lowell realized, but "something larger and quite different, other galaxies of stars."

## NOVAE IN SPIRALS

In 1917 this reborn idea of distant galaxies received new support when several spiral nebulae sprouted novae. In July of that year George Ritchey at Mount Wilson Observatory in Pasadena, California, detected a nova on a photographic plate of the spiral nebula NGC 6946 in Cepheus. Ritchey had largely built Mount Wilson's 60-inch telescope and was then at work on an even larger telescope, the mighty 100-inch. The nova Ritchey had discovered was a true nova, not a supernova, and it was faint, far fainter than the eye could see. Meanwhile, up the California coast at Lick Observatory, astronomer Heber Curtis had found two novae in M100 and another in NGC 4527.

After these discoveries, astronomers scrutinized photographic plates of other spirals, turning up additional novae. Curtis recognized the implications: "It seems to me that they furnish weighty evidence in favor of the well known 'island universe' theory of the spiral nebulae. . . . Though reasoning by analogy frequently leads to error, the occurrence of objects of the same type in the spirals would reasonably be expected, were these spirals in fact congeries of vast numbers of stars, like our own galaxy."

Of course, the novae in spirals differed in one important respect from their counterparts in the Milky Way: they looked fainter. But that was only because they were farther. From the difference in apparent brightness, Curtis estimated that the novae, and their spiral hosts, were millions of light-years away. Curtis also knew why the spiral nebulae avoided, or appeared to avoid, the disk of the Milky Way. Photographs of edge-on spiral galaxies showed their disks to be lined with dust. If the Milky Way's disk were similar, dust should clog the Galactic plane, blocking the view of other galaxies so that the spiral nebulae appeared only outside the Galactic plane.

However, just as the idea that the spirals lay outside the Milky Way was gathering momentum, two new developments—one right, the other wrong—were about to catapult some astronomers back to the view that the spirals lay within the Milky Way.

## MIGHTY MILKY WAY

Right things first. In 1917, Mount Wilson astronomer Harlow Shapley shook up our own Galaxy, declaring it some ten times larger than had been thought. When he was a child, Shapley's only contact with astronomy had come one summer when his father told him and his brother about an upcoming meteor shower. "So we lay down on our backs on a beautiful August night to wait for the first ones to come," Shapley wrote, "and we both went soundly to sleep and never saw those Perseids."

Shapley discovered astronomy by accident. He went to the University of Missouri at Columbia intending to study journalism; when he got there, though, he found that the journalism school had not yet been established. "I opened the catalogue of courses," he said. "The very first course offered was a-r-c-h-a-e-o-l-o-g-y, and I couldn't pronounce it! . . . I turned over a page and saw a-s-t-r-o-n-o-m-y; I could pronounce that—and here I am!" Another astronomer later said of Shapley, "I have never seen a quicker mind, a more agile sense of humor, or a more complete absence of what usually passes for humility."

Nor was there anything humble about Shapley's oversized model of the Galaxy. He based it on globular clusters, tight-packed conglomerates that harbor hundreds of thousands of stars and look like frosted lights adorning a Christmas tree. Through various techniques, Shapley tried to ascertain their distances and at first thought they were outside the Milky Way. Then, in 1917, he boldly reversed himself: not only did the globular clusters belong to the Milky Way, but they outlined its structure the way Christmas lights outline a Christmas tree; so deduce the structure of the globular cluster system and you deduce the size and shape of the Galaxy.

From Shapley's model, two startling conclusions followed. First, whereas most other astronomers had thought that the Sun and Earth occupied the center of the Galaxy, Shapley showed that they reside about halfway from the center to the edge of the disk. This is obvious, he said, because the globular clusters congregate on one side of the sky, around the constellation Sagittarius, the home of the Galactic center. In the same way, streetlights congregate around a city's downtown, so a suburbanite could locate downtown by finding the part of the sky with the most streetlight glow.

"After I saw that mankind was peripheral, it occurred to me that this had philosophical implications," Shapley later wrote. "If man had been found in the center, it would look sort of natural. We could say, 'Naturally we are in the center because we are God's children.' But here was an indication that we were perhaps incidental. We did not amount to so much."

Second, the Galaxy must be big, because most of the globular clusters were tens or hundreds of thousands of light-years distant. Shapley actually overestimated the Milky Way's size, thinking its disk some 300,000 light-years across, about two and a half times its true size. Shapley's model of the Milky Way was so big that it would have swallowed the spirals, which he believed were just a few thousand light-years away. He believed this because the 1885 "nova" in Andromeda would otherwise have been superluminous. He also believed it because of the work of another Mount Wilson astronomer, Dutch-born Adriaan van Maanen.

## IMPROPER MOTIONS

Shapley liked van Maanen. "Van Maanen was aggressive and he was sociable. He could go to a dinner and soon have the whole table laughing. He was a social success. People liked him—until he became a sort of playboy. . . . He was a charming person, a bachelor; he and I were pals of a sort—I don't know why, because I wasn't 'society' and he was. I suppose we got together because he was rather an alert-minded person and I liked his nonsense."

Van Maanen measured proper motions, the year-by-year change in positions that result as objects move through space. In particular, he wanted to see the spirals spin. Vesto Slipher, the astronomer who had earlier discovered the spirals' high speeds, had detected their spin via their Doppler shifts. As an edge-on spiral like the Sombrero Galaxy rotated, one side moved toward the Earth and the other side away, so Slipher's spectra showed that the Doppler shift changed from one side of the galaxy to the other.

If the spirals were nearby, van Maanen might also see them rotate—not through the Doppler shift, but through the proper motions of their spiral arms, as these whirled around the galaxy's center and changed position on photographic plates. In 1916 van Maanen claimed to detect such proper motions in the spiral arms of M101. Then, in the early 1920s, he repeated the claim for six other spirals, including M33, M51, and M81.

If van Maanen was right, the spiral nebulae must be nearby, because proper motion cannot be seen in far-off objects, just as a distant mountain looks stationary to a speeding motorist. Shapley and some other astronomers used this "fact" to support their contention that the spiral nebulae were not other galaxies. Most other astronomers, however, did not believe van Maanen's work. The spirals should be distant: they raced through space and glistened with numerous novae. Furthermore, in 1924 astronomers realized that van Maanen's results meant that the spirals spun in the opposite direction from what Slipher had found.

To resolve the debate over the nature of the spiral nebulae, astronomers needed a celestial yardstick to ascertain how far the spirals were. Fortunately, nature provided just such a yardstick: remarkable pulsating stars called Cepheids, which expand and contract like a human heart, would figure in the two greatest astronomical discoveries of the twentieth century.

## SACRED CEPHEIDS

Nothing warms the heart of an astronomer fathoming the depths of the cosmos on a freezing night like a Cepheid variable. These bright yellow stars pulsate with the precision of a Swiss watch in a way that reveals just how bright and distant they are. As a result, astronomers seeking the size and age of the universe have enshrined Cepheid variables in the cathedral of cosmic calibrators.

As Cepheids expand and contract, they brighten and fade, alerting astronomers to the pulsations and making these valuable stars easy to find. The first two Cepheids came to light back in 1784, long before anyone recognized their great utility in probing space. On the night of September 10, Edward Pigott, a prosperous English amateur astronomer, discovered that the light of the star Eta Aquilae was varying. When brightest, Eta Aquilae shone twice as brightly as it did when faintest. Pigott noted that the star brightened quickly but faded slowly—a Cepheid hallmark—and soon determined that it repeated this variation every seven days. A month later, on October 20, Pigott's friend, a deaf-mute twenty-year-old named John Goodricke, noticed fluctuations in the light of Delta Cephei, which showed the same fast-rise, slow-fade behavior of Eta Aquilae. Delta Cephei lay so far north that it never set, so Goodricke and Pigott could observe it every clear night. In Delta Cephei's honor, variable stars with similar properties were later called Cepheids.

To Pigott and Goodricke, the flickering Cepheids must have resembled cosmic fireflies, nighttime oddities, but little more. In 1907, however, Harvard astronomer Henrietta Leavitt, who was nearly as deaf as Goodricke, made a crucial discovery. Leavitt was investigating variable stars in the Magellanic Clouds. If Eta Aquilae and Delta Cephei were individual fireflies blinking on and off, the Magellanic Clouds hosted a firefly nest: Leavitt's 1907 paper listed 808 variable stars in the Large Magellanic Cloud and 969 in the Small Magellanic Cloud.

For 16 of the variables in the Small Magellanic Cloud, Leavitt was able to determine how long the stars took to wax and wane. "It is worthy of notice that . . . the brighter variables have the longer periods," she wrote. Worthy of notice, indeed—yet no one did take notice, probably because that single statement followed lengthy tables listing 1,777 stars. In 1912, Leavitt devoted an entire paper to the remarkable Cepheid period-luminosity relation. Because all of the Cepheids in the Small Magellanic Cloud lay at about the same distance from Earth, she noted, the Cepheids that *look* brighter must also *be* brighter—their *intrinsic* brightness, or luminosity, must be greater, for they emit more light into space. Thus a Cepheid with a ten-day pulsation period must emit more light than a Cepheid with a five-day pulsation period. As astronomers know today, this is because the bigger and brighter a Cepheid is, the longer it takes to pulsate—just as a large musical instrument, like a cello, resounds at a lower frequency than a small one, like a violin. Measuring a Cepheid's period, which is easy, thus reveals the Cepheid's intrinsic brightness; comparing the intrinsic with the apparent brightness yields the distance to the Cepheid—and its galaxy.

There is one catch, however. To know which luminosity goes with which period, astronomers must know the distances of a few Cepheids in the Milky Way. Then they would know the Cepheids' intrinsic brightnesses. Unfortunately, Cepheids are so rare that even the nearest, Polaris, has a distance of 430 light-years and a parallax too small for astronomers of Leavitt's day to have measured.

Nevertheless, this obstacle did not deter astronomers. In 1913, Danish astronomer Ejnar Hertzsprung used thirteen Cepheids—including Eta Aquilae, Delta Cephei, and Polaris—to derive an approximate distance from the stars' proper motions, since the farther the Cepheids were, the smaller should have been their proper motions. A few years later, Harlow Shapley deemed two of the Cepheids peculiar, then repeated Hertzsprung's procedure with the eleven remaining Cepheids. Shapley used the resulting Cepheid period-luminosity relation to ascertain distances to what he thought were Cepheids in the nearest globular star clusters, leading to his large model of the Galaxy.

As astronomers now know, this Cepheid period-luminosity relation was flawed—a flaw that would lay buried, go uncorrected, indeed be repeatedly "confirmed" over the next three decades. A Cepheid is actually three times more powerful than Hertzsprung and Shapley had thought. They erred primarily because dust in space dimmed the Cepheids, making them seem feebler than they really were; and because the proper motions were not that accurate. Even with this handicap, however, the Cepheids were the key that unlocked the spirals' true nature.

## CEPHEIDS IN ANDROMEDA

Edwin Hubble almost did not become an astronomer. In college he was a heavyweight boxer, and a sports promoter had wanted him to battle the world champion. Hubble declined. His father wanted him to study law, which Hubble did, but after his father's death, he switched to astronomy. In 1919 he reached Mount Wilson Observatory in Pasadena, California, home of the world's largest telescopes, a 60-inch and a brand-new 100-inch.

In late 1923, Hubble used the 100-inch telescope to hunt for novae in Andromeda, the same spiral nebula that had astonished Vesto Slipher a decade earlier with its high speed. In October, Hubble took several photographic plates that captured what he thought were three Andromedan novae. He marked each with an "N." To see how bright the novae had been on previous nights, Hubble examined older plates. As he traced the trail into the past, two of the novae obediently disappeared. But one remained, sometimes brighter, sometimes fainter, and Hubble realized that this "nova" was no nova at all but instead something far more precious: a Cepheid variable. On the plate, Hubble x-ed out the "N" and wrote "VAR!"

Now Hubble could determine Andromeda's distance—and at last settle the debate over the spiral nebulae. First, though, he had to find the Cepheid's period. If the period was short, then the Cepheid emitted little light and Andromeda was nearby; but if the period was long, the Cepheid shone profusely and Andromeda was distant. Hubble therefore examined the star's fluctuating brightness on dozens of plates that went back to 1909. On October 23, 1923, he determined the period: it was long—31.4 days—so Andromeda must be distant. Hubble estimated a distance of about 1 million light-years, shy of the actual number, but more than enough to place the Andromeda "nebula" far beyond the shores of the Milky Way.

The debate over the spiral nebulae ceased. The universe was vast, and other galaxies were strewn throughout space. Just as centuries earlier Nicolaus Copernicus had declared that the planets were other earths and Thomas Digges that the stars were other suns, so Edwin Hubble had proved that the spiral nebulae were other galaxies.

Yet there were no news conferences, no press releases, no front-page headlines to trumpet the twentieth century's second greatest astronomical discovery. Instead, Hubble kept the news to himself. He photographed Andromeda again, to follow the Cepheid, to confirm its period, to discover additional Cepheids; he also found Cepheids in Andromeda's neighbor, M33, which proved to be as distant. During 1924, Hubble's triumph leaked out to astronomers around the world. Thirteen months after the actual discovery, *The New York Times* finally reported the historic news—but only on page six, where Hubble's name was repeatedly misspelled "Hubbell."

Hubble had hesitated because of Adriaan van Maanen's faulty work, which had supposedly revealed proper motions in the spirals, demanding that they be nearby. After Hubble's discovery, however, even those astronomers who had believed van Maanen dismissed his work.

Ironically, Hubble's discovery could have been made long before 1923. After all, photographic plates had captured Andromeda's bright Cepheid back in 1909, two years after Henrietta Leavitt discovered the Cepheid period-luminosity relation. There is even a story that around 1920, Milton Humason marked Cepheids on plates of Andromeda that Shapley had taken. Humason had no formal training as an astronomer; in fact, he had dropped out of school at fourteen, after which he began his career at Mount Wilson—as a mule driver. Then he became a janitor. According to the story, after Humason returned the Andromeda plates to Shapley, Shapley rubbed out the marks, telling Humason why they could not be Cepheids.

If Hubble had done no more than establish the existence of other galaxies—a word he disliked, by the way; to him they were "extragalactic nebulae"—he would still have achieved great fame. But Hubble's greatest discovery, and the greatest astronomical discovery of the twentieth century, was yet to come, a discovery that involved the dynamics of the entire universe.

## THE EXPANDING UNIVERSE

To Albert Einstein, the universe was anything but dynamic. In 1915 he formulated his greatest triumph, the general theory of relativity. Despite its mathematical complexity, general relativity describes the most familiar force—gravity—but in a way quite different from Einstein's predecessor, Isaac Newton. Newton had envisioned gravity as a force that one mass exerts on another: the Earth attracts an apple; the Sun attracts the Earth. In contrast, Einstein said that mass curves space, and other masses follow this curved space. For example, the Sun curves space into the shape of a vortex, so the Earth and other planets revolve around it like water spiraling around a drain.

In 1917, however, when Einstein tried to apply his theory to the universe, he found that he had the same problem Newton did: his universe collapsed. Every star attracted every other, and the whole thing fell together in the middle. Newton had wriggled out of the problem by invoking "a continual miracle" to hold up the universe's weight. Since miracles carried less credibility in the twentieth century, Einstein introduced a term into his equations called the cosmological constant. This represented a repulsive force, like antigravity, except that instead of arising from mass, it arose from space. The universe could therefore be static: the outward force of the cosmological constant counterbalanced the inward force of gravity.

Einstein's universe was finite but unbounded. In this it resembled the Earth's surface, which has a finite amount of area but lacks an edge. Sailors never fall off the edge of the Earth's surface, but if they voyage long enough, they can return to their starting point from the opposite direction. Likewise, a light beam setting off into Einstein's universe would never encounter an edge to space, but the curvature of space would bend the light beam and return it to its starting point.

Soon after Einstein formulated his universe, Dutch astronomer Willem de Sitter found another solution to Einstein's equations of general relativity. Unlike Einstein's universe, de Sitter's was infinite. A light beam would never return to its starting point but instead travel forever farther into the void. De Sitter's model had a peculiar feature: it had no mass, and if mass were introduced into it, the universe became unstable. Although the true universe possessed mass, de Sitter argued that the actual density of mass in the universe was so low that his model might be correct.

De Sitter's model had another unusual feature, one that led to Hubble's discovery of the universe's expansion. At large distances, said de Sitter, time would appear to run more slowly, so an observer would see a distant clock run more slowly than it really does. This de Sitter effect would also redshift light. To see why, imagine a distant galaxy emitting yellow light. Yellow light waves oscillate 545 trillion times a second. Because of the de Sitter effect, those light waves would appear to oscillate less often, as red light does; thus, the distant galaxy would exhibit a redshift. Furthermore, the farther the galaxy, the more slowly time should appear to run and the greater its redshift should be. During the 1920s, several astronomers looked for a link between distance and redshift, but none found it, in part because no one had accurate distances to galaxies.

Enter Edwin Hubble and the Cepheids. Because of his discovery of Cepheids in the Andromeda Galaxy and elsewhere, Hubble now knew the distances of nearby galaxies; because of Vesto Slipher's measurements, Hubble also knew their redshifts.

In 1928, Hubble traveled to Holland. There he met de Sitter, who urged him to look for a distance-redshift relation. When Hubble returned home, he sought to establish the distances not just to nearby galaxies, like Andromeda and M33, where he could see Cepheids, but also to more distant galaxies, where he could see no Cepheids. For those galaxies, Hubble used other distance indicators, such as their brightest stars, whose luminosities he knew from observing similar stars in Cepheid-calibrated galaxies like Andromeda and M33. When he was done, Hubble had distances for twenty-four galaxies, out to a redshift of 1,090 kilometers per second.

His result was profound: the farther the galaxy, the greater was its redshift. In a 1929 paper modestly entitled "A Relation between Distance and Radial Velocity among Extra-Galactic Nebulae," Hubble touched on the de Sitter effect only in the final paragraph, and astronomers began to interpret the distance-redshift relation otherwise—as the sign of an expanding universe. Furthermore, it is not even the Doppler shift that produces the redshift. A Doppler shift arises when an object moves *through* space. But the redshifts that most galaxies exhibit arise from the expansion of space itself. As a light wave travels through space, that space expands and stretches the light's wavelength, shifting it redward. The farther a galaxy is from Earth, the greater the length of time its light has spent traveling through space, so the more the light wave has been redshifted. Thus, the farther a galaxy, the greater is its redshift, just as Hubble found.

Unfortunately, Hubble's discovery fit neither Einstein's universe nor de Sitter's: Einstein's had mass but no motion, and de Sitter's had motion but no mass. At the time, British astronomer Arthur Eddington pondered the problem. Eddington would later write, perhaps overconfidently, "I believe there are 15,747,724,136,275,002,577,605,653,961,181,555,468,044,717, 914,527,116,709,366,231,425,076,185,631,031,296 protons in the uni-

WHO DID WHAT WHEN

| When | Who | What |
|---|---|---|
| 1912 | Vesto Slipher | Measures first galactic velocity; the high speed indicates spiral nebulae are other galaxies. |
| 1914 | Vesto Slipher | Finds most galaxies have redshifts rather than blueshifts. |
| 1915 | Albert Einstein | Formulates general relativity. |
| 1917 | Albert Einstein | Applies general relativity to derive static model of universe. |
| 1917 | Willem de Sitter | Formulates empty universe model; predicts distance-redshift relation, from dilation of time rather than expansion of space. |
| 1922 | Aleksandr Friedmann | Formulates first model of explicitly expanding universe. |
| 1923 | Edwin Hubble | Discovers Cepheid in Andromeda and thereby Andromeda's distance, proving that other galaxies exist. |
| 1927 | Georges Lemaître | Formulates model for expanding universe. |
| 1929 | Edwin Hubble | Reports distance-redshift relation for galaxies, now interpreted as arising from the universe's expansion. |
| 1931 | Georges Lemaître | Formulates big bang model. |

verse, and the same number of electrons." Now, as he contemplated the universes of Einstein and de Sitter, he wished that Einstein's equations allowed a third solution, for a universe with both mass and motion.

Such a solution already existed. Ironically, it had been formulated in 1927 by a former Eddington student, Belgian priest Georges Lemaître. In 1922, five years before Lemaître, Russian meteorologist Aleksandr Friedmann had worked out much the same idea: an expanding, mass-filled universe that satisfied Einstein's equations of general relativity. Perhaps the turbulence Friedmann saw in the atmosphere inspired him to contemplate a dynamic model for the universe. Unfortunately, Friedmann died in 1925, so he never saw the verification of his idea. Nevertheless, with the work of Friedmann and Lemaître, and Hubble's distance-velocity relation, cosmologists had a new view of the universe. If extrapolated back in time, it implied that the cosmos was once smaller and denser, compressed into a tiny space. This "primeval atom," in Lemaître's words, or big bang, in modern terminology, burst forth and gave birth to the expanding universe.

Actually, long before Friedmann and Lemaître, the same poet who had contemplated the dark night sky—Edgar Allan Poe—managed to deduce much the same, starting with the observation that all matter attracts all other matter: "That each atom attracts—sympathizes with the most delicate movements of every other atom, and with each and with all at the same time, and forever, and according to a determinate law of which the complexity, even considered by itself solely, is utterly beyond the grasp of the imagination of man. . . . Does not so evident a brotherhood among the atoms point to a common parentage? Does not a sympathy so omniprevalent, so ineradicable, and so thoroughly irrespective, suggest a common paternity as its source? . . . In a word, is it not because the atoms were, at some remote epoch of time, even *more than together*—is it not because originally, and therefore normally, they were *One*—that now, in all circumstances—at all points—in all directions—by all modes of approach—in all relations and through all conditions—they struggle *back* to this absolutely, this irrelatively, this unconditionally *one*?"

In early 1931, Albert Einstein visited Mount Wilson Observatory and met Edwin Hubble. There Einstein threw out the cosmological constant, the "antigravity" force that had allowed the universe to be static. With the discovery that the universe was expanding, Einstein no longer needed the cosmological constant, and he later called it the biggest blunder of his life.

At Mount Wilson, Einstein's wife was shown the mammoth 100-inch telescope with which astronomers like Hubble were determining the universe's structure. She was not impressed. "Well, well," she reportedly said, "my husband does that on the back of an old envelope."

# FOUR

# BIG BANG BATTLES

THE first three decades of the twentieth century witnessed three great astronomical discoveries: the Sun resided not at the Galaxy's center but in its suburbs; other galaxies speckled space beyond our own; and the universe was expanding, pushing most of those galaxies away. The last two discoveries revolutionized cosmology, and in their wake two competing cosmological models—the big bang and the steady state—rose to battle.

The two rivals gave radically different visions of the universe. The big bang theory posited a flashy beginning to the universe, a fiery explosion that unleashed both matter and space itself. In contrast, the steady state cosmology offered the luxury of eternity, a cosmos with neither beginning nor end. Stakes in the battle between the big bang and the steady state theories were enormous: the outcome would affect humanity's perception of the entire universe.

Centuries earlier, astronomers had waged another great battle, after Nicolaus Copernicus challenged the Ptolemaic view that the Earth constituted the solar system's center. As in that epic battle, religious overtones pervaded the new cosmic struggle. Indeed, the Catholic Church again intervened, though this time on the winning side. Furthermore, each side's leading proponent was not only a brilliant scientist but also an excellent writer, authoring popular books and articles that allowed the public to watch the great cosmic debate.

## THE GENESIS OF GENESIS

Of the two cosmologies, the big bang theory arose first—from Georges Lemaître, the Belgian priest who in 1927 had proposed that the universe was expanding. Today cosmologists cite the universe's expansion as evidence favoring the big bang theory, but the expansion alone hardly proves it. In fact, when he wrote his 1927 paper, Lemaître believed that the universe had no beginning but instead had existed forever. Originally it had been a static universe, he said, like Albert Einstein's 1917 model. Then, for some reason, the universe began expanding. "It remains to find the cause of the expansion of the universe," he wrote.

Einstein didn't like Lemaître's expanding universe. "Your calculations are correct," Einstein said to Lemaître, "but your physical insight is abominable." Einstein had written much the same about Aleksandr Friedmann, the Russian meteorologist who had been the first to propose an explicitly expanding universe. (Willem de Sitter's earlier model also expanded, but he failed to recognize this.)

Nevertheless, after Edwin Hubble discovered that farther galaxies had greater redshifts, scientists recognized that the universe was expanding. Unlike Einstein, British astronomer Arthur Eddington had no problem with an expanding universe—he even wrote a book with that title—but he did worry about the universe's birth: "It has seemed to me that the most satisfactory theory would be one which made the beginning *not too unaesthetically abrupt.*" As Lemaître originally did, Eddington favored a universe that had existed forever. In 1927 he narrowly missed coining the name for what would emerge as the winning cosmology when he said, "As a scientist I simply do not believe that the present order of things started off with a bang."

Ironically, despite Eddington's misgivings about a birth for the universe, he planted the seed for the big bang model. In early 1931 he gave a lecture that Lemaître attended. The talk was not about the beginning of the universe but the end. Eddington discussed the second law of thermodynamics, a bleak law which states that entropy—the amount of disorder—continually increases, so the universe faces a "heat death" of complete disorder. If the universe is infinitely old, the amount of entropy should have risen to an infinite level—yet the actual universe retains so much order that life exists. It was a thermodynamic version of Olbers' paradox. Just as Olbers' paradox stated that an infinitely old universe should launch an infinite amount of light toward Earth, so Eddington's argument implied that an infinitely old universe should be infinitely chaotic. Yet it wasn't.

Lemaître saw a possible solution—the birth of the universe. In May 1931 he published a four-paragraph paper in *Nature* that enunciated what would later be called the big bang theory. Strangely, he made no mention of Hubble or the universe's expansion; indeed, he barely alluded to observations at all. Instead, he proceeded through thought and logic.

Paraphrasing the second law of thermodynamics, Lemaître began by stating that the number of quanta in the universe is ever increasing. Thus, he reasoned, if we went back in time, we would eventually reach an era when there was only one great quantum. "Now, in atomic processes, the notions of space and time are no more than statistical notions; they fade out when applied to individual phenomena involving but a small number of quanta. If the world had begun with a single quantum, the notions of space and time would altogether fail to have any meaning at the beginning; they would only begin to have a sensible meaning when the original quantum had been divided into a sufficient number of quanta. If this suggestion is correct, the beginning of the world happened a little before the beginning of space and time." At the start, said Lemaître, there was one massive but highly unstable atom, whose atomic weight equaled that of the entire universe. This unstable atom then split into smaller and smaller atoms. The *New York Times* headline read LE MAITRE SUGGESTS ONE, SINGLE, GREAT ATOM, EMBRACING ALL ENERGY, STARTED THE UNIVERSE.

It sounded like Genesis, and Lemaître's paper originally ended, "I think that everyone who believes in a supreme being supporting every being and every acting, believes also that God is essentially hidden and may be glad to see how present physics provides a veil hiding the creation." Before publishing the paper, though, Lemaître crossed out these words, probably because he did not want to mix science and religion.

In a lengthier paper later in 1931, Lemaître called the universe's first atom the "primeval atom." Elsewhere he wrote, "The evolution of the world can be compared to a display of fireworks that has just ended: some few red wisps, ashes and smoke. Standing on a well-chilled cinder [the Earth], we see the slow fading of the suns, and we try to recall the vanished brilliance of the origin of the worlds."

If Lemaître could have asked an all-knowing oracle just one question, it would have been: has the universe ever been at rest, or did the expansion start from the beginning? "But, I think, I would ask the oracle not to give the answer, in order that a subsequent generation would not be deprived of the pleasure of searching for and of finding the solution."

And search they did.

## THE UNIVERSE'S PREHISTORIC KITCHEN

This Belgian priest's unlikely successor, the man who would emerge as the leading advocate for the big bang, was Russian-born physicist George Gamow. As a child Gamow went to church and snuck a communion bread crumb home, where he examined it under a microscope. Although the Catholic Church claimed that the bread had become the flesh of Christ, Gamow found that it still looked like bread. "I think this was the experiment which made me a scientist," he said. Gamow liked to recount how God must reside nine light-years from Earth. In 1905, when Russia lost the Russo-Japanese War, Russian churches had prayed for God to punish the Japanese. The prayers, traveling at light speed, took nine years to reach God, then nine more years followed for the response, when God devastated Japan with the 1923 earthquake.

Gamow hated communism, which he escaped in 1933, and loved practical jokes. "On some occasion," recalled his former student Ralph Alpher, "he had a party at his home and invited the Russian scientific attaché—who came, despite the fact that Gamow was still under sentence of death from the Stalin regime. So the guy participated in the party, and Gamow snuck out and sent the Soviet limousine home. When the party was over, this guy came up, having had too much vodka, and he couldn't find the limousine. He rang Gamow's bell, and Gamow didn't answer; so the guy had to hike back to the embassy." It was two or three miles away.

In Russia, Gamow had learned cosmology in part from Aleksandr Friedmann, the meteorologist whose model of the expanding cosmos predated Lemaître's. But Gamow was a nuclear physicist, not an astronomer. Nuclear physics dictates which elements can fuse or fission into which others. Gamow was especially interested in the *origin* of the elements—hydrogen, helium, carbon, oxygen, gold, and all the rest. Where did they come from? Why were some elements, like hydrogen and helium, common, while others, like silver and gold, rare?

Gamow sought the answers to both questions in the big bang, in what he called the "'prehistoric' kitchen of the universe." During that primordial era, Gamow said, the universe was so hot and dense that nuclear transformations occurred as readily as boiling water cooks an egg. He believed that the nuclear fury just after the big bang cooked up an abundance pattern—much hydrogen and helium, little silver and gold—that still exists today, preserved in the stars and galaxies.

In order to probe the big bang's element-creating power, Gamow enlisted graduate student Ralph Alpher. Alpher had received a college scholarship to the Massachusetts Institute of Technology from an MIT alumni organization, which withdrew it upon learning that Alpher was Jewish. Alpher instead attended college and graduate school at George Washington University, where Gamow was stationed. In 1948, Alpher and Gamow wrote a paper describing how the hot, dense conditions following the big bang could have created all of the elements from hydrogen to uranium.

In this paper, Gamow played his most famous practical joke. He had noticed that Alpher's name and his own resembled the first and third letters of the Greek alphabet, alpha and gamma. Missing, though, was a name corresponding to the second Greek letter, beta. So Gamow made Cornell physicist Hans Bethe the paper's second author, even though Bethe had nothing to do with it. Appropriately, the paper appeared April 1.

Gamow also tried to persuade another scientist, Robert Herman, to change his name to Delter, to resemble the fourth Greek letter, delta, but Herman refused. Nevertheless, in an article the following year, Gamow referred to the paper as "Alpher, Bethe, Gamow, and Delter."

Later in 1948, Alpher and Herman teamed up to pen a crucial prediction. If the early universe forged the elements, its heat must still exist today, albeit thinned out by the universe's enormous expansion since then. Alpher and Herman pegged the present temperature of the universe at a mere 5 Kelvin, or −451 degrees Fahrenheit.

Anything with a temperature—a star, planet, or person—radiates energy. Thus, if Alpher and Herman were right, the entire universe should radiate. A star is so hot that it radiates light, while planets and people, being cooler, emit infrared radiation, which has less energy. If the universe's temperature was just a few Kelvin, it would give off electromagnetic radiation carrying even less energy—microwaves and radio waves.

Here was something: a property of the universe that observers might measure. At the time, cosmology suffered from a paucity of data, leaving theorists free to speculate about the nature of the universe. Indeed, a hard-nosed scientist could have said that in all of cosmology, there were only two real facts: the night is dark, and the farther a galaxy, the greater is its redshift. But Alpher and Herman had put their finger on an observational consequence of the big bang theory, one that could help confirm the cosmology.

"We expended a hell of a lot of energy giving talks about the work," said Alpher. "Nobody bit; nobody said it could be measured. We even, some years later, talked to [Allan] Sandage." Sandage had been Hubble's observing assistant, and he carried on Hubble's work after Hubble died, in 1953. "Well, he was the wrong guy to talk to about making radio-frequency observations, but he didn't see how it could be done either. And so over the period 1948 to 1955, we sort of gave up. Later we were severely criticized by some people for having not pursued it. Well, neither one of us was in a position to do so."

Nevertheless, Alpher and Herman continued to publish papers predicting the universe's temperature. In a popular 1952 book, *The Creation of the Universe*, Gamow himself did so, although at an inflated level of 50 Kelvin.

According to Alpher, there was a problem. "The cosmology was not believed," he said. "There were a lot of doubters and skeptics. The then-existing cosmological parameters did not give a very happy age for the universe. That was a *major* problem."

The universe's troublesome age followed from Hubble's observations. After discovering that farther galaxies had greater redshifts, Hubble worked out the universe's expansion rate, a number now called the Hubble constant. The Hubble constant is the redshift, expressed as a speed in kilometers per second, divided by the distance, in megaparsecs. (One megaparsec is 3.26 million light-years, a bit farther than the Andromeda Galaxy is now known to be.) Most of Hubble's redshifts were right, but his distances were much too small. He thought that the Andromeda Galaxy was under a million light-years away and the Virgo cluster just a few million farther. Since the Hubble constant is the redshift *divided* by the distance, his distance *under*estimates led him to *over*-estimate the Hubble constant. He obtained a Hubble constant of 530 kilometers per second per megaparsec. This said that for every megaparsec farther he looked, he thought the redshift increased by 530 kilometers per second.

Such a high Hubble constant meant that the universe expanded fast and was young, because a fast-expanding universe has taken little time to reach its present size. According to Hubble's work, the universe was only about 1.8 billion years old. But geologists were saying that the Earth was older; its age is now known to be 4.6 billion years, contradicting the youthful cosmic age estimate derived from big bang cosmology.

This contradiction argued against the big bang theory—and partially inspired a troika in England to devise a new cosmology, one that had neither beginning nor end: an everlasting "steady state" cosmos that expunged the deus ex machina of the big bang.

## THE STEADY STATE

Of the three steady state proponents, the one destined for greatest fame was Fred Hoyle, a maverick forever at war with the establishment. "The first priority among scientists," Hoyle wrote in his autobiography, "is not to be correct but for everybody to think in the same way."

Hoyle's rebellion against authority started early. He had mastered the multiplication tables at age four, so his teacher told him to learn the Roman numerals. "How could anybody be so daft as to write VIII for 8?" Hoyle wondered. Well, his teacher replied, Roman numerals were old and sometimes appeared in books. "This was more than I could reasonably stomach, and the day this outrage to the intelligence was perpetrated became my last at that particular school." Alas, the law compelled Hoyle to attend school. "I concluded that, unhappily, I'd been born into a world dominated by a rampaging monster called 'law,' that was both all-powerful and all-stupid." Later, he brought another teacher a flower with six petals, and she whacked his left ear—she had told the class the flower had only five petals. Later in life, Hoyle became deaf in his left ear.

In the 1940s, Hoyle met up with two scientists who had fled their native Austria, Hermann Bondi and Thomas Gold. Bondi would later coin the term that describes the dark night sky, *Olbers' paradox*; Gold would go on to recognize that pulsars were fast-spinning neutron stars.

All three scientists despised religion. "Religion," said Gold, "is a force that is driving us back to the misery of the Middle Ages. It creates artificial divisions: India against Pakistan; in Northern Ireland, the Catholics against the Protestants; and the Sunni Moslems against whatever other Moslems there are. Religion is a disaster." Bondi said his marriage to a woman of a different religious background has been successful because neither he nor she is in the least religious. And Hoyle once proposed a simple solution to the conflict in Ireland: jail every priest and clergyman. "Not all the religious quarrels I ever saw or read about," Hoyle wrote, "is worth the death of a single child."

Hoyle is no atheist, however. "I believe that the universe shows signs of logical structure," he said. "I mean, if you take the Dirac equation, add to it the Pauli principle, that controls everything all the way from the electron up to biology. How on Earth do you get a situation like that in a random universe? I have to think, I have to believe, that this has all been thought out; but I don't believe it's been thought out by the sort of gods that people talk about. It's got to be something much superior to what the Pope believes."

All three steady staters denied that their hostility to religion led them to develop the steady state cosmology. Yet they surely noticed that the big bang theory, formulated by a priest, echoed the words of Genesis. Indeed, Hoyle has occasionally attacked the big bang theory as religious fundamentalism.

It was Gold, however, who originated the steady state idea. "During the war we worked together," Gold wrote, "and also roomed together for some time. In the evening Fred would typically walk around and with great emphasis say: 'Well what could that Hubble observation mean? Find out what it could mean!' He would continue along this line, sometimes being rather repetitious, even aggravating, drumming away at particular points without any obvious purpose. At other times Fred would have Bondi sit cross-legged on the floor, then sit behind him in an armchair and kick him every five minutes to make him scribble faster, just as you might whip a horse. He would sit there and say: 'Now come on, do this, do that,' and Bondi would calculate at furious speeds, though *what* he was calculating was not always clear to him—as on the occasion when he asked Fred, 'Now, at this point do I multiply or divide by $10^{46}$?'"

Legend has it that Gold's steady state idea was inspired by a 1945 ghost movie starring Michael Redgrave, *Dead of Night*, which ended as it began. "That's probably not correct," said Gold, "but it's a good story. It's a movie that has no beginning and no end, very cleverly arranged. I think we saw that movie several months before, and after I proposed the steady state I said to them, 'Isn't that a bit like *Dead of Night*?'"

Gold's idea of a never-beginning and never-ending universe solved the problem vexing the big bang theory, a universe that seemed younger than the Earth. If the universe was infinitely old, as the steady state theory maintained, it could easily accommodate a planet whose age was several billion years.

Nevertheless, Gold's idea posed a new problem. Because the universe expands, it also thins out, as galaxies race away from one another. The universe's density of matter thus gets diluted, the way a drop of red dye loses its vigor as it expands in a pool of water. Therefore, if the universe was infinitely old, as Gold proposed, it should also be infinitely rarefied. Yet plenty of galaxies lay within range of telescopes. So Gold also proposed that particles pop into existence out of nothing, maintaining the universe's density of material.

At first, neither Bondi nor Hoyle thought much of Gold's proposal. Said Bondi, "Fred Hoyle and I said, 'What a crazy idea! We'll shoot it down before dinner.' But dinner was remarkably late that night."

Gold's idea of continuous creation seemed outrageous because it violated the law of conservation of mass—but then, so did the big bang itself, and far more recklessly. Furthermore, the rate of creation necessary to maintain the universe's density was so meager that it contradicted no experiment. In order to see even one atom pop into existence in a laboratory beaker, you'd need to observe it for billions of years.

Hoyle worked out one formulation of the steady state cosmology, but the first journal he submitted his work to rejected it—claiming a shortage of paper. "It seems unbelievable now," said Hoyle, "but that was the excuse they used. The referee couldn't fault the solution of the equations, so the secretary needed an excuse."

Bondi and Gold didn't like Hoyle's paper, either, finding it too specific. So they wrote a more philosophical approach, which invoked what they called the perfect cosmological principle. Decades earlier, Albert Einstein had implicitly formulated what others later called the cosmological principle. This states that over large distances, space is uniform. Although galaxies congregate here and there, over large distances, no matter where you are, the universe looks pretty much the same.

Bondi and Gold extended the cosmological principle to include time. Their *perfect* cosmological principle stated that not only was the universe uniform in space, but it was also uniform in time. No matter where or *when* you looked, the universe should look pretty much the same. The big bang theory violated the perfect cosmological principle: it said that at one time the entire cosmos had been compressed into a point.

Even as he helped formulate steady state cosmology, Hoyle was starting to deliver science to the public. In early 1950 the British Broadcasting Corporation planned to offer lectures on Saturday evenings. However, the original lecturer backed out, and the show's producer scrambled for a replacement. He asked Hoyle and then checked Hoyle's file at the BBC. It said: DO NOT USE THIS MAN.

The producer ignored the warning. Hoyle agreed to give the lectures for a practical reason—he needed the money. The five broadcasts proved far more lucrative than Hoyle had imagined, for they led to a handsome book deal, first in England and then in America. Hoyle and his wife, both from modest backgrounds, saved most of their newfound wealth; they did, however, buy their first refrigerator.

In his fifth and final BBC broadcast, Hoyle delved into cosmology—advocating the steady state theory—and then attacked religion, provoking protests from Christians. "It was a pity that Fred Hoyle spoilt his otherwise fine lectures by leaving the subject in which he is qualified, and ending with 'religion' in which he is manifestly not qualified," wrote one listener. Wrote another, "In the last of his interesting broadcast lectures Mr. Fred Hoyle remarks: 'It strikes me as very curious that the Christians should have so little to say about how they propose eternity should be spent'. Now, it strikes me as very curious that Mr. Hoyle should find it very curious that Christians are unable to describe the next life."

"Isn't it time we had some common sense?" demanded a bishop from Eastbourne. "We are told the universe ultimately comes out of nothing—something out of nothing. Isn't it far more probable that, as everything planned comes from a planner and everything we make comes from a maker, the whole creation comes from a Creator—it is His thought that thickens into created things. Painters do it in much the same way. Why not God? And really! Christianity is an 'escapism' and offers one nothing except everlasting frustration. Was anything so muddled and untrue ever written?" Angry Christians wanted to know what right Hoyle had to air his views, and the BBC dutifully reminded listeners that "this is a free country and the personal beliefs of all men of high intelligence deserve consideration."

During a BBC broadcast that aired the year before, in 1949, Hoyle first named the competing cosmology when he said, "These theories were based on the hypothesis that all the matter in the universe was created in one big bang at a particular time in the remote past."

Popular books claim that Hoyle coined *big bang* to slam the theory; but Hoyle disputed that. "The BBC was all radio in those days, and on radio, you have no visual aids, so it's essential to arrest the attention of the listener and to hold his comprehension by choosing striking words. There was no way in which I coined the phrase to be derogatory; I coined it to be striking, so that people would know the difference between the steady state model and the big bang model." Thus, radio was partially responsible for giving birth to one of cosmology's most colorful terms. If Hoyle had been doing television, he might not have felt the need to coin such a memorable phrase.

Over forty years later, *Sky and Telescope* hosted a contest that invited readers to rename the big bang. Science writer Timothy Ferris penned an editorial blasting the term as ugly and bellicose. His editorial claimed that the big bang was dark (in fact, it was extremely bright) and repeated the myth that Hoyle had coined the term to denigrate the theory. The magazine received some 13,000 entries from people proposing new names for the big bang, but the judges thought none surpassed the original.

Hoyle was not one of the judges. "It was something of a farce," he said. "I thought, 'If they can do it, good luck to them,' but I was very interested that they couldn't."

## COSMIC BATTLES

During the 1950s and 1960s, the big bang theory waged a far more important contest with the steady state theory. In their favor, big bang proponents had the simplest extrapolation from the observed expansion of the universe—that the universe had once been smaller and denser. Furthermore, the big bang theory had no need to postulate the continuing creation of matter, which struck many scientists as absurd. Aesthetically, the big bang cosmology also had its appeal, for the theory promised an exciting universe, one that had metamorphosed from a hot, bright, dense state at birth to the cool, dark, sparse one we inhabit today. Thus, if observers looked far into space, and therefore far back in time, they could glimpse a universe different from the present one. Furthermore, the big bang theory was more familiar to astronomers: it had been formulated in 1931, whereas the steady state theory didn't arise until 1948.

On the other side, though, steady state proponents could marshal a host of philosophical arguments. "I'm trained as a mathematician," said

Hoyle, "and the instinct of a mathematician is to broaden the possibilities. The crazy idea of physicists is to limit everything: 'You shan't think this, and you shan't think that.' The physicists want to stop you thinking about various possibilities, and astronomical observers are the same. They're negative, whereas mathematicians are positive. If you worked the way physicists and astronomers work, you'd still be with the integers. You'd never have got the rationals, you'd never have got the irrationals, you'd never have got the imaginary numbers. It's because mathematicians are always pressing for new possibilities that we get them."

The steady state theory accorded with uniformitarianism, a principle in geology which held that the key to the past was the present: what happened in the past can be deciphered by studying present processes, such as erosion. In the same way, the steady state theory postulated a universe whose laws never changed, so that scientists could test it by examining present processes. In contrast, the big bang theory, with its splashy initial explosion, resembled catastrophism, the then discredited geological principle which held that calamities governed Earth's history. Furthermore, the big bang theory pushed the key event—the big bang itself—out of sight, to high redshift, making the theory harder to confirm. Finally, the steady state universe was immortal, renewing itself by creating new matter and new galaxies, whereas the big bang universe was doomed: if it collapsed, it would end in a fiery "big crunch"; if it expanded forever, it would suffer a heat death of total disorder.

The battle between the big bang theory and the steady state theory reached the public as well, for both George Gamow and Fred Hoyle were outstanding popular writers. "People often ask me how I write books that are so successful," Gamow said in his autobiography. "Well, it is a deep secret, so deep that I do not know the answer myself!" On a personal level, however, the two rivals got along well. Hoyle said that one conference during the 1950s tried to pit the two cosmic antagonists against each other, but the attempt failed when they refused to fight.

The cosmological battle had religious overtones, especially with the resemblance between the big bang and Genesis. In 1951, Pope Pius XII—whose disdain for democracy, appeasement of the Nazis, and silence during the Holocaust induced one critic to brand him "Hitler's Pope"—endorsed big bang cosmology, saying that it proved God's existence. As *Time* quoted him, "Hence, creation took place in time. Therefore, there is a creator, therefore, God exists."

Gamow, ever the joker, seized on the pronouncement to bolster his cosmology, quoting the Pope in a scientific paper: "It can be considered now as an unquestionable truth that '*from one to ten thousand million years ago, the matter of the (known) spiral nebulae was compressed into a relatively restricted space, at the time the cosmic processes had their beginning*' and that during that stage '*the density, pressure, and temperature of matter must have reached absolutely enormous proportions*' since '*only under such conditions can we explain the formation of heavy nuclei and their relative frequency in the period system of elements.*'"

Steady state proponent Thomas Gold was less impressed. "Well, the Pope also endorsed the stationary Earth." Of course, that was a different Pope.

Even some big bang advocates criticized the Pope's intervention, notably the original big bang proponent, Catholic priest Georges Lemaître. Said one witness, "[I] can recall very vividly Lemaitre storming into class on his return from the Academy meeting in Rome, his usual jocularity entirely missing. He was emphatic in his insistence that the Big Bang model was still very tentative, and further that one could not exclude the possibility of a previous cosmic stage of contraction." Moreover, Lemaître thought it wrong to use cosmology to do theology. "As far as I can see, such a theory remains entirely outside any metaphysical or religious question," Lemaître wrote.

The Pope's pronouncement must have reaffirmed suspicions among steady state supporters that big bang cosmology was religiously inspired. In fact, however, the big bang's leading advocate of the time was not religious; Gamow had, after all, subjected a communion wafer to microscopic examination.

Nor was Ralph Alpher, who had given up religion as a child. "I don't know how you can be a scientist and not be an agnostic," he said. "You have no evidence for the existence of some extramundane entity. Gosh, I get more kick out of seeing a pretty sunset or listening to Beethoven's Ninth than I do sitting in a temple or a church or whatever, listening to somebody give me homilies. I think the world, the universe, is so beautiful, so awe-inspiring, that I just don't see the need for much else."

Some steady state supporters were themselves religious. They could view the creation of any matter—whether all at once, in the big bang, or little by little, in the steady state—as divine. Furthermore, because the steady state theory held that this creation was ongoing, it demonstrated God's continuing involvement with the universe, whereas the big bang could have arisen from a God who had created and then abandoned the universe.

If steady staters attacked the big bang theory as religious, big bangers struck back by slapping the steady state theory's country of origin. "It is not surprising that the steady-state theory is so popular in England," wrote Gamow, "not only because it was proposed by its three (native-born and imported) sons H. Bondi, T. Gold, and F. Hoyle, but also because it has ever been the policy of Great Britain to maintain the *status quo* in Europe."

In 1959, *Science News Letter* polled thirty-three astronomers. The big bang theory fared better, but it hardly won an overwhelming victory. Did the universe start in a big bang? Of those polled, 33 percent said yes, 36 percent said no, and the rest were undecided. Does the universe create matter continuously, as the steady state theory maintained? Of those polled, 24 percent said yes, and 55 percent said no.

The debate would be over by 1965. The first blow against the steady state theory had come in 1952, when astronomers kicked its original inspiration out from under it: the age conflict between the Earth and the universe.

## A UNIVERSAL AGE

When it was first proposed, the steady state theory offered a way out of the embarrassment of a universe younger than the Earth. "The timescale difficulty was very serious," said Bondi. "Hubble had enormous prestige, and his estimate of the inverse Hubble constant was 1.8 billion years. Everybody took that as absolutely the fact; so compared with the rocks of the Earth, there was a severe timescale difficulty."

The universe's age was based on its expansion rate—the Hubble constant, which Edwin Hubble had put at 530 kilometers per second per megaparsec. That Hubble constant, in turn, was based on Hubble's estimates of the distances to other galaxies.

The man who would prove those distances wrong was Walter Baade, a cheerful German-born astronomer who worked at the same Pasadena observatory as Hubble. "Hubble was the outstanding astronomer of the twentieth century," said Lick Observatory's Donald Osterbrock, who has written about Baade's life. "Baade was the second outstanding astronomer of the twentieth century. He was a fantastically smart and technically skilled astronomer and also had a wonderful personality—he got along very well with others."

Baade had long suspected that something was wrong with the galaxies. Using Hubble's distances, astronomers could translate the apparent sizes of other galaxies into true sizes, but our Galaxy turned out to be larger than every other. Furthermore, Hubble's distance to the Andromeda Galaxy implied that its globular star clusters shone with only a quarter of the brilliance of their counterparts in the Milky Way.

In 1943, Baade made his great discovery, one that would eventually undermine Hubble's distances: stars in our Galaxy and others fall into two populations. At that time, because of World War II, the lights over nearby Los Angeles were dimmed to thwart Japanese attacks on the city, so the sky was dark. With Mount Wilson Observatory's 100-inch telescope, Baade took aim at the Andromeda Galaxy and for the first time "resolved" its stars—that is, he glimpsed its individual stars. He also resolved stars in four galaxies that orbit Andromeda. Unlike the Andromeda spiral, its four attendants were round or egg-shaped galaxies called ellipticals.

"There can be no doubt," Baade wrote, "that, in dealing with galaxies, we have to distinguish two types of stellar populations." The first population, now known to be young, he called population I; it dominated Andromeda's disk and spiral arms, its brightest members being blue and white supergiant stars such as Rigel and Deneb in the Milky Way. The second population, now known to be old, Baade called population II; it dominated Andromeda's central bulge as well as globular star clusters and elliptical galaxies, its brightest members being red and orange giant stars. Baade thus recognized that spiral galaxies, such as the Milky Way and Andromeda, hosted stars of both populations I and II, while elliptical galaxies had stars of only population II.

Baade's discovery spelled trouble for the distances Hubble had determined to other galaxies. These distances were based on Cepheids, the pulsating yellow supergiant stars that serve as cosmic yardsticks: the longer a Cepheid takes to pulsate, the bigger it is and the brighter it shines. In order to use Cepheids to measure distances, astronomers had to know which luminosity went with which pulsation period. No Cepheid lay close enough to Earth for astronomers to measure its distance directly, so first Ejnar Hertzsprung in 1913 and then Harlow Shapley in 1917 attempted to estimate the Cepheids' distances and luminosities from their proper motions. The farther the Cepheids were, the smaller these motions look, just as a distant airplane appears to move more slowly than one nearby.

Shapley had used what he thought were Cepheids in globular star clusters to help him formulate his "big Galaxy" model for the Milky Way, the one that put the Sun well away from the Galactic center. The trouble was, most Cepheids near the Sun were population I stars, whereas globular clusters contained only population II stars. Even in 1940, before Baade's discovery of the two stellar populations, another Mount Wilson astronomer, Alfred Joy, had noted peculiarities in a "Cepheid" that belonged to the globular cluster M3. Might the "Cepheids" in globular clusters obey a different period-luminosity relation than true Cepheids?

One way to answer this question was to examine another distance indicator that sprinkled globular clusters—RR Lyrae stars, which pulsate as Cepheids do but faster, typically in less than a day. RR Lyrae stars are fainter than Cepheids but easy to use, because unlike Cepheids they all have nearly the same luminosity. Astronomers know this because all RR Lyrae stars in a particular cluster look equally bright.

Not only do RR Lyrae stars throng globular clusters, but they also speckle the Galaxy at large. In fact, RR Lyrae itself, the brightest representative, is a so-called field star, unaffiliated with any globular cluster. The star's variability had been discovered in 1899 by Harvard astronomer Williamina Fleming. Astronomers could determine the intrinsic brightnesses of field RR Lyrae stars in the same way as they had the Cepheids': from their proper motions. The luminosity thereby deduced for the RR Lyrae stars yielded distances to the globular clusters that agreed with the distances the "Cepheids" there gave, so all seemed well.

With Mount Wilson's 100-inch telescope, Hubble and Baade could see Cepheids in the Andromeda Galaxy, but not RR Lyrae stars, which were too faint. In 1948, however, the great 200-inch telescope at Palomar Observatory went into action. If Andromeda were really just a million light-years away, as Hubble had said, the new telescope should reveal the galaxy's RR Lyrae stars.

Baade failed to find them; so Andromeda had to be farther than Hubble had said. "Already the very first [photographic] plates indicated that the accepted form of the [Cepheid] period-luminosity relation did not represent the true situation," Baade wrote. "If it had, the [RR Lyrae] variables of the Andromeda nebula should have appeared at the limiting magnitude of the plates. Instead, the brightest stars of the population II appeared at about this magnitude." From studies of globular clusters, Baade knew that the brightest population II stars shone about four times more brightly than RR Lyrae stars. This allowed him to deduce Andromeda's approximate distance: some 2 million light-years, twice as far as Hubble had said. Baade's doubling of the distance meant that Andromeda was twice as large as had been thought, making it larger than the Milky Way—our Galaxy was no longer the largest in the cosmos. Also, doubling Andromeda's distance boosted the luminosity of its globular clusters up to that of the Milky Way's globulars.

Most importantly, Baade doubled the estimated size and age of the entire universe. That's because Hubble had used Andromeda as a stepping stone to more distant galaxies in which Cepheids could not be seen. According to Baade, the universe was some 3.6 billion years old, lessening the age conflict between the Earth and the cosmos, and removing an argument against the big bang theory.

Baade announced the historic find in 1952, at a conference in Rome. Ironically, the man Baade appointed to record his talk was none other than the big bang's biggest rival, Fred Hoyle. After his talk, Baade was elated to hear South African astronomer Andrew Thackeray confirm it. Thackeray and a colleague had discovered RR Lyrae stars in NGC 121, a globular cluster in one of the Milky Way's satellite galaxies, the Small Magellanic Cloud. But NGC 121's RR Lyrae stars looked four times fainter than had been expected, which meant the Small Magellanic Cloud was twice as far as the flawed Cepheid period-luminosity relation had indicated.

How had this Cepheid error occurred? In the 1910s, when Hertzsprung and Shapley had used a handful of Cepheids near the Sun to calibrate the Cepheid period-luminosity relation, the astronomers had been forced to choose stars near the Milky Way's plane, because that's where population I stars, like Cepheids, reside. But the Galactic plane harbored gas and dust, which dimmed the Cepheids. In addition, the Cepheids' proper motions were not that accurate. To make matters worse, the incorrect Cepheid relation was then "confirmed" by RR Lyrae stars in the Milky Way. Unlike Cepheids, RR Lyrae stars do *not* suffer much dimming by gas and dust. That's because these stars are old, members of population II, so most lie far above and below the Galactic plane.

To astronomers, Baade's greatest discovery was his recognition of the two stellar populations; but to the public, his doubling of the size and age of the universe was far more dramatic. However, the unsuspecting Baade was about to be sabotaged and this latter work plagiarized, by none other than Harlow Shapley. Back in the 1910s, as he had constructed his radical new model for the Milky Way, Shapley had pioneered the Cepheid period-luminosity relation, but he now refused to hear of any problems with it. Even after Baade's discovery of the two stellar populations, and of differences between population I and population II Cepheids, Shapley had insisted that all Cepheids obeyed the same period-luminosity relation.

A few months after Baade's announcement, however, Shapley reversed himself—but now said that *he*, not Baade, had doubled the size and age of the universe. The media hailed Shapley's "discovery." A WIDER AND AN OLDER UNIVERSE, reported *The New York Times*; DOUBLE THE UNIVERSE, said *Time*. Shapley's name appeared prominently; Baade's was completely absent.

"I was Baade's student," said Allan Sandage, "and I saw Baade's reaction directly. You have to know Walter Baade. He was not a volatile man. He was outgoing, jovial, loved to talk to people, was congenial. But he was incensed. It was a scandal—an absolute, complete scandal. Exactly the same thing that Shapley had done to Hubble." In 1929, a few months after Hubble announced his greatest discovery—that farther galaxies exhibit greater redshifts—Shapley reported that *he* had discovered it. One of the few times that Sandage ever saw the reserved Hubble get emotional was when Sandage had said something good about Shapley.

Fortunately for Baade, astronomers recognized Shapley's plagiarism for what it was. They knew it was Baade who had halved the Hubble constant and doubled the size and age of the universe. During the rest of the 1950s, Sandage revised the Hubble constant further downward, further boosting the universe's estimated age. He and his colleagues realized that earlier astronomers had mismeasured the brightnesses of the stars, including the Cepheids, in other galaxies. These stars were fainter and their galaxies therefore farther than had been thought. Using the correct brightnesses lowered the Hubble constant.

So did another discovery. In most galaxies beyond Andromeda, Hubble could not see Cepheids, so to estimate those galaxies' distances he had used their brightest stars instead. Sandage discovered that what Hubble had thought were the brightest stars in distant galaxies were actually large regions of ionized hydrogen gas lit by many bright stars; they therefore far outshone solitary stars and were farther than Hubble had thought.

By 1958, Sandage estimated that the Hubble constant lay between 50 and 100, five to ten times lower than Hubble's value. If the universe had expanded uniformly since its birth—neither accelerating nor decelerating—this range meant that it was 10 to 20 billion years old, more than old enough to accommodate the 4.6-billion-year-old Earth. Thus, the big bang theory's age conflict with the Earth evaporated.

## ELLIPTICAL SEASONS

With Walter Baade's 1952 doubling of the age and distance scale, and the elimination of the age conflict between the Earth and the universe, the big bang theory's rival—the steady state theory—lost part of its impetus. Furthermore, it was suffering age problems of its own, problems that stemmed from the universe's chief building blocks, the galaxies.

"Steady state was never a viable theory," said Sandage. "There was no evidence of the formation of galaxies over time, which steady state required." For that reason, Sandage said, he and fellow observers in California, such as Hubble and Baade, never seriously considered the theory. "Bondi pronounced in England that he was coming to the United States to sell the steady state," said Sandage, "and the Pasadena astronomers were essentially appalled, because they believed that there was nothing to sell."

Although the steady state cosmology predicted that galaxies ranged from very young to very old, it also predicted that the *average* galaxy was young. That's because continuous creation of matter was giving birth to new galaxies, while the expansion of the universe swept older ones away. The steady state theory predicted that the average galaxy should be only a third as old as big bang supporters said the entire universe was. But astronomers could find no young galaxies.

The steady state theory also had problems with the oldest galaxies, the ellipticals. In 1943, Baade recognized that elliptical galaxies consisted of population II stars, the older stellar population. Furthermore, elliptical galaxies all had the same yellow-orange color, suggesting they all had the same age. That's because a galaxy gets redder as it ages. This color change has nothing to do with the redshift. Instead, it arises from the galaxy's stars themselves. A young galaxy shines with stars of all colors—blue, white, yellow, orange, red—so its overall color is white. But blue stars don't live long. If a galaxy gives birth to no new stars, all of its blue stars die and the galaxy turns yellow. White stars don't live long, either; when they die, the galaxy turns yellow-orange.

Most elliptical galaxies no longer give birth to new stars, so they are yellow-orange. Moreover—and this posed trouble for the steady state theory—elliptical galaxies are *uniformly* yellow-orange, suggesting that they all have the same old age. In the steady state picture, these galaxies should have formed over an eternity of time. They should have a range of ages and therefore of colors: some should be yellow, others yellow-orange, orange, orange-red, and red. In contrast, the big bang theory can explain the uniform color and old age of elliptical galaxies by saying that they arose right after the big bang, like flowers blossoming after a spring storm.

In 1952, steady state theorist Hermann Bondi hardly charmed astronomical observers when he published a paper stating that when observations contradict a well-developed theory, one should believe the *theory*, not the observations. "It really hit the observers here," said Sandage, "because Bondi said, 'Look, I'm coming to America to tell the observers at Mount Wilson and Palomar where they have misinterpreted what they have observed. The steady state theory is a well-developed theory, and if the California astronomers' observations contradict it, too bad for the observations.'" In short, said Sandage, Bondi's paper meant that observers like him didn't know what they were doing.

"I know," said Bondi, laughing. "I caused a lot of annoyance with that paper. What I did was to go through the literature and look at how often things had been proved wrong, and showed that it was just as likely with observational material as with theoretical material. I just wanted to show that observations weren't immune from being disproved."

## COUNTING ON RADIO

Another problem was emerging for the steady state theory, one involving a new form of radiation: radio waves. Astronomers had first detected celestial radio waves in the 1930s. During the 1940s and 1950s, scientists such as British astronomer Martin Ryle pioneered new ways to build radio telescopes, detecting numerous sources of radio waves. But where were these radio sources located? Were they stars inside the Milky Way, or distant galaxies far beyond?

Ryle—and most other astronomers—thought they were stars, radio stars, within the Galaxy. "Here is an example of what seems to be general practice in astronomy," Hoyle wrote: "When two alternatives are available, choose the more trivial." If the radio sources lay within the Milky Way, as Ryle believed, they would be close and thus weak. But if the radio sources lay millions or even billions of light-years distant, they had to be extremely strong in order to be detected.

Nevertheless, Thomas Gold, one of the original steady state proponents, thought that the radio sources were located beyond the Milky Way. For one thing, despite Ryle's contention that the radio sources were stars, no radio source had been identified with any star besides the Sun. In fact, one radio source coincided with the Crab Nebula, a supernova remnant in our Galaxy; another with M87, a giant elliptical galaxy in the Virgo cluster; and yet another, Centaurus A, with a disturbed galaxy in the constellation Centaurus.

Ryle angrily dismissed Gold as a mere theorist who didn't know what he was talking about. Ryle believed the radio sources were dim stars, smaller and cooler than the Sun, that spewed giant, radio-emitting flares. To support his argument, Ryle said that the positions of radio sources did not correlate with those of the three hundred nearest galaxies, implying that the radio emissions which seemed to be coming from M87 and Centaurus A arose not from those galaxies but from radio stars that happened to lie in front of them. Furthermore, Ryle noted, although he had picked up radio signals from four nearby galaxies, their emissions were so weak that they would have gone undetected if they had been much more distant.

Key evidence for Gold's side and against Ryle's came in 1952, at the same Rome conference and from the same astronomer who announced the doubling of the size and age of the universe. "In the large antechamber to the conference room," said Gold, "one was milling around like one usually does, and Walter Baade was there. He said, 'Tommy! Come over here! Look what we've got!'" New, more precise radio observations had pinpointed the position of a radio source in the constellation Cygnus, and Baade showed Gold new optical photographs of this region. They didn't show a star. Instead, they showed two galaxies smashing into each other. Furthermore, the galaxies had large redshifts, placing them hundreds of millions of light-years beyond the Milky Way, which meant they must be radio powerhouses. Thus, Gold was right, Ryle wrong.

Gold continued the story. "Then Ryle comes into the room. Baade shouts, 'Martin! Come over here! Have a look at what we've found!' Ryle comes and looks with a very stern face at the photographs, does not say a word, throws himself on a nearby couch—face down, buried in his hands—and weeps."

But Ryle would have his revenge. Ironically, once the radio sources were known to be extragalactic, Ryle could use them to attack the steady state theory. It worked like this. If the universe followed the steady state cosmology—unchanging in space and time—then any type of celestial object should be just as prevalent now as in the past. Since as astronomers look farther away, they see further back in time, the steady state universe predicts that over large distances any type of celestial object should be evenly distributed in space. In contrast, the big bang theory holds that the universe was once denser; thus, the different conditions that prevailed long ago might have given rise to objects which are no longer so common.

The radio sources offered a key test for the steady state cosmology. Was their distribution with distance—and thus with time—uniform? The crucial number turned out to be –1.5. That was the most negative slope that a uniform population of radio sources should have if their number *N* was plotted on a logarithmic graph against their strength *S*. Anything more negative than –1.5—for example, –1.8, –2.0, or –3.0—would indicate that the sources were more common in the past, contradicting the steady state tenet.

According to all three steady staters, Ryle set out to disprove their cosmology in order to get even with Gold, the astronomer who had humiliated him. "Establishments in all realms of life do best when nothing is happening," Hoyle wrote. "Great reputations are made over the supposed virtue of doing nothing, which is just as true in cosmology as elsewhere. It is a situation well summed up by a remark of Johann Sebastian Bach. Following a visit from a supposedly famous violinist, Bach wrote to a friend: 'He played well enough, but only from known music.' Throughout my troubles with Ryle, it always seemed to me that the cosmological work of his group was rather like the playing of Bach's visitor."

In 1955, Ryle reported results from the 2C survey, the second Cambridge investigation of radio sources. The slope on the log *N*-log *S* plot was –3.0, devastating for the steady state theory. However, Bondi, Gold, and Hoyle refused to believe it, saying the data were poor. In fact, the data *were* poor. Three years later, Ryle and his colleagues completed a new survey—the 3C—and found little correspondence between 2C sources and 3C sources, implying that most of the former had been spurious. Yet the new, 3C survey also had bad news for the steady state theory: the slope on the log *N*-log *S* plot was around –2.0, still disagreeing with the steady state prediction of –1.5.

The end came in 1961. In his autobiography, Hoyle recounted a phone call from the Mullard Company, which had donated money to Ryle's group. "A polite voice informed me that, during the coming week, Professor Ryle would be announcing new, hitherto undisclosed results that I might find of interest and asked if my wife and I would care to accept an invitation to be present. So it came about that, in the afternoon a few days later, I turned up with my wife at the Mullard headquarters in London. A smartly dressed Mullardman of about my own age led us into a modest-sized hall in which a number of media representatives were assembled. We were escorted by our host to the front row, where my wife was bowed into a seat. Then I was led on to a raised dais and bowed into a chair, not so comfortable as the one my wife had just been given. The smartly dressed man then withdrew, leaving me to gaze down on the media representatives. The rest of the stage decor consisted of a blackboard on an easel, a lowered screen for slides, and, I believe, a lectern.

"So what was I to think about as I sat there under the bright lights? It needed no great gift of prophecy to foretell that what I was about to hear would have something to do with the log *N*-log *S* business. But was I being uncharitable in thinking that the new results Ryle would shortly be announcing were adverse to my position? Surely, if they were adverse, I would hardly have been set up so blatantly. Surely, it must mean that Ryle was about to announce results in consonance with the steady-state theory, ending with a handsome apology for his previously misleading reports. So, I set about composing an equally handsome reply in my mind.

"A curtain parted, and Ryle entered. The Mullardman made a short introduction, and, pretty soon, Ryle had launched not into the promised statement but into a lecture. I was well used to its form, so I sat there, hardly listening, becoming more and more convinced that, incredible as it might seem, I really had been set up. The results involved the sources of what were now called the 4C survey. The 4C survey contained more sources than before, Ryle explained, greatly reducing statistical fluctuations. Yet the slope of –1.8 had been maintaining, showing that the steady-state theory was wrong, and would Professor Hoyle care to comment? The media leant forward in anticipation."

That evening, newsagents in London hawked their wares by shouting, "The Bible was right!" "For the next week," Hoyle wrote, "my children were ragged about it at school. The telephone rang incessantly. I just let it ring, but my wife, fearing something had happened to the children, always answered, fending off the callers."

For this and other work, Ryle later received the Nobel prize. Hoyle was once asked, "Why is it, Fred, that you can't get along with Martin Ryle?" Hoyle was silent; a cascade of possible replies went through his head. Finally he answered: "It must be because I have no sense of humor."

Two years later, in 1963, the steady state theory suffered a similar blow when astronomers realized that peculiar starlike radio sources—dubbed quasars a year later—lay at enormous distances. Most quasars are billions of light-years distant, so they existed billions of years ago. None does today. That quasars were once common and now aren't contradicted the steady state tenet that the universe remains the same.

## HELIUM

The second lightest element in the cosmos also proved too much for the steady state theory to handle. Originally, in the 1940s, George Gamow, Ralph Alpher, and Robert Herman had looked to the big bang as the source of all the elements from hydrogen to uranium. But they had run into trouble. In their scheme, the first isotope was hydrogen-1, the simplest and lightest, which has mass number 1—one particle in its nucleus. This built up to hydrogen-2, or deuterium, which has mass number 2; then to helium-3, which has mass number 3; and then to ultrastable helium-4, with mass number 4.

Beyond this, however, lay two chasms: one at mass number 5 and another at mass number 8. Neither mass number corresponds to a stable element. Thus, the big bang can't easily create elements heavier than helium.

During the 1950s, astronomers recognized that most of these heavy elements arose in stars: the oxygen we breathe, the calcium in our bones, the iron in our blood. Astronomers had found that old stars bore fewer heavy elements than young stars, so the old stars had forged heavy elements and cast them into the Galaxy, enriching the material that gave birth to younger stars. If instead the big bang had created the heavy elements, then every star should have received an equal share. Moreover, astronomers detected a short-lived element, technetium, on red giant stars. Technetium, atomic number 43, is radioactive and decays after just a few million years. Thus, if the big bang had created technetium, *billions* of years ago, it should have vanished, so the red giant stars themselves must have produced the element.

Even in 1946, before he developed the steady state cosmology, Hoyle had proposed that stars—especially supernova explosions—had created the elements. This idea culminated in a mammoth 1957 paper that Hoyle wrote with Margaret Burbidge, Geoffrey Burbidge, and William Fowler. Today astronomers can say which types of stars created which elements. For example, the oxygen you breathe was created inside massive stars and then released into the Galaxy when they exploded, whereas the iron in your blood was created *during* the explosions, which fried lighter elements into heavier ones.

Because Hoyle advocated a stellar origin for the elements, and because he was the steady state theory's best-known advocate, many people incorrectly believe that his cosmology motivated his work on the origin of the elements. After all, steady state cosmologists did not believe in a big bang, so they had to find another site in which to create the elements. Sloppy science writers like to state that an incorrect idea, the steady state cosmology, inspired a correct one, the stellar origin of the elements.

"That's absolute rubbish," said Hoyle. "This had nothing to do with cosmology at all. It was [J. Robert] Oppenheimer, with whom I never got on very well, who spread the canard that the reason I got onto this theory was to support the steady state cosmology. Anybody who looked at the dates in the literature could see it was rubbish: my first paper on the synthesis of the elements came in 1946, whereas the steady state cosmology didn't come till 1948."

Even though stars do create heavy elements, they fail to explain one *light* element: helium. Strangely, this is the element that most stars, including the Sun, manufacture. From the start, helium had a celestial reputation, for it had first been seen not on the Earth but on the Sun, when astronomers in 1868 detected a mysterious feature in the solar spectrum. Scientists named the element after the Greek word for the Sun, *hēlios*. Not until 1895 did they detect it on Earth, in a mineral bearing uranium, the *heaviest* naturally occurring element.

The elusiveness of helium on Earth stems in part from its extreme terrestrial scarcity: it makes up only 1/200,000 of the Earth's air. The element is rare because it is so light that it escapes the Earth's gravity and drifts into space. Furthermore, it is inert, so molecules don't lock it up, the way water captures the even lighter element hydrogen. The small supply of helium on Earth comes from the decay of radioactive elements such as uranium.

Although helium is rare on Earth, it abounds in space. It is the second most abundant element, after hydrogen. About one atom in eleven is helium. This high helium abundance posed trouble for steady state cosmology. Try though they might, stars can't manufacture this much helium. Only the big bang can. Furthermore, whereas the abundances of heavy elements like calcium and iron increase greatly from old stars to young, indicating that stars created these elements, the helium abundance varies much less, suggesting that most helium arose before the stars did—in the big bang. Ironically, this scheme for helium production was first worked out by big bang opponent Fred Hoyle and by Roger Tayler, who published their work in 1964.

By then, the steady state cosmology was already dying. It took just one more blow to kill it altogether.

## AFTERGLOW

In 1948, Alpher and Herman had predicted that because the big bang had been hot, the universe still had a temperature, about 5 Kelvin. Any object with this temperature radiates microwaves and radio waves, so this afterglow of the big bang should pervade the entire cosmos, emerging from all directions.

In the 1960s, theorists at Princeton University, unaware of Alpher and Herman's work, repeated it, reached the same conclusion, and set about trying to detect the radiation. However, two scientists at nearby Bell Labs beat them to it. Arno Penzias and Robert Wilson weren't cosmologists seeking to distinguish between two rival cosmologies. Instead, they were trying to use a radio antenna to study the heavens. But a persistent hiss interfered. Even after eliminating all obvious sources of noise, their antenna suffered from the hiss, no matter where it pointed.

In 1965, word of their trouble reached the Princeton scientists, who realized the hiss was the big bang's afterglow. SIGNALS IMPLY A 'BIG BANG' UNIVERSE, said the front-page headline in *The New York*

*Times*. Subsequent work put the universe's temperature at 2.7 Kelvin, or –455 degrees Fahrenheit—some 55 degrees Fahrenheit colder than Pluto. For this discovery, Penzias and Wilson later won the Nobel prize.

Despite the discovery's vindication of big bang cosmology, Alpher, Herman, and Gamow were anything but pleased. "That whole business was a nightmare," said Alpher, "all the junk that went on during this period. Scholarship among scientists is pretty lousy." No one—not the Princeton theorists, not Penzias and Wilson, not *The New York Times*—mentioned that Alpher, Herman, and Gamow had been the first to predict the cosmic microwave background, even though Alpher and Herman had published their prediction in such widely read journals as *Nature* and *The Physical Review* and Gamow had done so in a popular book. Instead, the Princeton theorists claimed credit for the successful prediction.

"If I lose a nickel," said Gamow, "and someone finds a nickel, I can't prove that it's my nickel. Still, I lost a nickel just where they found one." Actually, Gamow could prove that he had lost this particular nickel, and he wrote a letter of protest citing his book and the many papers his group had published. "Thus, you see," he concluded, "the world did not start with almighty Dicke." Robert Dicke was one of the offending Princeton theorists.

Even two years after the 1965 discovery, two other Princeton scientists—James Peebles and David Wilkinson—published a *Scientific American* article that glorified the Princeton work and mentioned Alpher, Herman, and Gamow only briefly, on the second-to-last page. There the article said the old work "implied" that the universe had a temperature. In fact, the work had explicitly stated this. Only in the 1970s did Alpher, Herman, and Gamow begin to receive credit for their work, but by then Gamow was dead.

Alpher admitted, "We too probably can be accused of poor scholarship." Long before Penzias and Wilson's discovery of the cosmic microwave background, and even before Alpher and Herman predicted it, a Canadian astronomer had detected it. In 1941, Andrew McKellar analyzed spectral lines that another astronomer, Walter Adams, had observed. These spectral lines arose from the simple molecule CN, which exists in the space between the stars. Like all else in the universe, the interstellar CN molecules get warmed by the cosmic microwave background. McKellar found that the CN molecules had a temperature of

2.3 Kelvin. McKellar had published this work, albeit in an obscure place, and Alpher, Herman, and Gamow had missed it.

Ironically, one astronomer *did* know McKellar's work—but it was the wrong astronomer: Fred Hoyle, arch foe of the big bang. Hoyle had mentioned McKellar's work in a 1949 paper. Hoyle knew this work because it had once saved a paper of his from being rejected.

In 1956, Hoyle even discussed the matter with Gamow, who was predicting a much higher temperature for the cosmic microwave background. Gamow gave Hoyle a ride in his new luxury car. "I recall George driving me around in the white Cadillac," said Hoyle, "explaining his conviction that the Universe must have a microwave background, and I recall my telling George that it was impossible for the Universe to have a microwave background with a temperature as high as he was claiming, because observations of the CH and CN radicals by Andrew McKellar had set an upper limit of 3 K for any such background. Whether it was the too-great comfort of the Cadillac, or because George wanted a temperature higher than 3 K, whereas I wanted a temperature of zero K, we missed the chance of spotting the discovery made nine years later by Arno Penzias and Bob Wilson."

## AFTERMATH

The cosmic microwave background was the big bang's afterglow, and its discovery left the steady staters reeling. Later in 1965, Hoyle himself conceded defeat, penning an article in *Nature* calling it and the high helium abundance clear signs of the big bang. Perhaps because of Hoyle's concession, and because the discovery was dramatic and easy to explain, popular writers often say that the cosmic microwave background disproved the steady state theory. In fact, the problems that had accumulated before then—the ages of the galaxies, the radio counts, the high helium abundance—probably played a greater role; furthermore, many observational astronomers, such as Hubble, Baade, and Sandage, had always thought the steady state theory a joke.

"It is widely believed that the existence of the microwave background killed the 'steady-state' cosmology," Hoyle wrote, "but what really killed the steady-state theory was psychology. Tommy Gold and Hermann Bondi had for long urged the predictability of their version of the theory as its outstanding virtue, and yet here, in the microwave background, was an important phenomenon which it had not predicted. Bad." For Hoyle, who had known McKellar's work, things were worse: he had failed to seize the advantage. "For many years this knocked the stuffing out of me."

Hoyle recovered, however, and soon returned to his old ways, attacking the big bang and religion—in fact, attacking the big bang *as* religion. "I have always thought it curious," he wrote in 1982, "that, while most scientists claim to eschew religion, it actually dominates their thoughts more than it does the clergy. The passionate frenzy with which the big-bang cosmology is clutched to the corporate scientific bosom evidently arises from a deep-rooted attachment to the first page of Genesis, religious fundamentalism at its strongest."

Another member of the steady state triumvirate, Thomas Gold, is equally unrepentant. "I can't really see anything wrong with the steady state theory," said Gold, who nevertheless acknowledged that most every other astronomer gave up on it decades ago. "I'm not deflected by the numbers of people who believe in one thing or another. Science does not proceed by Gallup poll."

Of the three original steady staters, only Hermann Bondi has abandoned the theory. "I had said, 'If there was a big bang, show me some fossils of it,'" said Bondi. "Helium was the first fossil. Very early on, I saw the large abundance of helium as an Achilles' heel of the steady state theory."

Today, nearly all cosmologists work within the framework of big bang cosmology. Depending on one's point of view, those who cling to the steady state theory are brave mavericks challenging an ossified scientific establishment; recalcitrant reactionaries refusing to accept new data; or just plain sore losers. Whatever the case, other cosmologists ignore their work, convinced that the secrets of the universe lie in its fiery birth: the big bang.

# FIVE

# THE BIG BANG FOR BEGINNERS

COSMOLOGY has rocketed ahead since Edgar Allan Poe explained the darkness of night. During the twentieth century, astronomers progressed from the discovery of other galaxies and of the expanding universe to the extraordinary conclusion that all the matter in those galaxies—as well as the space between them—had once been compressed into a region tinier than a pinhead, one that burst forth in the fiery explosion Fred Hoyle dubbed the big bang. To probe the universe further, astronomers and cosmologists must know more—about the nature of galaxies, the expansion of space that carries these galaxies away, and the three cosmological parameters that dictate the universe's size, shape, age, and fate.

## CELESTIAL CITIES

Few sights rival the grandeur of a giant spiral galaxy as it twirls through space. This glowing cosmic whirlpool sports spiral arms that adorn themselves with knots of blue supergiants, pulsating yellow Cepheids, magenta gas clouds, and the flash of an occasional supernova. Nestled amongst these luminaries shine hundreds of billions of lesser suns—red, orange, yellow, white, and blue—each bound to the galaxy by gravity, each contributing its light, mass, and beauty to the metropolis.

These stunning starlit cities are the building blocks of the universe.

They house most of its stars, so they emit most of its light. Thus, observers can *see* galaxies, a simple but practical fact that makes them crucial to cosmology. Galaxies mark the universe's expansion, just as rocking nighttime buoys indicate waves of dark water. Galaxies harbor much, perhaps most, of the universe's mass. Since mass exerts gravitational pull, galaxies try to slow the universe's expansion. And observations of remote galaxies—5 to 10 billion light-years from the Milky Way—can reveal whether the universe's expansion is speeding up or slowing down.

That such convenient probes exist is no coincidence, because astronomers would not have arisen without them. The universe could have spread its material uniformly rather than clump it into galaxies, but then life would not have developed to contemplate the consequences. Giant galaxies like the Milky Way and Andromeda are vital for life. Stars within galaxies forge the heavy elements life needs—the oxygen we breathe, the calcium in our bones, the iron in our blood—but in order to do any good, *galaxies* must retain these heavy elements and recycle them into new generations of stars and planets. So crucial is this process that even galaxies much smaller than the Milky Way will not do. Small galaxies lack the star power to create large quantities of heavy elements. They also fail to summon the gravitational strength to hold on to such elements when exploding stars blast them into space. As a consequence, dwarf galaxies possess a pittance of the Milky Way's heavy element treasure. If giant galaxies did not exist, neither would we.

Galaxies come in different types. In 1926, three years after proving the galactic nature of Andromeda, Edwin Hubble published his scheme for classifying galaxies. The bright galaxies he saw fell into three main types: ellipticals, spirals, and irregulars.

Hubble's first galaxy type, the ellipticals, look like smooth eggs or balls, their light sharply concentrated at the center. Some elliptical galaxies are giants, like M87 in the Virgo cluster, while others are lesser entities, like M32, an attendant of Andromeda. Even in nearby M32, Hubble failed to find stars, sparking speculation that elliptical galaxies consisted of gas, the chief substance out of which stars form. Actually, as astronomers know today, just the opposite is true: elliptical galaxies have stars but almost no gas, so they don't give birth to any new suns. As a result, the typical elliptical hosts only old stars. Hubble failed to see them because even the most luminous lack the luster of bright young stars.

The old stars color their elliptical homes yellow-orange. All of the short-lived blue and white supergiant stars in elliptical galaxies have died, so yellow, orange, and red giant stars, which evolve from longer-lived stars like the Sun, dominate an elliptical galaxy's light and color.

When you think of galaxies, however, you probably don't think of ellipticals. You probably envision Hubble's second galaxy type, the majestic spirals. Our Galaxy is a spiral, as are the Andromeda Galaxy and the aptly named Whirlpool Galaxy, M51. Whereas elliptical galaxies range from large to small, most spirals are large, because only a large galaxy can whip up the spiral pattern.

Rotation, in fact, distinguishes spiral galaxies from giant elliptical galaxies. Spirals spin fast. In the next hour, the Sun and Earth will speed through half a million miles of space as they race around the Milky Way's center, but the Galaxy is so huge that even at that speed, completing a single orbit takes 230 million years. In contrast, a giant elliptical galaxy does not rotate. Instead, its stars dive into and out of the galaxy's center.

Spirals differ in another way. Whereas an elliptical galaxy consists of old stars, the gas and dust that line a spiral galaxy's arms give birth to new stars, so its members span the eons from young to old. Spiral galaxies therefore have a slower metabolism than elliptical galaxies. Elliptical galaxies converted all their gas and dust into stars early in life, whereas spirals did so gradually. If intelligent life is common throughout the universe, the oldest and wisest races probably inhabit elliptical galaxies, since most of the oldest and wisest stars do. For this reason, any intergalactic travelers may choose to avoid spirals because of their barbaric inhabitants, just as world travelers tend to avoid uncivilized reaches of the globe.

Because of the spirals' youthful stellar population, their arms sparkle with blue and white supergiants like Rigel and Deneb, making the galaxies bluer than ellipticals. Spiral galaxies also have Cepheids, which are young stars, so spirals—but not ellipticals—boast these bright distance markers. Elliptical galaxies have RR Lyrae stars, the distance indicators that shine more feebly than Cepheids and so can be seen only in nearby galaxies.

Ellipticals and spirals predominated among the bright galaxies that Hubble studied. A few galaxies, though, were too amorphous to fit either category, so Hubble called them irregular. Most irregular galaxies are smaller than most spirals. The nearest examples, the Large and Small Magellanic Clouds, orbit the Milky Way. The Large Magellanic

Cloud is a tenth as luminous as the Galaxy, and most other irregular galaxies are smaller and dimmer still. Like spiral galaxies, irregular galaxies typically have gas and dust that create new stars, so they have Cepheids. Indeed, by studying the Cepheids in the Small Magellanic Cloud, Henrietta Leavitt discovered the relationship between a Cepheid's pulsation period and its luminosity. Irregular galaxies might have achieved spiral status if they had been born bigger, with enough spin to generate a spiral.

How common are the three galactic types? If astronomers examined only the brightest and easiest-to-see galaxies, such as those Hubble studied, the spirals would win. They make up more than two thirds of all prominent galaxies, with ellipticals accounting for most of the rest. Consider, for example, the great catalogue of bright objects that French comet hunter Charles Messier—the "M" in M51 and M87—compiled in 1781. Of his 103 objects, 33 are now known to be galaxies, of which 23 are spiral, 9 are elliptical, and only 1 is irregular. Most of these galaxies shine about as brightly as the Milky Way. This analysis suggests that most galaxies are spirals, and that the Milky Way is an average galaxy.

Both conclusions are wrong. Sampling galaxies by choosing those which exceed some threshold of brightness is akin to sampling the human race by including only people taller than seven feet. You'd miss most of the population. Far better to sample everyone, tall or short, who happens to live on your street. Galactically speaking, that means choosing all galaxies, bright or faint, within a certain distance.

The nearest galaxies, including the Milky Way, belong to a small gathering called the Local Group. When Hubble wrote his 1936 book *The Realm of the Nebulae*, he listed only 10 Local Group members, but today astronomers know 36. Most lie within 3 million light-years of the Milky Way, all within 4 million. Despite their proximity, only 3 of the 36 Local Group galaxies make Messier's list—the Andromeda Galaxy (M31) and two of its satellites, M32 and M33. Thus, the typical galaxy in the Local Group, and presumably the universe, doesn't match the luminosity of the galaxies Messier saw. Most Messier galaxies therefore outshine most Local Group galaxies, just as most people on today's front page are more famous than most people who live on your street.

Furthermore, whereas the Messier galaxies made the Milky Way seem ordinary, the Local Group galaxies reveal that it is actually extraordinary. Of the 36 galaxies in the Local Group, only one—Andromeda—

outshines the Milky Way. All the other Local Group galaxies—97 percent—are dimmer, and most are much dimmer. Even this percentage understates the case, because in the Local Group dwell other galaxies too dim to be seen. The Milky Way therefore outpowers most other galaxies.

Prominent galaxies like Messier's can lead astronomers astray in another way. Over two thirds of the Messier galaxies are spiral, while only the three largest Local Group members are—Andromeda, the Milky Way, and M33. That is just 8 percent of the total. And only four Local Group galaxies are elliptical, the other main Messier type. Instead, most Local Group galaxies are irregular or dwarf spheroidal, the latter a type that did not make Messier's list and that Hubble did not know about when he first classified galaxies. Dwarf spheroidal galaxies are dim and diffuse ellipticals. The faintest harbor just a few million stars—1/100,000 of the Milky Way's total—and can't even muster the luster of the Galaxy's single brightest star. Yet most galaxies in the Local Group, and presumably the universe, are dwarf spheroidal.

Galaxies do not go through the dark alone, because most belong to small gatherings called groups. The typical group harbors a few bright galaxies and a few dozen dim ones. As in the Local Group, the brightest galaxies are usually spiral, and the members' gravity holds the group together. Nearby groups include the Sculptor, Maffei, and M81 groups.

Galactic gatherings far grander than groups are called galaxy clusters. Clusters are crammed with hundreds or thousands of galaxies and bear many of the giant elliptical galaxies of the universe. The nearest galaxy cluster, the Virgo cluster, hosts the best-known giant elliptical, M87. Indeed, the Virgo cluster's presence explains why over half of the Messier galaxies congregate in the constellation Virgo and its neighbor, Coma Berenices.

Groups team up with other groups and clusters to form superclusters. The Local Group resides on the outskirts of a supercluster that is centered on the Virgo cluster. The Local Supercluster stretches across some 100 million light-years of space.

By good luck, when astronomers peer into the Local Supercluster's heart, they look nearly perpendicular to the Milky Way's dust-clogged disk. The gas and dust in the Galactic disk cloak one fifth of the extragalactic sky, so hordes of uncharted galaxies, some nearby, must lurk behind the murk. Astronomers' view of the universe therefore resembles that of a bumbling geographer who failed to spot

North America, whose land area contributes about a fifth to the world's total.

## THE OBSERVABLE UNIVERSE VERSUS THE ACTUAL UNIVERSE

*On the hill we viewed the silence of the valley*
*Called to witness cycles only of the past*

—YES ("CLOSE TO THE EDGE")

Astronomers never see the universe as it is, only as it was. To observe planets, stars, and galaxies, astronomers must detect the light these objects emit, reflect, or deflect. This light may be visible or invisible, taking the form of radio waves, infrared radiation, or x-rays, but it does not travel infinitely fast. Thus, astronomers can see only what has already happened, not what is presently happening, just as a newspaper photograph shows only the past, never the present.

For nearby celestial objects, the difference between the observable universe and the actual universe is so minuscule it hardly matters. The nearest celestial body, the Moon, is just 1 light-second away, so observers see it as it looked 1 second ago, during which time it has hardly changed. The Sun and its planets look as they were minutes or hours ago, and most of the brightest stars as they were dozens, hundreds, or thousands of years ago. For example, the light now reaching Earth from Orion's brightest star, Rigel, set out long before Christopher Columbus did. Nevertheless, this time is small compared with Rigel's total lifetime of many million years, so chances are Rigel is still there. Outside the Milky Way, the time delay grows, but even for nearby galaxies, it is no great concern. For example, during the 2.4 million years that the Andromeda Galaxy's light has taken to reach Earth, thousands of its stars have exploded, but the galaxy harbors hundreds of billions more, so the loss barely diminishes Andromeda's appearance.

However, when venturing farther afield, gazing at galaxies *billions* of light-years distant, astronomers may be observing galaxies that have vanished. For example, all quasars are remote, so all existed long ago. None is nearby, suggesting that none exists now, anywhere in the actual universe. But astronomers still see them, in the observable universe, the finite speed of light having preserved their memories.

Light therefore wields a double sword. If light were infinitely fast, astronomers could see everything, but not "everywhen." They could see regions of the universe where the laws of physics might differ from those near Earth, but they would struggle to reconstruct the universe's past, since they would never know of quasars and other extinct celestial species. Unfortunately, nature failed to arrange for galaxies to emit two types of radiation: one traveling at finite speed, to show the past; the other traveling at infinite speed, to show the present and total reality.

Astronomers see each galaxy at the place and time appropriate to it. They see a galaxy 1 million light-years distant as it was 1 million years ago, no sooner and no later; a galaxy 2 million light-years distant as it was 2 million years ago—again, no sooner and no later—and so on. Each galaxy exists at earlier and later times, too, but terrestrial observers cannot see them then. Analogously, a student sees his teachers as they were when he was in their classes, no sooner and no later. Thus, the student saw his first-grade teacher when he was in first grade, his second-grade teacher when he was in second grade, and so on. The teachers looked different before he saw them and look different today, yet he remembers them as he experienced them in each particular grade. And his younger sister, who had the same teachers, would see each at a different time than he did, just as an astronomer in another galaxy would see the same galaxies at a different time than terrestrial astronomers do.

By the way, when astronomers say that an event happened in a particular year—for example, that a star exploded in the Andromeda Galaxy in 1885—the year refers to when astronomers on Earth saw it. Nonastronomers sometimes complain that this date is fictitious, since the star in Andromeda actually exploded more than 2 million years earlier. Unfortunately, citing the actual year would require knowing Andromeda's exact distance. As a result, if the estimated distance changed, so would the stated year of the supernova. In fact, when astronomers saw the 1885 supernova, they thought Andromeda was not more than a few thousand light-years from Earth, which would have caused them to assign the supernova an erroneous year.

Light's finite speed restricts observers not just temporally but also spatially. As Edgar Allan Poe recognized when he explained Olbers' paradox of the dark night sky, astronomers can see only as far as the universe is old. If the universe is 15 billion years old, then the farthest galaxies they can see have a distance of 15 billion light-years. More distant galaxies may exist, but their light hasn't had time to reach the

Earth. When journalists talk about the "edge" of the universe, this is what they mean—the edge of the *observable* universe, imposed by the twin finitudes of the universe's age and the speed of light. Thus, the edge of the universe is an illusion, like the end of a rainbow. A rainbow may look as though it ends over there, by the red barn; but if you were in the red barn, it would look as though the rainbow's end were elsewhere. Cosmologists favor a less flamboyant term for the edge of the observable universe: the horizon. Just as a sailor on Earth can't see beyond the terrestrial horizon, so even the most powerful telescope can't peer beyond the cosmic horizon.

In a sense, ancient people who thought the Earth was the center of the universe were right. Earth is indeed the center of the *observable* universe, surrounded by spherical shells of space that carry information from different epochs, with the edge marked by the universe's creation, some 15 billion light-years away. That expanse sounds more impressive than it is, however. If you wanted to represent the Local Group of galaxies with the period at the end of this sentence—about 1/100 inch across—then to map the observable universe, you'd need a sheet of paper only 30 inches across: the entire observable universe would fit on a card table.

The leap to largeness occurs *within* the Milky Way. Because light is so swift, speeding through space at 670 million miles per hour, a single light-year is immense. If you shrank the Milky Way so that the Sun were just 1/100 inch across, then to map the entire Galaxy, including its dark halo, you'd need a sheet of paper that stretched far beyond the Moon. So in a sense, the Galaxy is big, but the observable universe—all space within the horizon—is not.

The horizon recedes every year, because each year astronomers can see one light-year farther out. Thus, the observable universe gets bigger and bigger. This increase has nothing to do with the universe's expansion. Even if the universe were static, astronomers would see more each year, as light from previously unseen objects finally made it to Earth.

Beyond the horizon, the universe may differ from that within. For example, all stars and galaxies in the observable universe seem to be made of matter, not antimatter. If they weren't, huge explosions would erupt where the two types touched. Beyond the horizon, however, stars and galaxies could consist of antimatter. In these distant reaches, the physical constants—the speed of light $c$, the charge of an electron $e$, the gravitational constant $G$, Planck's constant $h$, Boltzmann's constant $k$—might even differ from terrestrial values. Perhaps huge stretches of space harbor no galaxies at all, or realms where the laws of physics break down. From their limited vantage point, terrestrial astronomers would never know, just as, centuries earlier, Hawaiians did not know of blizzards, nor Alaskans of palm trees.

## EXPANDING SPACE

As Edwin Hubble discovered, the universe is expanding—the actual universe, not just the observable one. Space opens up between galaxies, hurling them away from one another. And the farther apart two galaxies are, the more space exists between them to open up, and the faster they race apart. Thus, as Hubble found, the farther a galaxy from the Milky Way, the greater is its redshift.

Contrary to popular belief, this redshift does *not* arise from the Doppler shift. A Doppler shift results when a galaxy moves *through* space. But these redshifts are cosmological: they arise from the expansion of space. As a light wave travels through the fabric of space, space expands and stretches the light wave, increasing its wavelength and producing a redshift, since red light has a longer wavelength than blue. The farther a galaxy is from the Milky Way, the more space the galaxy's light has traveled through en route to Earth, so the more stretched out, or redshifted, the light has become. Thus, the farther a galaxy from the Milky Way, the greater is its redshift. The expansion of space therefore causes both the galaxy's redshift and its movement away from Earth. Yet no galaxy is rushing through space. In the same way, if the Earth began to expand, New York and San Francisco would move apart, yet each would retain its present longitude and latitude.

Because nearly all galaxies move away from ours, and because the farther they are, the faster they move away, our Galaxy might seem the center of the universe—the site of an ancient explosion that cast these galaxies away, with the fastest having traveled the farthest. But observers in other galaxies would see the same thing: most other galaxies moving away from theirs, with those farther moving away faster. To understand why, compress the space of the universe into a two-dimensional sheet and wrap it around a round balloon. Each galaxy is a coin glued to the balloon's skin. Light from other galaxies travels only on the skin of the balloon, curving as necessary to stay on the balloon's skin. By rules of the analogy, astronomers can detect such light, but they cannot observe the vast space outside the balloon nor the hollow within.

Now blow up the balloon. Point to one coin and call it the Milky Way. The farther a galaxy is from the Milky Way, the faster it moves away from it. It matters not which coin you point to—Hubble's distance-redshift relation holds for them all.

Furthermore, the big bang was *not* the explosion of matter into preexisting space. Instead, it was the creation of both the matter *and* the space, which has expanded ever since and today carries most galaxies away from most others. In like fashion, as the balloon expanded, its skin stretched, carrying the coins away from one another.

Nonastronomers often ask where the center of the universe is. Astronomers see no center within the space they observe, just as no point on the balloon's skin is the balloon's center. True, the balloon does have a center, inside the hollow, but by rules of the analogy, astronomers cannot observe anything but the balloon's skin. Therefore, the actual universe may or may not have a center; if it does, it lies outside the space that astronomers can observe.

In the same way, the actual universe has no known edge, just as the balloon's skin has none. A spacecraft could travel forever through the universe and never meet an edge, just as an ant could crawl forever around the balloon's skin and never encounter a barrier.

Although the universe is expanding, no one knows what if anything it expands into. In order to know, astronomers would have to observe outside the space of our universe, just as observers on the balloon would have to observe outside the skin of the balloon, which the rules of the analogy forbid. So astronomers are stuck. Perhaps the universe is expanding into nothing, or into God's living room, or into some mad physicist's laboratory. Take your pick.

For the same reason, no one can say whether other universes, perhaps arising from other big bangs, exist. Just as other balloons could exist, so could other universes, but unless they intersected ours, they would presumably remain unknown to astronomers. As a result, any statement about other universes is speculative. We see only the skin of our own "balloon"—and the horizon prevents us from seeing most of that.

Admission of ignorance should also greet the popular question, "What happened before the big bang?" For what it's worth, the standard answer is that you aren't supposed to ask: there was no time before the big bang, just as there is no point north of the north pole, because time was created in the big bang. However, this answer is suspect, because known physics breaks down at the high temperatures and densities that prevailed just after the big bang. If known physics can't state what was going on $10^{-90}$ second *after* the big bang, why believe what someone says about what happened, or did not happen, *before*?

Because the universe expands, some people mistakenly think that *every*thing expands—that the Sun's planets are moving away, that the other stars are moving away, that all other galaxies are moving away. Gravity, however, can overwhelm the expansion of space. The attractive gravitational pull of the material within galaxies holds these star cities together, so galaxies do not expand, nor do their stars, planets, and solar systems. In the same way, the coins on the balloon's skin did not expand when the balloon did.

Even the nearest galaxies do not, as a whole, move away from the Milky Way. That's because they belong to the Local Group, whose members gravitationally anchor one another. Thus, an astronomer observing only Local Group galaxies would never know that the universe expands.

The universe's expansion appears just beyond the Local Group's frontiers, at a distance of 4 million light-years from the Local Group's center of mass. The center of mass lies between Andromeda and the Milky Way, because these two galaxies contain nearly all of the Local Group's mass. Andromeda has about 60 percent more mass than the Milky Way, so the center of mass lies closer to the former—900,000 light-years from Andromeda, 1.5 million light-years from the Milky Way.

The Local Group moves away from other nearby groups as well as from the Virgo cluster, even though they all belong to the Local Supercluster. They feel one another's gravitational pull, which slows, but does not prevent, the expansion of space between them.

Because most galaxies belong to groups and clusters, and because galaxies move around within these groups and clusters, galaxies can collide, even though the universe at large expands. Right now, the Andromeda Galaxy is moving toward the Milky Way, because the two giants feel each other's gravitational tug. Billions of years in the future, Andromeda may smash into the Milky Way, merging the two into an even greater galaxy.

## LIGHT'S WEARY WAY?

The cosmological redshift arises because light waves get stretched by the expansion of space. However, every cosmologist who gives public talks has undoubtedly been asked about a competing explanation. Why can't light lose energy as it races through space, growing more and more tired, like a runner completing a marathon? Red light is weaker than blue, so this energy loss might cause a redshift. The farther a galaxy from the Milky Way, the longer the light has traveled, so the weaker and redder it has gotten. Thus, according to this tired-light theory, the farther the galaxy, the greater should be its redshift, just as Hubble found.

The tired-light theory is not new. It was first proposed by maverick scientist Fritz Zwicky in 1929, a few months after Hubble discovered the distance-redshift relation. But two observations rule it out. First, astronomers see that exploding stars in distant galaxies brighten and fade more slowly than those nearby. This time dilation arises from the expansion of space. To see how, imagine that a star in a far-off galaxy emits one pulse of light toward Earth on January 1 and a second pulse on February 1. Initially, the two pulses are separated by a distance of one light-month. As they travel toward Earth, though, the space between them expands, perhaps doubling; so astronomers receive them two months apart. In the tired-light theory, this should not happen—the pulses of light weaken but do not separate. In fact, astronomers do observe that distant supernovae wax and wane more slowly than nearby ones, agreeing with the idea that space expands and contradicting the tired-light theory.

Second, the tired-light theory disagrees with the observed spectrum of the cosmic microwave background, the big bang's afterglow. This has a specific shape which physicists call a blackbody: it is most intense at one particular wavelength, falls off slowly at longer wavelengths, but rapidly at shorter wavelengths. The universe's expansion degrades the cosmic microwave background's spectrum, stretching it to longer wavelengths, but in a way that preserves the blackbody shape. In contrast, the tired-light theory predicts that as the light composing the cosmic microwave background loses energy, the spectrum ceases to remain a blackbody, contrary to observations.

## THE HERE AND NOW VERSUS THE THERE AND THEN

The expansion of space complicates the meaning of galactic distances, causing them to grow. When astronomers state that a galaxy is, for example, 8 billion light-years from Earth, is that how far the galaxy was when it emitted its light toward the Earth, or is that how far the galaxy is now?

Actually, neither. That distance gives the lookback time, the length of time the light took to travel from the galaxy to Earth. The lookback distance conveniently reveals that astronomers see the galaxy as it looked 8 billion years ago.

Because the universe is expanding, however, the galaxy must have been closer than the lookback distance when it emitted its light toward the Earth. Suppose this past distance was only 6 billion light-years. As the light left the galaxy, the 6 billion light-years of space grew, so the light resembled a runner racing on a track that kept stretching, with the finish line farther than when he set out. That's why the light required 8 billion years, not 6 billion, to reach the Milky Way. And today, the galaxy is farther than the lookback distance—let's say 12 billion light-years. Cosmologists often call the present distance the proper distance.

The galaxy's present distance, 12 billion light-years, is twice its past distance, 6 billion light-years. Thus, since the galaxy emitted that light, the universe's size has doubled, doubling the light's wavelength. For example, the galaxy's light had a spectral feature due to calcium at a wavelength near 4,000 angstroms, which is purple. The redshift doubled this wavelength, so astronomers on Earth would see the calcium line near 8,000 angstroms, beyond the red region of the spectrum, in the infrared.

A doubling corresponds to a 100 percent increase, so astronomers say the galaxy has a redshift of 1. In a galaxy at a redshift of 2, the wavelength has tripled (a 200 percent increase); in a galaxy at redshift 3, the wavelength has quadrupled (a 300 percent increase); and so on. The Virgo cluster, heart of the Local Supercluster, has a redshift of only 0.004. The far grander Coma cluster, five to six times farther, has a redshift of 0.02, so its spectral features are shifted by 2 percent.

Now for something bizarre. As you would expect, the greater a galaxy's redshift, the greater are its lookback distance and its present distance. But this is not necessarily true of the past distance. When a high-redshift galaxy emitted its light toward Earth, the galaxy may have been quite close to the Milky Way. For example, a galaxy at redshift 10 may have emitted its light when it was only 3 billion light-years from the Milky Way, whereas a redshift 1 galaxy may have done so when it was 6 billion light-years from the Milky Way. Still, the redshift 10 galaxy's light took longer to get here, which is why its light got so redshifted.

DISTANCE AND REDSHIFT*

| REDSHIFT | PAST DISTANCE | **LOOKBACK DISTANCE** | PRESENT DISTANCE |
|---|---|---|---|
| | | (BILLIONS OF LIGHT-YEARS) | |
| 0.0 | 0.0 | **0.0** | 0.0 |
| 0.1 | 1.3 | **1.4** | 1.5 |
| 0.2 | 2.4 | **2.6** | 2.9 |
| 0.3 | 3.2 | **3.7** | 4.2 |
| 0.4 | 3.9 | **4.6** | 5.4 |
| 0.5 | 4.4 | **5.4** | 6.6 |
| 0.6 | 4.8 | **6.1** | 7.8 |
| 0.7 | 5.2 | **6.8** | 8.8 |
| 0.8 | 5.4 | **7.4** | 9.8 |
| 0.9 | 5.6 | **7.9** | 10.7 |
| 1.0 | 5.8 | **8.3** | 11.6 |
| 1.1 | 5.9 | **8.7** | 12.4 |
| 1.2 | 6.0 | **9.1** | 13.2 |
| 1.3 | 6.1 | **9.4** | 14.0 |
| 1.4 | 6.1 | **9.7** | 14.7 |
| 1.5 | 6.1 | **10.0** | 15.3 |
| 1.6 | 6.1 | **10.2** | 16.0 |
| 2.0 | 6.1 | **11.0** | 18.2 |
| 3.0 | 5.6 | **12.2** | 22.3 |
| 4.0 | 5.0 | **12.9** | 25.2 |
| 5.0 | 4.6 | **13.3** | 27.3 |
| 6.0 | 4.1 | **13.5** | 29.0 |
| 7.0 | 3.8 | **13.7** | 30.3 |
| 8.0 | 3.5 | **13.8** | 31.4 |
| 9.0 | 3.2 | **13.9** | 32.3 |
| 10.0 | 3.0 | **14.0** | 33.1 |
| 15.0 | 2.2 | **14.2** | 36.0 |
| 20.0 | 1.8 | **14.3** | 37.7 |
| 40.0 | 1.0 | **14.4** | 41.1 |
| 100.0 | 0.4 | **14.5** | 44.2 |
| 1000.0 | 0.05 | **14.5** | 47.7 |

***If the Hubble constant is 65, omega is 0.3, and lambda is 0.7, yielding an age for the universe of 14.5 billion years.***

The exact distances—past, present, and lookback—that correspond to a particular redshift depend on how fast the universe expands and on how much the expansion has slowed down or sped up. The table gives these distances for one choice of cosmological parameters. The central distance column—the lookback distance—is in boldface, to emphasize that this is the distance astronomers normally mean.

Some of the numbers in the table's last column, giving the present distances, may make you queasy. Light travels at the speed of one light-year per year, and Einstein's special theory of relativity suggests that nothing can travel faster. If all matter was together during the universe's birth, 10 to 15 billion years ago, how can some things have already gotten 20, 30, even 40 billion light-years distant?

The answer lies in the expansion of space. Special relativity prohibits faster-than-light travel *through* space. But high-redshift galaxies do not race through space. Instead, space swells up between Earth and these distant galaxies, so much so that a redshift 100 object is now over 40 billion light-years distant—but it has not traveled through 40 billion light-years of space. And the most far-off objects are so remote their light has yet to reach the Earth, so astronomers can't even see them.

## THREE NUMBERS, ONE UNIVERSE

Born in a big bang some 15 billion years ago, space has been expanding ever since, flinging most galaxies away from most others. Their gravity tries to stop their flight, while empty space may oppose their effort by accelerating the very expansion that catapults them away.

In order to quantify the expanding universe, cosmologists employ three parameters: the Hubble constant, which says how fast the universe presently expands; omega, which denotes the density of matter in the universe and exerts a gravitational force that tries to stop the expansion; and lambda, the cosmological constant, which exerts a repulsive force and tries to speed up the expansion. Together, the numerical values of these three parameters dictate the size, shape, age, and fate of the cosmos.

THREE COSMOLOGICAL PARAMETERS

| Parameter | Symbol | Likely Range | Units | What It Represents | What It Does | Affects Universe's Age? | Affects Universe's Fate? |
|---|---|---|---|---|---|---|---|
| Hubble constant | H or $H_0$ | 40 to 90 | km/sec/Mpc | Universe's expansion rate | — | Yes | No |
| Omega | $\Omega$ or $\Omega_M$ | 0.01 to 1 | dimensionless | Universe's mass density | Slows universe's expansion | Yes | Yes |
| Lambda | $\lambda$ or $\Omega_\Lambda$ | 0 to 1 | dimensionless | Repulsive energy of empty space | Accelerates the expansion | Yes | Yes |

## THE HUBBLE CONSTANT

Because Edwin Hubble discovered the universe's expansion, his name is attached to its rate of expansion. The Hubble constant is the most basic cosmological parameter, one so crucial that astronomers and cosmologists have fought over it for decades. If astronomers could determine the Hubble constant's numerical value, it would state how much galaxy redshifts (in kilometers per second) increase for every megaparsec farther an astronomer looks (one megaparsec is 3.26 million light-years). For example, if the Hubble constant is 50 kilometers per second per megaparsec, then on average a galaxy one megaparsec farther than another will recede 50 kilometers per second faster. But if the Hubble constant is 100, then the farther galaxy will be speeding away 100 kilometers per second faster than the nearer one.

The Hubble constant anchors the distance scale for the universe. It translates galaxy redshifts, which modern astronomers find easy to measure, into hard-fought galaxy distances. For example, suppose astronomers wanted to know how far it is to the Coma cluster of galaxies. They already know that the expansion of space causes the Coma cluster to speed away from the Galaxy at 7,000 kilometers per second. If the Hubble constant is 50 kilometers per second per megaparsec, then that redshift means Coma is 7,000/50 = 140 megaparsecs, or 460 million light-years, distant. But if the Hubble constant is twice as high—100—then Coma is only half as far, 230 million light-years.

Distance, in turn, affects other key galactic quantities. Suppose astronomers wanted to compare a Coma galaxy with the Milky Way, to see which galaxy emits more light. Using various instruments, they can measure how much light they receive from the Coma galaxy. But to translate that into the amount of light the galaxy actually emits, astronomers must know its distance. Looks can deceive: a weak neighbor galaxy can dazzle observers because of its proximity, just as a nearby candle can appear to outshine a distant streetlight. In fact, the power, or luminosity, that astronomers deduce for the galaxy depends on the *square* of the distance they estimate. So an astronomer using a Hubble constant of 50 would conclude that the Coma galaxy is twice as far, twice as large, and four times more luminous than would a rival astronomer using a Hubble constant of 100.

The Hubble constant interests cosmologists in large part because it indicates approximately how old the universe is. The higher the Hubble constant, the faster the universe expands, so the younger it must be, because a fast-expanding universe has taken less time to attain its present size. For example, if the Hubble constant is 100, then the universe is only half as old as if the Hubble constant is 50.

The Hubble constant yields a useful number called the Hubble time—the age the universe would have if it had expanded at a steady rate since its birth. A Hubble constant of 50 gives a Hubble time of around 20 billion years; a Hubble constant of 65, a Hubble time of around 15 billion years; and a Hubble constant of 100, a Hubble time of around 10 billion years. These are not, however, exact ages for the universe, because its expansion has not been steady since its birth. Nevertheless, they are close enough to cause trouble. After all, when Edwin Hubble found the Hubble constant to be 530, it meant the Hubble time was a mere 1.8 billion years—and partially inspired the big bang's biggest competitor, the steady state cosmology.

## OMEGA

The universe has matter—a good thing, for a universe without matter might also be without life. Matter exerts gravitational force, so the combined tug of the universe's many galaxies might slow its expansion. The denser the universe, the more the expansion brakes.

The universe's density also affects its size, shape, and destiny. If the universe is sufficiently dense, the galaxies will so slow the expansion that it will halt and then reverse. The universe will start to collapse, ending its days in a fiery big crunch. If instead the universe is sufficiently tenuous, the galaxies will slow but never stop the expansion, and the universe will live forever.

To distinguish between these two outcomes, cosmologists designate the universe's density with the final letter of the Greek alphabet, omega ($\Omega$). If *omega* has an ominous ring, it should: it dictates the universe's fate, be it fire or ice, mortal or immortal.

If omega exceeds 1, the universe is so dense that it will eventually collapse. If omega is less than 1, the universe is so tenuous that it will expand forever. A universe with omega precisely equal to 1 is right at the critical density: add a single atom and it collapses. An omega of 0.2, for example, means the universe has only 20 percent of the density needed to cause collapse, whereas an omega of 2 means the universe has twice the needed amount.

It does not take much to collapse the universe. The critical density is only around $10^{-29}$ grams per cubic centimeter—100 billion million billion times less dense than terrestrial air, and nearly a trillion times less dense than even the Moon's extremely tenuous atmosphere. The critical density's exact value depends on the square of the Hubble constant. The higher the Hubble constant, the faster the universe expands, so the denser the universe must be to halt its expansion.

Every day, stars lower the density by converting matter into energy. Could they reduce the universe's density below the critical value and thereby alter the universe's fate? Or could scientists who found that they lived in a universe destined to collapse save the cosmos by annihilating matter themselves? No—the energy that emerges from the destruction of matter slows the universe's expansion just as much as the matter that perished. So if the universe is going to collapse, there's nothing you can do about it. Conversely, no terrorist could ever convert energy into matter and make an expanding universe collapse.

If no force but gravity affects the universe's expansion, cosmologists can use omega alone to distinguish three possible universes: if omega is greater than 1, the closed universe; if omega equals 1, the flat universe; and if omega is less than 1, the open universe.

OMEGA AND THE UNIVERSE
(WITH LAMBDA = 0)

| UNIVERSE | DENSITY | OMEGA | MASS | SPACE | FATE | LIFETIME |
|---|---|---|---|---|---|---|
| Closed | High | More than 1 | Finite | Finite | Collapses | Finite |
| Flat | Moderate | 1 | Infinite | Infinite | Expands forever | Infinite |
| Open | Low | 0 to almost 1 | Infinite | Infinite | Expands forever | Infinite |

## THE CLOSED UNIVERSE: FROM CREATION TO CREMATION

Robert Frost predicted that the world would end in fire, not ice, so he must have thought that omega exceeded 1. If so, the universe will someday stop expanding. Then it will start to shrink—slowly at first, then faster and faster. Nearby galaxy groups will begin to exhibit blueshifts. Distant galaxies, though, will still exhibit redshifts, for their light set out during the era of expansion and en route to Earth passed through more expanding than contracting space. As time goes on, the sphere of blueshifted galaxies will grow, and galaxies will smash into one another. The cosmic microwave background, cold today, will heat up so much that the entire sky will begin to glow—first red, then orange, yellow, white, and blue. Less than a year before the apocalypse, the sky's extreme heat will trigger runaway nuclear reactions in stars, causing them to explode; then, mere seconds before the end, the heat will decimate the very nuclei the exploding stars spewed into space. Ultimately, everything—all light, matter, heat, and space—will crash together in a tumultuous big crunch.

A universe this dense has a distinct shape. According to both Newton and Einstein, mass causes gravity. Newton said that every mass exerts an attractive force on every other mass. Einstein, however, said that mass curves space, the way a bowling ball causes a mattress it rests on to sag. As a result, other masses—and light itself—follow curved paths through space, as would a marble shot near the bowling ball. For example, whereas Newton said that the Earth orbits the Sun because the Sun attracts the Earth, Einstein said that the Sun curves space and that the Earth revolves around it like water spiraling around a drain.

According to Einstein, if omega exceeds 1, the cosmic matter density curves space so much that it closes on itself. As the name implies, a closed universe has a finite amount of space and matter. Yet it has no edge. In the same way, the Earth's surface has a finite amount of area, but no edge to fall off of. A light beam could shoot through a closed universe, gravity gradually bending its path, until it returned to its starting point—just as sailors could sail around the world, the Earth's curvature gradually bending their path, until they returned to their home port.

In a closed universe, parallel lines eventually converge, just as terrestrial lines of longitude converge at the North and South Poles. For the same reason, the three angles of a triangle add up to more than 180 degrees. On Earth, you could trace out such a triangle by starting at the North Pole and proceeding straight down to the equator; making a 90-degree turn and following the equator for a thousand miles; then making another 90-degree turn, back north, until reaching the North Pole, where you would make an angle of about 15 degrees with your starting path. Your total triangular trek would add up to more than 180 degrees, allowing you to deduce that the Earth is round.

The specifics of a closed universe's future depend on both the Hubble constant and omega. For example, if the Hubble constant is 55 and omega is 2, then the universe is 10 billion years old and will reach its maximum size 46 billion years from now. The big crunch will follow 56 billion years later, so the universe's total lifetime, from big bang to big crunch, is 112 billion years.

After collapsing, a universe might bounce back, with a new big bang giving birth to a new universe, possibly with different physical laws and constants. Our own universe could be such a phoenix universe, born from the ashes of a previous one. And that previous universe may itself have arisen the same way, and so on—the string of past universes could extend infinitely far back in time. However, some cosmologists have argued against this scenario, saying that radiation from previous universes would overpower the cosmic microwave background, and that the amount of entropy would increase with each new universe, until the universe became too disordered for life. Such objections may be irrelevant, though, for they follow from the laws of physics, which may break down each time the universe collapses.

If the universe is closed, astronomers can presently see only a fraction of it, because the horizon is smaller than the universe. But the horizon will keep growing, even as the universe's expansion slows, so they will see more and more. When the universe peaks in size, astronomers will finally see all the way to the other side of the universe—the equivalent of Santa Claus, at the North Pole, glimpsing penguins at the South Pole. As the universe begins to shrink, the horizon will keep growing. Ultimately, just as the big crunch destroys everything, fireproof astronomers will be able to see all the way around the universe, allowing them to view the backs of their heads.

## THE FLAT UNIVERSE: EUCLID'S PARADISE

If fiery universal collapse doesn't appeal to you, and if you'd prefer that parallel lines stay parallel and that a triangle's angles add up to 180 degrees, you might enjoy a universe with omega precisely equal to 1. Then the cosmic matter density exactly equals the critical density.

Cosmologists call such a universe flat. That's because the matter density is neither so great as to curve space into a closed shape nor so small as to curve it the other way. Such space has no curvature—it is flat, like a tabletop. That's why parallel lines stay parallel, just as Euclid said they would, and the angles of a triangle add up to 180 degrees. A light beam shot into space would never return. Here and there it would get deflected one way or another by the gravity of various galaxies, but its overall path would be straight.

Unlike a closed universe, a flat universe has infinite space and mass. Thus, a flat universe is more massive than a closed universe, even though it has a lower mass *density*. And in cosmology, density is destiny.

A flat universe expands forever. The gravitational pull of all its matter is just barely insufficient to reverse the expansion. If the universe is flat, galaxy groups and clusters will continually recede from the Local Group, and the universe will grow colder and darker, as stars burn out and as the expansion of space further attenuates the heat of the cosmic microwave background. T. S. Eliot once said that the world would end with a whimper, not a bang. So he must have thought that omega was 1 or less.

The expansion of a flat universe with omega equal to 1 forever slows but never reverses. According to the equations, after an infinite amount of time, the universe expands infinitely slowly, which is another way of saying that it stops expanding—and recalls the adage, "Infinity is where things that don't happen, happen." At any *finite* time, a flat universe with an omega of 1 is expanding; it's just doing so more slowly than yesterday. And if you wait long enough, its expansion rate, or Hubble constant, will dip below any positive number you name—20, 10, even 0.00000001 kilometer per second per megaparsec—but it never reaches 0.

A flat universe with omega equal to 1 expands at such a rate that when it gets eight times older, most other galaxies will be four times farther from the Milky Way. Since eight is greater than four, observers will see more and more of the universe: the horizon expands faster than the universe, so if observers wait long enough, they can see any particular galaxy they want to. Even then, however, because a flat universe is infinite, an infinite number of other galaxies lie beyond the horizon and therefore are invisible. In a flat universe with omega equal to 1, the present distance to the horizon is thrice its lookback distance. Thus, if the universe is 12 billion years old, so that the farthest observable objects have a lookback distance of 12 billion light-years, they are presently separated from the Milky Way by 36 billion light-years.

The age of a flat universe with omega equal to 1 is easy to calculate. It is simply two thirds of the Hubble time. For example, if the Hubble constant is 65, then the Hubble time is 15 billion years, and the age of the universe is two thirds of that, or 10 billion years.

## THE OPEN UNIVERSE: BORN INFINITE

If omega is less than 1, then the universe will also expand forever. However, its matter density is so low that the universe assumes a distinct shape—open. Whereas a closed universe resembles the surface of a globe, and a flat universe a tabletop, an open universe resembles a saddle. Draw parallel lines on a saddle and they diverge; draw a triangle and its angles add up to less than 180 degrees. So if two light beams set off on parallel paths, they eventually skirt away from each other.

Like a flat universe, an open universe is infinite. Strangely, the equations say that at birth both flat and open universes had zero size, but an instant afterward they were already infinitely large.

An open universe expands faster than a flat universe. Recall that when a flat universe gets eight times older, it gets only four times larger, no matter what its age, because the universe's matter density continually restrains the expansion. In contrast, in an open universe, after an infinite amount of time, matter is so spread out that it can no longer brake the expansion. Thus, when an old open universe gets eight times older, it also gets eight times larger.

As in a flat universe, if observers wait long enough, they can see any particular galaxy; but because the universe is infinite, an infinite number of galaxies always lie out of sight. So if astronomers want to see all the universe's galaxies, they are out of luck, but if they only want to see their best friend's galaxy, they will eventually succeed, no matter how far it is.

An open universe is older than a flat universe with the same Hubble constant. That's because the universe's age depends not just on the Hubble constant but also on omega. An open universe with no matter—omega equal to 0—expands forever at the same rate, because no matter slows the expansion. Such a universe is exactly as old as the Hubble time. Because the actual universe has matter, however, the gravitational pull of that matter tries to slow the expansion. In the past, therefore, the universe should have expanded faster. If so, the universe is younger than the Hubble time, because a fast-expanding universe has taken less time to reach its present size.

The greater omega, the faster the universe expanded in the past, so the younger is the universe, and the more its age differs from the Hubble time. For example, if the universe has only 10 percent of the matter density needed to cause it to collapse—that is, an omega of 0.1—then its age is 90 percent of the Hubble time, and if omega is 0.2, its age is 85 percent of the Hubble time. As omega rises, the universe gets younger and younger. A flat universe—omega of 1—has an age that is two thirds (67 percent) of the Hubble time, and if omega is 2, then the universe's age would be only 57 percent of the Hubble time. For a Hubble constant of 65, this translates into an age for the universe of just 8.6 billion years—younger than the oldest stars are thought to be.

## LAMBDA: A RUNAWAY UNIVERSE?

The third cosmological parameter, designated by the Greek letter lambda (λ), is the wild card. Everyone knows that the Hubble constant and omega exist, because the universe both expands and harbors matter. Lambda, however, is another story. It indicates the power of the cosmological constant, what Einstein called his biggest blunder.

The cosmological constant represents the repulsive energy of empty space. Logically, it might seem that empty space should possess no energy whatsoever. If so, the cosmological constant does not exist and the numerical value of lambda is 0. But if empty space does exert a repulsive force, that force tries to accelerate the universe's expansion.

Cosmologists quantify lambda as follows. Einstein said that mass and energy are equivalent, so in principle the energy density that the cosmological constant represents could be transformed into a mass density. This would correspond to some numerical value of omega and is what cosmologists call lambda. For example, if omega is 0.2 and lambda is 0.4, then the cosmological constant is twice as strong as omega.

Lambda is omega's enemy. Whereas omega represents the attractive force of gravity and tries to pull the universe back together, lambda represents a repulsive force that tries to tear it apart. And time is on lambda's side. As the universe expands, more and more empty space opens up, which powers lambda. Furthermore, lambda itself accelerates this expansion, thereby creating more empty space, which feeds its own power. In contrast, over time omega weakens, because the galaxies get farther apart, and their gravitational tugs on one another grow ever more feeble. A lambda-dominated universe therefore becomes a runaway, getting ever larger ever faster.

Obviously, lambda can drastically affect the universe's fate. If lambda does not exist, then that fate is sealed by omega. For example, if omega is 2, the universe is so dense that it will reverse its expansion and collapse. But even a small lambda—just 0.05—can eventually overwhelm this omega, stave off the collapse, and incite a runaway expansion. Lambda therefore saves the universe from collapse. The price: an accelerated expansion and an even faster frosty fate.

Like omega, lambda alters the geometry of space, but not in the way you might think. When only omega ruled the cosmos, geometry governed our fate: a closed universe collapses, a flat or open universe expands forever. With lambda, however, a closed universe can also expand forever. Strangely, even though lambda opposes omega, they both affect the geometry of space in the same way—because lambda's energy curves space just as omega's matter does. Without lambda, the dividing line between an open and closed universe is at omega equal to 1. With lambda, the dividing line is at omega *plus* lambda equal to 1. For example, a universe with an omega of 2 and a lambda of 0 is closed and will collapse; a universe with an omega of 2 and a lambda of 0.05 is also closed, with finite space and mass, but it will expand forever. A universe with an omega of 0.2 and a lambda of 0 is open and will expand forever; a universe with an omega of 0.2 and a lambda of 0.8 will also expand forever but is flat.

Lambda increases age estimates for the universe. If lambda exists, the universe was expanding more slowly in the past, when the universe was smaller and lambda was too weak to give the expansion much of a push. Because a slow-expanding universe has taken more time to reach its present size, the universe could be older than the Hubble time. For example, if the Hubble constant is 65, omega is 0.3, and lambda is 0, the universe is 12.2 billion years old. But if lambda is instead 0.7, the universe is 2.3 billion years older—14.5 billion years old.

Because the three cosmological parameters—the Hubble constant, omega, and lambda—govern the universe, cosmologists have expended enormous effort measuring them, achieving only partial success. The Hubble constant is difficult to determine because it requires knowing the distances of distant galaxies, those whose redshifts probe the universe's expansion. Lambda, deriving its power from empty space, is even worse, for in order to see its effects, astronomers must peer across billions of light-years of space.

Omega might seem easier. After all, astronomers can see light from the galaxies that contribute to the universe's matter density; therefore, a galactic census might decide whether the universe is dense enough to collapse. However, even as astronomers reveled in the 1965 detection of the big bang's afterglow, they were about to discover that the universe was far more complex and less cooperative than they had thought, for the beautiful starlit cities that speckled space held a dark secret—one that might alter the fate of the entire universe.

# SIX

# DARK MATTERS

FLY over a city at night and picture the giant galaxy below: strands of streetlights, glowing orange, purple, and yellow, interspersed with traffic lights flashing red and green, peer over countless headlights heading down highways. Engulfing the bustling city and its radiant suburbs, however, lurk vast expanses of dark countryside, spotted only here and there by the occasional rickety streetlamp lighting the occasional dirt road.

During the 1970s astronomers discovered that the beautiful starlit cities they had been studying resembled their terrestrial counterparts—each spiral was embedded in an enormous dark halo, lit only here and there by the occasional star or star cluster, yet outweighing the luminous galaxy many times over. All this matter matters: dark galactic halos contribute mightily to the universe's matter density, omega, which determines whether the cosmos faces eternal expansion or eventual collapse.

If light traced mass, cosmic prognosticators would have it easy, for they could infer omega simply by adding up all the light they see. It's not much, so they would conclude that the universe will expand forever. For example, if every Sun's worth of visible light corresponded to the Sun's worth of mass, omega would be a mere 0.001, far shy of the value of over 1 needed to cause cosmic collapse.

Contrary to popular belief, however, the Sun is no average star. Most other stars weigh less and shine *much* less. Faint, cool stars called red dwarfs throng the Sun's vicinity, but they glow so feebly that the unaided eye can't see a single one. The Sun's nearest stellar neighbor, dim Proxima Centauri, has 1/10 of the Sun's mass but emits only 1/18,000 as much visible light. Thus, it would take 18,000 red dwarf stars like Proxima to match the Sun's luster, but their combined mass would top 1,800 Suns. If such dim suns generated all of the universe's light, omega would catapult from 0.001 to 1.8—more than enough to halt the expansion.

Unfortunately, dark matter eludes astronomers' telescopes, which are set up to detect light. Light is elitist, casting some objects into prominence while condemning others to obscurity. Astronomers probing dark matter must therefore summon a more egalitarian power: gravity, which weighs and sways all matter, light or dark. Dark matter may not glitter through telescopes, but its gravity tugs luminous objects that do, steering stars, gas, and even entire galaxies through space.

Great astronomical discoveries can sometimes be tied to a specific person, a specific event, even a specific date. When Italian astronomer Galileo Galilei turned his crude telescope on the glowing band of light called the Milky Way, he immediately saw that it was "nothing but a congeries of innumerable stars." When, one February afternoon, Lowell Observatory astronomer Clyde Tombaugh spotted a faint speck flittering on a photographic plate and exclaimed to himself "That's it!", he had discovered the solar system's farthest planet, Pluto.

Dark galactic halos, however, emerged from the astronomical gloom much less triumphantly. "In romantic views, you go back and you say, 'Oh—there was one event,'" said Princeton University's Jeremiah Ostriker. "But if you look at the evidence, it was accumulating slowly over time, then finally became overwhelming, and people shifted, with a sudden lurch. There was no very strong opposition to massive halos; it just took a while for people to be convinced."

## THE ANDROMEDA STRAIN

Popular accounts usually associate the discovery of dark halos with one optical astronomer, Vera Rubin at the Carnegie Institution of Washington. "Everybody was studying centers of galaxies," she said, "but I had always been interested in how galaxies *ended*. One of the reasons I was interested in it, even as a graduate student, was that nobody seemed to be studying it."

Rubin, who earned her doctorate under big bang proponent George Gamow, has encountered few obstacles in being an astronomer. "The obstacles were all in *becoming* an astronomer," she said. "There were very few places where a woman could study. My high school physics teacher said, 'You should do okay as long as you stay away from science.' When I was an undergraduate, I wrote a postcard to Princeton, asking for a catalogue of the graduate school, and I got back a fancy letter saying that inasmuch as they didn't accept women, they wouldn't send me a catalogue." One college asked what else she liked to do. "I told them, 'Well, I do a little painting,' and the woman interviewing me said, 'Have you ever considered a career where you paint pictures of astronomical objects?' That became a tag line in my family: for many years, whenever anything went wrong for anyone, we said, 'Have you ever considered a career in which you paint pictures of astronomical objects?'"

In 1963 astronomers discovered that quasars had enormous redshifts, unleashing a stampede of observers to study the far-off objects. Rubin's Carnegie colleague, W. Kent Ford, had developed an image tube that picked up faint light. Since quasars were dim—even the brightest barely outshone Pluto—Rubin and Ford had the ideal instrument with which to observe them.

However, Rubin found the quasar work distasteful. "I would get calls from very, very important people, very nice people, asking me if I had a redshift of thus and so; and if I did, they wouldn't observe it, and if I didn't, they would." Often Rubin's redshift was sufficiently uncertain that she wanted to confirm it before announcing it, so if she said yes to the caller, she might release bad data, whereas if she said no, she would relinquish her stake in the quasar's spectrum, allowing others to publish it before she did. "Finally," she said, "after about two years, I just decided I was going to get out of the quasar world and find something to do that no one would care about."

She turned her attention to the rotation of the most important galaxy outside our own, Andromeda, 2.4 million light-years away, the largest member of the Local Group, a beautiful spiral like the Milky Way. As do the terrestrial hurricanes they resemble, spiral galaxies spin fast. Only in the twentieth century, however, did astronomers detect this rotation—by exploiting the Doppler shift. Relative to its center, one edge of a spinning spiral swings toward Earth and the other away, so the approaching edge exhibits a blueshift and the other edge a redshift. In 1913, Lowell Observatory astronomer Vesto Slipher—who had earlier discovered Andromeda's high speed—used the Doppler shift to uncover the Sombrero Galaxy's rotation, and German astronomer Max Wolf discovered another spiral's spin, M81. The following decade, astronomers found that the Milky Way itself rotated. An observer north of the Galaxy's disk would see most of the Milky Way's stars, including the Sun, racing clockwise around the Galaxy's center at 220 kilometers per second, or half a million miles per hour.

From 1966 to 1968, Rubin and Ford traced the Andromeda Galaxy's rotation, observing the galaxy from snowy peaks in Arizona. "The temperature was below zero Fahrenheit," said Rubin. "One telescope was a telescope where the dome rolled off, and we were standing out in the cold. I think it's the only time when Kent Ford and I actually took turns at the telescope. Instead of us both being there, one was inside warming up, the other was outside working."

Rubin and Ford wanted to see how fast Andromeda's stars revolved around its center. In the solar system, the planets closest to the Sun race around it fastest. For example, Mercury dashes ten times faster than Pluto, because Mercury feels the Sun's gravity more intensely and must move faster to keep from falling in. Thus, the planets resemble lazy employees: the closer the boss is, the harder they work. The Andromeda Galaxy should show similar behavior, with stars closer to its center revolving faster.

In Andromeda, Rubin and Ford observed not stars but star-forming clouds similar to the Milky Way's own Orion Nebula, a cauldron of magenta gas and hot blue stars. Such stellar nurseries line Andromeda's spiral arms, consist mostly of hydrogen gas, and bear newborn stars so hot that they ionize the hydrogen, stripping off its electron. After the electron rejoins the hydrogen atom, it can radiate red light with a wavelength of 6,563 angstroms. This red spectral line outshines any from Andromeda's stars, making it easier to observe. By measuring its Doppler shift, Rubin and Ford could determine how fast an ionized hydrogen region revolved around the galaxy's center.

Altogether they observed sixty-seven regions of ionized hydrogen across Andromeda's disk. "We weren't even sure it would work," said Rubin. "We'd expose for an hour or two on a spot where we saw nothing—because the telescopes were just not equipped to be able to see anything that faint." Ford's image tube, though, captured the red spectral line. "It was grand," said Rubin. "I really enjoyed the observing, all the spectra looked beautiful, and I just thought this was a great program, because we were getting so much interesting science." Plus, no one was phoning her for redshifts.

In the end, as they reported in 1970, Rubin and Ford tracked Andromeda's rotation from nearly one edge of its disk to the other—a span of almost 4 degrees, which at Andromeda's distance translates into 160,000 light-years, or 80,000 light-years on either side of Andromeda's center. The galaxy's velocities, however, failed to match the declining pattern that prevailed in the solar system. Instead, after rising to 270 kilometers per second, Andromeda's velocities fell only slightly, then held steady all the way to the disk's edge. In other words, much of the rotation curve was flat. It was as if Pluto sped around the Sun as fast as the Earth.

Yet nowhere in their paper did Rubin and Ford make much of this oddity. "No, I don't regret it," said Rubin. "That's how science goes. I really believe that if you're going to be a happy scientist, you have to enjoy every day's work. Science to me is really a process—it's the process of my life, essentially. Deciding what to observe is perhaps the most important thing I do, and then laying out the observing program I find incredibly fun and delightful, and I enjoy the observing, and I even enjoy the measuring. I can't second-guess and say I should have understood it then. Making a discovery is great, but you invest a lot of yourself in this, and you might *not* discover anything."

## HEAVY IMPLICATIONS

On the other side of the world, however, another astronomer had already recognized the startling implications—not of the Andromeda work, which was just going to press, but of similar rotational behavior in two other spiral galaxies.

"It was really just a throwaway comment," said Kenneth Freeman of Australian National University. "I'd completely forgotten that I had ever said anything like that myself." In fact, his remarks were tucked in a brief appendix to a 1970 paper on spiral galaxies.

For other reasons, Freeman's paper attracted great attention. "The main thing in the paper," he said, "was that a spiral galaxy is a spiral galaxy. When you look at a bunch of different galaxies, you find that the disk's central surface brightness is pretty much the same. That was a big surprise." Surface brightness indicates how concentrated light is. For example, a lightbulb has a higher surface brightness than the lamp shade it illuminates. In the disk of a spiral galaxy, the surface brightness is greatest at the center, then plummets exponentially in all directions from the center to the edge.

Freeman wondered what the exponential light distribution implied for the spirals' rotation. However, mass, not light, dictates rotational behavior, because mass alone exerts the gravitational force that makes stars revolve around a galaxy's center. In the solar system, nearly all mass sits at the center, in the Sun, so planetary velocities decline with distance from the center. In a galaxy, though, mass is spread out, complicating the rotation curve. Freeman calculated that the rotation curve should first rise with distance from the center, then fall—*if* light meant mass.

"Then I got interested to see how that compared with observation," said Freeman. "Now there were a few galaxies around at that time that had surface photometry, so we knew where the rotation curve *should* peak." Despite such data, Freeman couldn't say how much mass lay beyond the last point in a galaxy's rotation curve. "You can only say what the mass distribution is like *inside* where you've got rotation data," he said. "If you've got data out to some point, you can't say *anything* about the mass distribution farther out."

This peculiar handicap stems from an equally peculiar property of gravity: only material interior to a star's orbit affects its speed; exterior material exerts no net force. In the same way, people on the Earth's surface feel the Earth's gravitational pull because all of the planet's mass lies closer to the Earth's center than they do. But any people at the Earth's center would feel no force, because the gravitational pull from the Earth's material in one direction would cancel the pull from material in the opposite direction. And people halfway between the center and the surface would feel only the force from the material inside their radius, none from the shell beyond. The upshot: to probe a dark halo, astronomers must measure velocities of objects *within* or *beyond* the dark halo. Only when the observations have reached the mass distribution's end will the velocities decline with distance as they do in the solar system.

Fortunately, in many spiral galaxies, neutral atomic hydrogen gas skirts far beyond the edge of the star-filled disk. Unlike the red *ionized* hydrogen that Rubin and Ford studied, neutral hydrogen emits no visible light. It does, however, send out radio waves that are 21 centimeters long. Just as Rubin and Ford measured velocities by observing the Doppler shift of the red 6,563-angstrom line, so radio astronomers had deduced velocities from the Doppler shift of the 21-centimeter line.

The radio data Freeman examined came from two nearby spiral galaxies: M33, the Local Group's third largest member, after Andromeda and the Milky Way; and NGC 300, M33's twin and a member of the Sculptor group, 7 million light-years away. For both galaxies, the radio rotation curves failed to fall with distance from the center. Instead, both had rotation curves that were flat, just as Rubin and Ford had found for Andromeda. The clouds of neutral hydrogen gas beyond the starry disks were moving just as fast as the stars closer in.

In his paper, Freeman wrote, "These data have relatively low spatial resolution; if they are correct, then there must be in these galaxies additional matter which is undetected, either optically or at 21 cm. Its mass must be at least as large as the mass of the detected galaxy, and its distribution must be quite different from the exponential distribution which holds for the optical galaxy."

Dark matter had begun to come out of the dark.

## VOICE OF ANDROMEDA

At that time, however, radio telescopes were primitive. For this reason, Freeman's statements about M33 and NGC 300 began with phrases like, "*If* the data are correct . . ." In the end, most of the data did turn out to be correct, but scientists had legitimate concerns. For one thing, radio telescopes of that era viewed galaxies only fuzzily. For another, if a radio telescope pointed in one direction, it picked up signals from other directions, too. In like manner, if you tried to tape a singer singing middle C at the edge of a choir singing high E, your tape would pick up not just the middle C but the high E, too. Thus, radio astronomers worried that the high velocities they measured beyond a disk's starry edge might actually be coming from material closer in.

"In those days," said Morton Roberts at the National Radio Astronomy Observatory in Charlottesville, Virginia, "you tried to find a big galaxy." The bigger the galaxy looked, the better a radio telescope could scrutinize it. "So you sat down with a list of the biggest galaxies in the northern hemisphere, and you marched through them. Working for you was the fact that the hydrogen went out much farther than the light."

The galaxy with the largest apparent size in the northern sky was the same one that Rubin and Ford had chosen: Andromeda. "We went well beyond the optical limit," said Roberts, who observed the galaxy for several years. "The radio data said there was just no light, no optical light, but there was sure something holding up that rotation curve."

Whereas Rubin and Ford had measured Andromeda's rotation curve out to 80,000 light-years from the galaxy's center, the radio observations of Roberts and his colleague Robert Whitehurst reached a distance of 105,000 light-years. In Andromeda's stellar disk, the radio velocities confirmed the optical ones. Beyond that disk, to Roberts' surprise, the rotation curve remained flat—all the way out to 105,000 light-years, the last data point.

"I can tell you this," said Roberts. "It was not received well when I spoke about it at colloquia or meetings. This was in the early 1970s, before it was published. There were a lot of raised eyebrows." In particular, two radio astronomers in England claimed from their observations that Andromeda's rotation curve declined.

A flat rotation curve has a simple interpretation. It means that the mass inside a certain distance of the galaxy's center is proportional to that distance. For example, if Andromeda has 180 billion solar masses within 50,000 light-years of its center, then it has twice this mass, or 360 billion solar masses, within twice this distance, 100,000 light-years. And if the flat rotation curve held out to 200,000 light-years, then inside that distance the galaxy would have 720 billion solar masses. The *density* is decreasing—as the inverse square of distance from the galactic center—but because so much volume exists at large distances, the total mass in the outer regions exceeds that closer in.

This mass emitted no light. Roberts and Whitehurst suggested that it consisted of large numbers of dim red dwarf stars, similar to Proxima Centauri.

## DARK ANGELS

As optical and radio astronomers were uncovering Andromeda's odd rotation, two theorists at Princeton were finding other peculiarities closer to home—in the Milky Way Galaxy. "I had been working on rotating stars," said Jeremiah Ostriker, "and I found out something very curious. A spherical star is, of course, quite stable: the Sun is quite stable. But if you make the star rotate faster and faster, when it gets to some critical point, then it becomes unstable and fissions into a binary star. And you can think of that as the shape of the star—that if the shape of the star is as flattened as approximately 2 to 1 by rotation, the amount of rotation is sufficient so that the star would be happier as a binary.

"Then I realized, if you look at pictures of edge-on spiral galaxies, they're *much* flatter than that. Their length to their height may be 50 to 1." That's flatter than a dime. "So I got to worrying: how can ordinary spirals be stable?"

Ostriker and his colleague James Peebles used a computer to simulate a galaxy like the Milky Way. They watched its stars revolve around its center and interact with one another via gravity.

"The galaxy was violently unstable," said Ostriker. "We set it out looking like the picture, and it changed radically. It broke into pieces. So the computer calculations that we did indicated the same conclusion for a galaxy as I had found for stars: our Galaxy couldn't be as flat as it looked. So we said, 'Well, what else could there be? What other element?' We knew that there was a halo of ordinary stars, the spherical component of our Galaxy, so we said, 'Let's up the mass in that component and see if the halo tends to stabilize it.'"

It did. Like a protective blanket, the halo's gravity calmed the disk—but only if the halo's mass equaled or exceeded the disk's.

"Then you say, 'What about the solar system? Isn't that a flat, rotating thing?' But most of the mass is in the center, in the Sun, and the Sun is round. So the solar system is not a counterexample; it's an example: any rotating system in which most of the mass is in a round component is all right.

"So we then said that it's likely that the spherical component of our Galaxy, even though it's not bright, has at least as much mass as the flat component. There weren't observations to support it, but there weren't any observations to contradict it, either—because we simply didn't know what mass was in the two components." Ostriker and Peebles published their work in 1973.

The next year, Ostriker, Peebles, and Israeli scientist Amos Yahil marshaled a variety of evidence to make an even more audacious claim. "We tried to put together *all* the evidence and show it was all consistent," said Ostriker. "We were looking at all the dynamical measures, using gravity to weigh everything. We were saying that it all hung together and that you'd have to look for a model of the universe in which there was much more dark matter than stars." A team of Estonian scientists independently reached a similar conclusion.

Whereas the 1973 paper concluded that the halo must be at least as massive as the disk in order to stabilize it, the 1974 paper claimed that the halo's mass was ten times greater. This meant that giant spirals like the Milky Way and Andromeda harbored roughly a trillion times more mass than the Sun. This conclusion followed not only from the then-existing rotation curves but also from other measures that probed even farther, such as the motions of galaxies in groups.

Most of the Milky Way's light streams out of its disk, which harbors

THE MILKY WAY'S TWO HALOS

| | Stellar Halo | Dark Halo |
|---|---|---|
| Prominent components | Most globular clusters; many RR Lyrae stars | Unknown |
| Prime location | Inside solar orbit | Outside solar orbit |
| Density distribution (r = distance from Galactic center) | $r^{-3.5}$ | $r^{-2}$ |
| Visual luminosity (Sun = 1) | 1 billion | Near 0 |
| Mass (Sun = 1) | 3 billion | 1 trillion |
| Number of stars | 10 billion | Unknown |
| Typical metallicity (Sun = 1) | 0.02 | Unknown |

the spiral arms, but astronomers have long known about the Galaxy's rounder components, too: the Galactic bulge, which occupies the central region of the Milky Way like the yolk in a poached egg; and the stellar halo, a tenuous assortment of farther-flung stars that rise above and below the disk. The disk generates some 15 billion solar luminosities of visible light, the bulge about 4 billion, and the stellar halo another 1 billion or so. The stellar halo's stars are so old they formed before other stars had a chance to enrich the Galaxy with heavy elements such as oxygen and iron. As a consequence, halo stars were born from metal-poor material and so are metal-poor themselves. The stellar halo's most prominent members are metal-poor globular clusters, such as M13 in the constellation Hercules. But for every star in a globular cluster, the halo sports a hundred other stars outside them. Some halo stars shine in the Sun's vicinity—the nearest is just 13 light-years away—and a few even reside far beyond the edge of the disk, but most halo stars huddle closer to the Galactic center, lying interior to the Sun's orbit around the Galaxy. This is why globular clusters congregate around Sagittarius, home of the Galactic center. The flat rotation curves, however, implied large quantities of dark material in the *outer* parts of spiral galaxies. In the Milky Way, that meant beyond the Sun's orbit. So: one Galaxy, two halos—and much confusion among astronomers.

"The recognizable spatial and population components that make up the Milky Way have acquired a set of names that have become alarmingly obscure and confusing," wrote Berkeley astronomer Ivan King some years later. "It can be argued that the misnaming began when flat rotation curves in galaxies were discovered, since these showed that galaxies must contain an extended component that emits little or no light. This 'massive halo' conflicted in name with the classical stellar halo, since the two differ both in physical nature and in spatial distribution. Some misguided individuals then chose to distinguish the stellar halo by calling it the 'spheroid.' Unfortunately, that word was already being used to designate the central bulge of external galaxies that resemble the Milky Way. It has taken only a few years of carelessness to extend the term 'spheroid' to include in a single mélange the dark massive halo, the low-metal-abundance stellar halo, and the central bulge, which is dominated by stars of high metal abundance." Better, then, to distinguish the classical halo from the newer one by calling the first the stellar halo and the second the dark halo.

This radical revision in the structure of our Galaxy and others had cosmological consequences. Ostriker, Peebles, and Yahil said that omega, the universe's matter density, must be around 0.2—not enough to stop the universe's expansion, but more than enough so that most material in the universe hid in the dark countryside surrounding the starry cities that populated the cosmos.

## As the Galaxies Turn

During this time, Vera Rubin, who had measured Andromeda's optical rotation curve, was studying other subjects. In the mid-1970s, however, she returned to observe additional spiral galaxies. Despite the previous work, she didn't expect their rotation curves to be flat. "I must have seen Ostriker and Peebles' paper," said Rubin, "but I wasn't doing it because of that. The honest truth is, I just wanted to know how they rotated. I wanted to understand why spirals looked the way they did and why they came in different classes. It was *not* done to see if the rotation curves were flat. I wasn't smart enough to say that I was going to do that."

Using powerful new telescopes in Arizona and Chile, Rubin and her colleagues observed the rotation curves of several galaxies a night. "By the end of the first night," she said, "I knew that something very funny was going on: each one looked exactly like every other one." The rotation curves were all flat. "And actually the first idea that came to me, in the darkroom, was not even dark matter," she said. "My first idea was that there was some kind of a feedback mechanism—that if the stars got too slow, they got speeded up; that if they got too fast, they got slowed down—because the curves looked so exceedingly flat. And it probably was not till the second night, we immediately all thought of dark matter." The study, published in

1978, examined eleven spiral galaxies, including the galaxy in which astronomers had first detected rotation, the Sombrero. The paper's conclusion: "All rotation curves are approximately flat."

That same year, Albert Bosma, a radio astronomer in Holland, completed a thesis in which he observed dozens of spiral galaxies. "Bosma's influence was enormous," said Kenneth Freeman. "His thesis really took the world by storm. It was the real clincher. At that point, there was absolutely no question at all that there was dark material in most of these spirals."

Bosma used the new Westerbork array of radio telescopes, which yielded sharp views of other galaxies. Because he observed the neutral hydrogen gas that danced beyond the periphery of the stellar disks, he could measure galactic rotation to great distances from their centers. He wrote, "The most striking feature of the rotation curves . . . is that [the rotation velocity] declines very slowly, if at all."

## FIRST SIGNS IN THE DARK

Although dark halos came of age during the 1970s, the first clues actually emerged decades earlier. "I always think it's interesting when the evidence is around and people don't notice it," said Ostriker. "There was a very long period when people thought the stars were fixed, even though supernovae were seen—they just were ignored. That's what happened here: the evidence gradually accumulated, from radio rotation curves, optical rotation curves, the timing argument."

The first strong evidence for dark matter came in the 1930s, when maverick Caltech astronomer Fritz Zwicky sensed something amiss with galaxy clusters. Many of Zwicky's colleagues, however, sensed something amiss with him. Shortly before his death, Zwicky self-published a galaxy catalogue that opened with "A Reminder to the High Priests of American Astronomy and to their Sycophants" and went on to say "[M]ost astronomers were and still are convinced that the redshift observed in the spectra of the distant galaxies is a clear proof for the expansion of the universe, some scatter-brains among them [i.e., Allan Sandage] even claiming to know how the rate of this expansion has been changing in time . . . The naivety of some of the theoreticians, at all times, is really appalling. As a shining example of a most deluded individual we need only quote the high pope of American Astronomy, one Henry Norris Russell . . . Secondly, the most renowned observational astronomers in the 1930's also made claims that now have been proved to be completely erroneous. This retarded real progress in astronomy by several decades since the said observers had a monopoly on the use of the large reflectors of the Mount Wilson and Palomar Observatories, and inasmuch they kept out all dissenters. I myself was allowed the use of the 100-inch telescope only in 1948, after I was fifty years of age, and of the 200-inch telescope on Palomar Mountain only after I was 54 years old . . . E. P. Hubble, W. Baade and the sycophants among their young assistants were thus in a position to doctor their observational data, to hide their shortcomings and to make the majority of the astronomers accept and believe in some of their most prejudicial and erroneous presentations and interpretations of facts. Thus it was the fate of astronomy, as that of so many other disciplines and projects of man, to be again and again thrown for a loop by some moguls of the respective hierarchies. To this the useless trash in the bulging astronomical journals furnishes vivid testimony. . . . [E]xcept for some outstanding exceptions like George Ellery Hale, the members of the hierarchy in American Astronomy have no love for any of the lone wolves who are not fawners and apple polishers . . . Today's sycophants and plain thieves seem to be free, in American Astronomy in particular, to appropriate discoveries and inventions made by lone wolves and non-conformists, for whom there is never any appeal to the hierarchies and for whom even the public Press is closed, because of censoring committees within the scientific institutions. . . . A. Sandage of the Mt. Wilson Observatory in May of 1965 attempted one of the most astounding feats of plagiarism . . . In sharp contrast to their ready and uncritical acceptance of all sorts of childish phantasies and stolen ideas, the Editors of the Astrophysical Journal exhibited an almost unbelievable lack of tolerance and good judgement by rejecting my first comprehensive and observationally well documented article on compact galaxies." Zwicky reprinted the rejection letter for all to see.

Zwicky could pen calmer words, too. In 1933 he reported that the galaxies in the Coma cluster moved so fast that they should escape its grasp. Said Ostriker, "The galaxy clusters weigh everything that's in them—whether it's attached to the galaxies or in between the galaxies or whatever. It doesn't matter what you think your weight is—when you get on the scale, it tells you what your weight is. And Zwicky was weighing everything there." To hold on to the speeding galaxies, said Zwicky, the Coma cluster must have hundreds of times more matter than met the eye. Three years later, Mount Wilson astronomer Sinclair Smith reported that the nearer Virgo cluster had the same problem and proposed "a great mass of internebular [i.e., intergalactic] material within the cluster."

Shortly afterward, an astronomer in California, Horace Babcock, measured the Andromeda Galaxy's rotation curve. He found that it was rising out to his last measured point, now known to be 70,000 light-years from Andromeda's center. Most of his measurements, however, clustered around Andromeda's center, and those at the largest distances differ from modern data by more than a hundred kilometers per second. Still, Babcock's data agree with modern values out to 50,000 light-years. Furthermore, he recognized what the rotation curve meant: that Andromeda's outer regions harbored unseen material.

In 1959, Frank Kahn of the University of Manchester and Lodewijk Woltjer in Holland proposed the so-called timing argument, which weighed both the Milky Way and Andromeda. "Scientifically, I think the timing argument is a very important thing," said Freeman. "It gives us the best estimate that we've got of the dark matter content of our own Galaxy. But historically, I don't think it had much influence at all. At the time, I don't think it was really appreciated, because the answer was just so way out. I think people thought, 'Okay, something's missing in the physics,' so they discounted it. If their paper had come in the mid-1970s, after Ostriker and Peebles, then I think it might have been a different story." In fact, the 1974 paper by Ostriker and colleagues cited Kahn and Woltjer's work as evidence for dark halos.

Kahn and Woltjer's timing argument exploited the Andromeda Galaxy's motion through space. Although the universe expands and most galaxies move away from ours, Andromeda does not. Kahn and Woltjer assumed that shortly after the big bang, it did move away, but the two galaxies' gravitational pull caused them to fall back toward each other. To achieve this reversal during the lifetime of the universe, each galaxy must have at least a trillion times more mass than the Sun.

Other nearby galaxies also weigh the Milky Way. Several galaxies orbit the Milky Way, the brightest being the Magellanic Clouds, the farthest being the Leo I galaxy, 740,000 light-years distant, and a few globular star clusters also reside beyond the edge of the Galactic disk. These objects move fast, so to hold on to them the Galaxy must have roughly a trillion solar masses, agreeing with Kahn and Woltjer's timing argument.

In 1962, Vera Rubin studied blue stars in the Milky Way that shone so brightly they could be seen across thousands of light-years of space. She and her colleagues used the stars' velocities to trace the Galaxy's rotation curve. The conclusion: beyond the Sun's distance from the Galactic center, "the stellar curve is flat, and does not decrease as is expected."

Rubin had to fight to get the paper published—not because of its science, but because it included the names of her students. "I got a call from the editor," she said. "I still remember where I was standing—he called me at my house—and I really was angry enough to say that I would withdraw the paper if he didn't put the students' names on it." The editor backed down, and the paper appeared, complete with the students' names.

By the late 1970s, with evidence from rotation curves, galaxy clusters, satellite galaxies, and the timing argument, astronomers were confronting the realization that most material in the universe was dark, beyond the reach of telescopes. "All through the history of astronomy," said Rubin, "we were studying light, and we were assuming that in studying light we were studying mass. It was such a fundamental assumption that I think no one really thought about it. It takes a while to get used to the concept that there may be matter for which radiation is not one of its properties."

Unfortunately, although gravity's egalitarian nature unveiled dark matter's presence, it said nothing about dark matter's composition, because gravity didn't care: the force affected everything, whether blue supergiants or red dwarfs, black holes or dust grains. Astronomers did know that dark matter couldn't be bright stars like the Sun; otherwise, telescopes would have seen them. Still, that left open a myriad of other possibilities: dim stars, such as red dwarfs, white dwarfs, brown dwarfs, black holes, and neutron stars; substellar objects, such as planets, comets, and asteroids; invisible gas; and subatomic particles.

There's one other possibility, too: trouble with Newton's law of gravity. "I would be ill at ease if Newton didn't hold on those scales," said Morton Roberts, "but by God, if you've got an accelerating universe these days, why not change the form of Newton's equations at large distances?" Deductions of dark matter depend on assuming that Newton's law of gravity remains valid over thousands, even millions, of light-years. If that assumption is wrong, then there might be no dark matter at all.

Few astronomers, though, wanted to explain their observations by altering the basic laws of physics. And just as astronomers were uncovering dark matter's ubiquity, others were gaining clues to its nature by examining an unlikely epoch: the era, mere minutes after the big bang, when creation's fiery aftermath forged the hydrogen and helium that still pervade the cosmos today.

# SEVEN

# ASHES OF CREATION

*In the beginning is the future*

—YES ("THE CALLING")

ONE second ago, the big bang burst forth. What triggered it no one knows, but now heat a thousand times hotter than the Sun's center sears the entire cosmos. Because of the intense heat, light floods all space, spanning the spectrum from gamma rays to radio waves, with every hue of the rainbow in between. A few brave protons and neutrons swirl about, dodging the photons, smashing into one another, struggling to construct elements that the merciless photons tear apart. As space expands, the heat diminishes, the photons weaken, and the universe's first nuclear reactor springs into action, cooking up the three lightest elements, hydrogen, helium, and lithium. Although nearly lost in all the light—photons outnumber protons over a billion to one—these three lightweight elements emerge from the primordial fireball, ready to build the material world.

"The light elements are the only window on the early evolution of the universe," said Gary Steigman of Ohio State University, who has spent decades divining their meaning. "If we look at the abundances of those nuclei left over from the big bang, we can learn about conditions when the universe was less than a half hour old." One of those conditions was the universe's matter density, the cosmological parameter omega; so astronomers who measure the primordial abundances of hydrogen, helium, and lithium might also foretell the universe's fate.

## PRIMEVAL ALCHEMY

Surprisingly, the extremes during the first few minutes of the universe's life simplify the calculations. "Because the early universe was so dense and so hot, reactions occurred very, very rapidly," said Steigman. "What that means is that various processes came to equilibrium, so it's much easier to calculate what you expect. It makes the predictions very clean, so people tend to take them more seriously—because there are fewer loopholes."

One second after creation, the universe's protons and neutrons—the building blocks of atomic nuclei—were crowded together over a billion billion billion times more than they are now. By terrestrial standards, this density was *low*, a mere 1/100 that of water. As the universe expanded, it thinned further. When four seconds old, the universe's proton-neutron soup was already fluffier than air. Cosmically, these are mammoth densities. They seem ordinary to us only because we inhabit an extraordinary place—a planet, a dense oasis of matter in a sparse cosmic desert. Because of the low density, each collision involved only two particles. Three-particle nuclear reactions were as rare in the rarefied early universe as a three-car wreck on a quiet street. Neglecting three-particle collisions further simplifies the calculations.

The primordial nuclear reactor, however, had a defect. "Unlike a normal nuclear reactor," said Steigman, "this one was expanding and cooling—so the energy went down, the density went down, and as a result you had only a short interval of time in which you could have transformed nuclei." That interval lasted just a few minutes. When the universe was 1 second old, though, its high temperature—10 billion Kelvin, or 18 billion degrees Fahrenheit—prevented nuclear reactions, because the gamma rays spawned by the heat ripped apart any compound nuclei that tried to form.

All atomic nuclei have protons, which are positively charged, and all nuclei but hydrogen-1 also have neutrons, which carry no charge. Protons and neutrons far outweigh the electrons that, in full-fledged atoms, dance around them. The early universe had more protons than neutrons. As time went on, protons gained greater ground, because after about ten minutes free neutrons decay into protons, electrons, and neutrinos. In contrast, neutrons that are bound into nuclei survive. Thus, once the big bang exploded, the neutrons felt the clock ticking. If the universe didn't expand fast enough, the temperature wouldn't fall soon enough to allow nuclear reactions; therefore, the neutrons would fail to get swept into the sheltering nuclei and would perish, leaving only protons and electrons. Since protons are really hydrogen-1 nuclei, the big bang's sole production would have been unadulterated hydrogen.

Fortunately for cosmologists, actual primordial nucleosynthesis offered more variety. It began in earnest about a minute and a half after the big bang. By that time, the temperature had fallen to 1 billion Kelvin, dwindling the number of gamma rays so that compound nuclei could survive. The first arose when protons smashed into neutrons to make hydrogen-2, also known as deuterium:

proton + neutron —> $^2$H.

Deuterium is a hydrogen isotope. Whereas normal hydrogen—hydrogen-1—has just a proton in its nucleus, deuterium has a proton plus a neutron. The number of protons, not neutrons, determines the element.

Deuterium is fragile. As soon as it formed, protons, neutrons, and other deuterium nuclei converted it into nuclei with mass 3, hydrogen-3 and helium-3:

$^2$H + neutron —> $^3$H

$^2$H + $^2$H —> $^3$H + proton

$^2$H + proton —> $^3$He

$^2$H + $^2$H —> $^3$He + neutron.

Hydrogen-3, also known as tritium, is another hydrogen isotope, this time with one proton and *two* neutrons. Unlike deuterium, it is radioactive, so any that survived primordial nucleosynthesis vanished long before astronomers could observe it.

In contrast, helium-3, with two protons and one neutron, is stable. It's tougher than deuterium, but it didn't last long, either, because it and hydrogen-3 became helium-4, the most tightly bound light nucleus:

$^3$H + proton —> $^4$He

$^3$H + $^2$H —> $^4$He + neutron

$^3$He + neutron —> $^4$He

$^3$He + $^2$H —> $^4$He + proton

$^3$He + $^3$He —> $^4$He + proton + proton.

The denser the universe, and the greater omega, the more these reactions occurred, destroying deuterium and helium-3, and creating helium-4.

After helium-4 took the stage, primordial nucleosynthesis neared its finale. Helium-4 could meet a proton or neutron, but the resulting nucleus had mass 5, and all mass 5 nuclei decay into lighter particles. Two helium-4 nuclei could make a mass 8 nucleus, but this also disintegrates. Because of these obstacles, the big bang failed to forge the carbon-12 that upholds terrestrial life, the nitrogen-14 that fills Earth's air, and the oxygen-16 we breathe. We are mostly stardust, not big-bangdust.

Nevertheless, collisions between helium-4 and hydrogen-3 (tritium) did yield a little lithium-7:

$$^{4}\text{He} + {}^{3}\text{H} \longrightarrow {}^{7}\text{Li}.$$

Protons split lithium-7 apart, and the denser the universe, the more protons abounded, so the less lithium-7 sprang forth. However, if the density was higher, helium-4 could smash into helium-3 to create hardier beryllium-7,

$$^{4}\text{He} + {}^{3}\text{He} \longrightarrow {}^{7}\text{Be},$$

which decayed by capturing an electron and forming lithium-7:

$$^{7}\text{Be} + \text{electron} \longrightarrow {}^{7}\text{Li}.$$

This decay occurred after the universe cooled and shut down its nuclear reactor, so the lithium-7 survived. Thus, lithium-7's behavior with omega is complicated: at low omega, the abundance of lithium is high; at somewhat higher omega, lithium's abundance is lower; but at still higher omega, the abundance is high again.

Because the quantity of each nucleus that emerged from the fiery maelstrom depended on the universe's density, their abundances can reveal omega. Unfortunately, deuterium, helium-3, and lithium-7 swim in a vast sea of hydrogen-1 and helium-4, which together accounted for over 99.9 percent of all big bang nuclei. Furthermore, their present abundances don't equal their primordial ones. During the 15 billion years since the big bang, stars have altered the universe's original composition, fusing the light elements into heavy ones, such as carbon and oxygen, that life needs. Thus, astronomers who probe the big bang are themselves contaminants of the untainted universe they seek to study.

THE LIGHTEST ELEMENTS

| Atomic Number | Isotope Name | Isotope Symbol | Made by Big Bang? | Abundance as Omega Rises | Abundance as Galaxies Age |
|---|---|---|---|---|---|
| 1 | Hydrogen-1 | $^{1}H$ | Yes | ↓ | ↓ |
| 1 | Hydrogen-2 (deuterium) | $^{2}H$ | Yes | ↓ | ↓ |
| 2 | Helium-3 | $^{3}He$ | Yes | ↓ | ? |
| 2 | Helium-4 | $^{4}He$ | Yes | ↑ | ↑ |
| 3 | Lithium-6 | $^{6}Li$ | No | — | ↑ |
| 3 | Lithium-7 | $^{7}Li$ | Yes | ↓↑ | ↑ |
| 4 | Beryllium-9 | $^{9}Be$ | No | — | ↑ |
| 5 | Boron-10 | $^{10}B$ | No | — | ↑ |
| 5 | Boron-11 | $^{11}B$ | No | — | ↑ |
| 6 | Carbon-12 | $^{12}C$ | No | — | ↑ |

To start such a study, astronomers must know each light element's life story. For fragile deuterium, that's easy: its abundance has declined, because stars destroy it. Helium-3 has led a more complicated life. Some stars make it, others destroy it, so no one knows whether its abundance has risen or fallen. Its big brother, helium-4, has definitely gained market share since the creation, as stars have transformed hydrogen-1 into helium-4. The heaviest big bang nucleus, lithium-7, is fragile, so stars usually destroy it; however, particles called cosmic rays race through space and bust interstellar atoms into lighter ones, elevating lithium's abundance. With helium-3 out of the picture, big bang detectives hunt for three light nuclei: in order of abundance, helium-4, deuterium, and lithium-7. Each nucleus took cosmic sleuths to a different locale—from miniature galaxies, to interstellar material, to the Milky Way's senior citizens.

## HIGH ON HELIUM

The big bang kicked out more helium-4 than any other compound nucleus. Indeed, the large helium abundance in the universe had argued for the big bang theory over the steady state. However, in the late 1950s, after scientists showed that stars had forged every element from carbon to uranium, some astronomers began to see stars as the source of all helium, too. It made sense: every main-sequence star, including the Sun, powers itself by burning hydrogen into helium at its center.

Some evidence even supported this view. If stars were the sole helium creators, the Galaxy should have formed from pure hydrogen, and the Galaxy's oldest stars—those in the stellar halo—should preserve this pristine material, modified only slightly by their own nuclear activities. Because the Sun resides in the Galactic disk, though, nearby halo stars are scarce. Nevertheless, a few halo stars dive through the solar neighborhood, their odd motions and low metallicities giving them away. If astronomers could measure the helium content of these interlopers, they could determine the primordial abundance of helium.

Unfortunately, the element refuses to cooperate. As a noble gas, helium guards its electrons so jealously that only hot, blue stars shake them loose. Therefore, only blue stars show prominent spectral lines of helium—but blue stars don't live long, and the halo is so old that its original blue stars have all died.

Still, even in a cool star, helium affects the total luminosity and temperature, and a 1962 study of halo star luminosities and temperatures suggested that they lacked helium. Also, some halo stars have entered a blue phase, giving their surfaces the energy to excite spectral lines of helium. These were weak, again suggesting little halo helium.

Other halo studies, however, contradicted this work. For example, RR Lyrae stars, the old pulsators that litter globular clusters and the halo at large, had colors indicating plentiful helium. Plentiful helium also appeared around one halo star that astronomers had caught dying. When a star like the Sun dies, it sheds its atmosphere and exposes its core, whose heat sets the cast-off atmosphere aglow. This glowing atmosphere is called a planetary nebula, and the core's heat can excite spectral lines of helium in the planetary nebula's gas. In 1963 astronomers observed a planetary nebula in M15, a globular cluster in the constellation Pegasus. Although M15's metal abundance was less than a hundredth solar, its planetary nebula harbored lots of helium. This discovery suggested to Fred Hoyle and Roger Tayler that the helium had formed before the Galaxy did, possibly in the big bang, and it was they who in 1964 worked out the details of primordial nucleosynthesis. Trouble was, the high helium levels in M15's planetary nebula might have originated in the dying star rather than the big bang. However, later work showed that high helium levels did explain the luminosities and temperatures of other stars in globular clusters.

Any remaining helium debate ended in 1971, when two British-born astronomers, Leonard Searle and Wallace Sargent, found huge quantities of the element—not in the Milky Way's halo, but in two odd galaxies 40 million light-years beyond.

Sargent had gotten interested in astronomy as a child, by reading books. "Then when I was fifteen," he said, "I heard Fred Hoyle give lectures on the BBC called 'The Nature of the Universe.' The idea that you knew what the temperature and density were at the center of the Sun came as a hell of a shock. At the age of fifteen, that sort of thing seemed beyond knowledge. It was not just the amazing numbers, but the fact that you could know them at all.

"Of course, Fred used the opportunity to describe the steady state theory, and I was quite attracted to it at the time. And also, he caught me at an age when I was trying to shed the Church of England religion that I'd been brought up in. His last lecture was a sort of antireligious lecture, and from that point on, I dropped any belief in organized religion—and got into trouble at school as a result."

After reaching Caltech, Sargent began observing galaxies catalogued by Fritz Zwicky, the same maverick astronomer who had detected dark matter in the Coma galaxy cluster—and who had antagonized most of his colleagues. "I got along with him very well," said Sargent, "partly because we British like eccentrics and deliberately try to exaggerate our own behavior, partly as a means of concealing our true thoughts. He was an extraordinarily creative and able scientist. He had a gift for finding things, which was highly unusual. You could be close to him for years and never figure out how he did it."

Some of Zwicky's galaxies were stellar breeding grounds like the Orion Nebula—only all alone, adrift in the intergalactic void, unanchored to any larger galaxy. Like the Orion Nebula, they bore hydrogen gas that had been ionized by the hot, blue stars they had just given birth to. These so-called extragalactic H II regions—"H II" is astrospeak for ionized hydrogen—were minuscule, dozens of times fainter than the Large Magellanic Cloud, hundreds of times fainter than the Milky Way. If the Milky Way was a mighty metropolis, extragalactic H II regions were small towns off in the country.

Meanwhile, one of Sargent's colleagues, Leonard Searle, had been observing H II regions in nearby spiral galaxies. Searle's spectra indicated that H II regions on the periphery of spiral galaxies had less oxygen than those more centrally located.

"We compared our spectra," said Sargent, "and he noticed that two of the galaxies that I'd studied out of Zwicky's list were in fact H II regions similar to but more extreme than the ones on the outsides of nearby spiral galaxies." The two galaxies were named I Zwicky 18 and II Zwicky 40, in the constellations Ursa Major and Orion, respectively. Using the Palomar 200-inch telescope, Searle and Sargent measured the galaxies' supply of oxygen and neon.

What they found startled Sargent. "No other galaxies had been discovered with that low an abundance," he said. "There was no particular reason to suspect, as we now do, that dwarf galaxies tend to get through their supply of gas more slowly than bigger galaxies." Life in small galaxies, as in small towns, proceeds leisurely. Since small galaxies form stars from their gas slowly, they also enrich themselves with elements like oxygen and neon slowly. Said Sargent, "In the case of I Zwicky 18 and II Zwicky 40, they are parsimonious in their use of their gas, for reasons which are not known. And curiously enough, I Zwicky 18 still holds the record for the lowest oxygen abundance of any known extragalactic H II region."

Searle and Sargent pegged that oxygen abundance at only 1/14 solar. Later work lowered that to a more extreme 1/50 solar. The neon abundance was also low. Thus, like the Milky Way's halo, the two dwarf galaxies nearly matched the big bang's composition—but unlike the halo, they also had hot, blue stars that should excite spectral lines of helium in their gas.

Searle and Sargent searched for those helium lines, and found them strong: the galaxies boasted about as much helium as the Sun and the Orion Nebula. Stars couldn't have made the helium, since they'd made so little else. Instead, the helium must have come straight from the big bang. Modern work puts the primordial helium abundance at around 24 percent by mass, which means that about one in every dozen big bang nuclei was helium. The Sun is 28 percent helium by mass, so even most of its helium originated in the big bang.

"In retrospect," said Sargent, "I think we made a mistake in not trumpeting this conclusion more loudly. I mean, it's in the paper, and we are credited with that discovery by the more discerning students of history, but nevertheless we didn't make as big a deal of it as we might have done." The main body of the nine-page paper mentioned the cosmological implication only in the last paragraph. "It was an understated way of doing things," said Sargent, "which partly arose from the fact that both Searle and I are British. In our minds, making an understated thing like that is stronger than shouting it from the rooftops.

"It reminded me, and maybe we had it in mind, of the last sentence in Watson and Crick's paper about the structure of DNA. The last sentence says something like, 'It has not escaped our attention that the structure that we have proposed provides a natural copying mechanism for the hereditary material.' The 'It has not escaped our attention' makes the claim and stops anybody else making the claim, but they don't go into any detail. And the last paragraph of our paper stops anybody else from making that claim, but it's not very explicit about what it is we're claiming!"

Although the high primordial helium-4 abundance certified the universe's fiery birth, it revealed little about its eventual fate—because the element's abundance depended more on the neutron-to-proton ratio during primordial nucleosynthesis than on the universe's matter density, omega. To unveil omega, astronomers required a much touchier big bang offspring.

## THE BOOK OF DEUTERONOMY

Deuterium leads a rough life. The big bang barely spat it out, and every star is its enemy. At the Sun's center, deuterium forms when two protons smack together, but this reaction typically takes billions of years. In contrast, deuterium's demise comes just a second later. That's how fast a proton transforms deuterium into helium-3.

"The unique thing about the big bang," said Donald York of the University of Chicago, "was that you had only a few seconds to make the deuterium. By the time more collisions would have turned it into helium-3, the universe had expanded and cooled, so there wasn't enough energy to have fusion occur anymore—and the deuterium survived."

This deuterium still exists—not in stars but in the cold space between them, and even right here on Earth. Drink a cup of water and you ingest over a billion trillion deuterium atoms from the big bang. That's because, for every 3,000 molecules of $H_2O$ on Earth, there's 1 molecule of HDO.

How much deuterium the big bang fireball tossed out depended on the universe's density during primordial nucleosynthesis. If the universe is a featherweight, with an omega of just 0.01, then some deuterium should have blazed out of the inferno; but if omega is 1, then less than a millionth as much deuterium should have survived. Water's high deuterium-to-hydrogen ratio—0.00015, or $1.5 \times 10^{-4}$—implied a lightweight universe, one that would expand forever.

In the early 1970s, however, scientists began to suspect that the celestial deuterium-to-hydrogen ratio was ten times lower, implying a denser universe, albeit one that would still expand forever. Strangely, they reached this conclusion by looking not at deuterium but at helium-3. The Sun's outer layers have burned their original deuterium to helium-3 but are too cool to burn that. Thus, the present helium-3 abundance on the Sun's surface lends clues to its original deuterium abundance. The solar surface shoots out a wind of particles that implant themselves on asteroids and the Moon. In 1971, by measuring the helium-3 abundance in small asteroids that had fallen to Earth as meteorites and in foils that the Apollo astronauts had unfurled on the Moon, scientists estimated the Sun's original deuterium-to-hydrogen ratio.

Still, these measurements relied on helium-3 rather than deuterium itself. To observe that, astronomers had to look to the tenuous scattering of atoms that dance between the deuterium-eating stars. Interstellar atoms absorb certain wavelengths of light, imprinting dark lines onto the spectra of stars they wander in front of. Like many interstellar atoms, however, deuterium has spectral lines at far ultraviolet wavelengths, which astronomers on Earth can't see. To study these interstellar atoms, NASA launched an ultraviolet telescope aboard a satellite named Copernicus.

In 1973 this satellite detected interstellar deuterium. "It was an accident," said York, who was then at Princeton, which had built the ultraviolet telescope. "You know, when you're young, you have to do something to identify yourself. I was the young guy on the team, and the senior people had picked topics that were the obvious things to do, such as the nature of molecular hydrogen in the interstellar medium. So if I wanted to do anything on my own, I had to pick something that was out of the way. I picked an obscure topic, which was looking at all the very brightest stars in which no interstellar lines had ever been seen."

York chose to observe Beta Centauri, a blue giant star hundreds of light-years beyond its famous cousin, Alpha. Blue stars like Beta Centauri emit copious amounts of ultraviolet light. To record Beta Centauri's ultraviolet spectrum, the satellite took twenty days.

"We made it a habit to go down and thumb through the spectra that came in every single day," said York, who examined the spectra with Princeton astronomer Jack Rogerson. "The spectra were plotted out on a piece of graph paper. And Jack went down one day and noticed that, beside all the hydrogen lines, just to the left, there was a little absorption line, and he immediately realized it was deuterium."

Another Princeton astronomer, Lyman Spitzer, recognized the cosmological implication. "The fact that we saw the deuterium so easily," said York, "clearly implied that the universe had a very low density." The deuterium-to-hydrogen ratio that Rogerson and York found, $1.4 \times 10^{-5}$, was ten times less than water's, just as the helium-3 work had suggested. Rogerson and York concluded that the universe had no more than a few percent of the matter density needed to collapse. Around the same time, other scientists detected the molecule HD on Jupiter, from which they derived a similar deuterium-to-hydrogen ratio.

Given the importance of the result, York said he wished that he'd had the foresight to search for the deuterium rather than to stumble across it. "Virtually the day after we discovered it," he said, "I got a letter from a group in France, led by Alfred Vidal-Madjar, asking if they could look for it. I'm very good friends with that guy now, because we said, 'Sure, come on and join.'"

## THE LITHIUM PLATEAU

Even after the sightings of primordial helium and deuterium, the heaviest and rarest big bang nucleus eluded detection. That was hardly surprising: primordial lithium's abundance relative to hydrogen was predicted to be a mere $10^{-9}$ to $10^{-10}$. If the world's people were big bang nuclei, then the vast majority would be hydrogen-1 or helium-4, a few hundred thousand would be deuterium or helium-3, and only one or two would be lithium-7.

Nevertheless, in 1981 two French astronomers observing ancient halo stars discovered big bang lithium—by accident. "We were interested in *aluminum*," said François Spite. "We were interested in very old stars, because it's interesting to understand how the metals were built little by little in the Galaxy. The relative proportions of the abundances of the metals are different in old stars than in young stars, because the metals in old stars were built by primitive supernovae, which are different from modern supernovae."

Spite had gotten into astronomy through physics. "I have always been interested in physics," he said. "In my youth, I was fascinated to understand how things work—how the telephone works, how the light shines, how the water flows from the tap." In college, he took physics. "Astronomy was present everywhere. Any extreme phenomenon was illustrated by astronomy: high density was illustrated by white dwarfs, and relativity was illustrated by astronomy." Spite's work on the aluminum abundance of halo stars was about to lead him to the most extreme phenomenon imaginable: the big bang.

However, searching halo stars for the big bang's heaviest element seemed pointless. "If we had made an application for telescope time to observe lithium," said Spite, "we probably would have been turned down, being told that it is impossible and stupid and you will find nothing—so it's just a waste of telescope time. Groombridge 1830 had been observed because it was bright, and there was no lithium." Groombridge 1830 is a nearby main-sequence star in the halo; it's yellow, like the Sun, lies just thirty light-years from Earth, and is visible through binoculars, south of the Big Dipper. Another bright halo star, the giant HD 122563 in the constellation Boötes, also lacked lithium. "So two old stars had no lithium," said Spite, "and people concluded that there was no lithium in old stars."

Lithium abounds most in young stars, so much so that astronomers once thought that these stars manufactured it. They now realize that at least some of it arises in the space between the stars, when cosmic rays split heavier elements into lighter ones, like lithium. As newborn stars form from this lithium-laden material, they inherit large quantities, but their heat soon destroys the fragile element. The Sun has depleted nearly all of its supply. Meteorites, which preserve a record of the Sun's natal material, have a lithium abundance over a hundred times the present Sun's. And halo stars, being much older than the Sun, should have annihilated all the lithium they acquired from the big bang, thereby burying a key clue to the universe's birth.

The element that Spite and his wife, Monique, were observing in halo stars, aluminum, had little in common with lithium, but both elements happened to have spectral lines at red wavelengths. In the same way, two words with little in common—such as *appealing* and *appalling*—can fall on the same page of the dictionary. By looking up one word, you chance upon the other; by observing aluminum, the Spites might see lithium.

Using the new Canada-France-Hawaii telescope atop Mauna Kea, the Spites scrutinized a slew of yellow main-sequence halo stars. "It was a big surprise," he said. "While observing in this spectral region, we found, unexpectedly, a lithium line." After the first detection, the Spites found lithium in every halo star they studied—except Groombridge 1830. Furthermore, even though the halo stars spanned metallicities from 1/12 to 1/250 the Sun's, all but Groombridge 1830 had nearly the *same* lithium-to-hydrogen ratio.

"The simplest explanation," said Spite, "is that this is the big bang abundance, nearly unchanged." A uniform abundance signifies an origin in the big bang, which conferred on all stars equal quantities of its progeny. In contrast, if the lithium had been made along with the heavier elements, then the more metal-rich halo stars should have had more lithium. And the stars hadn't destroyed any of their original lithium, because stars of different masses and metallicities should have destroyed the element at different rates.

But how did such old stars manage to retain their full lithium supply, when the more youthful Sun has already depleted its? The answer lies in the halo's low metal abundance. Metals absorb light, so light has a hard time leaving a metal-rich star like the Sun. As a result, the Sun's outer layers bubble and boil, transporting heat by convection; any lithium that once sat on the Sun's surface got dragged inside, where other processes dragged it deeper still, and the heat destroyed it. In contrast, light readily escapes from the more pristine halo stars, so they have thinner convective zones and preserved their surface lithium. Only Groombridge 1830 had destroyed all its lithium, because it was cooler than the Spites' other stars, and cool stars have deeper convective zones.

As did deuterium, the lithium abundance that the Spites measured pointed to a low-density, ever-expanding universe. "The quantitative result that they found was very interesting," said Gary Steigman. "The lithium abundance was low enough to give you a nearly unique value for the density, but high enough to be consistent with the predictions. That consistency need not have been the case, so many people took this as one of the striking successes of primordial nucleosynthesis." Unlike the other big bang nuclei, lithium-7's abundance first falls with omega, then rises. At very low omega, the lithium-to-hydrogen ratio is predicted to be $10^{-9}$. As omega rises, that ratio falls to $10^{-10}$. Then, as omega rises further, the ratio rises again, back up to $10^{-9}$. The actual ratio that the Spites measured—around $10^{-10}$—was near the predicted minimum, thereby constraining omega to lie within a narrow range.

Unfortunately, complications set in. Because omega is the ratio of the universe's actual density to the critical density, and because the critical density depends on the square of the Hubble constant (see Table 2 in the Appendix), the omega that scientists derive from the light elements depends on the inverse square of the Hubble constant. The lower the Hubble constant, the greater is the omega that the light-element abundances imply.

Nevertheless, for most measured values of the Hubble constant, this omega falls between 0.01 and 0.10. "This was a number bigger than what we inferred from what's shining in the universe," said Steigman. The light-element abundances therefore confirmed the presence of dark matter.

Still, the omega that cosmologists derived from the light-element abundances carried a catch: it measured only the density of ordinary matter, consisting of protons and neutrons, which participated in the primordial nucleosynthesis. Because physicists call protons and neutrons "baryons," they call ordinary matter "baryonic." Any *non*baryonic matter did not participate in primordial nucleosynthesis and thus did not affect the light-element abundances. If this nonbaryonic matter exists, it is exotic, differing from the stuff of our world and our bodies; nonetheless, it would also carry weight and contribute to omega. Thus, the omega derived from the light-element abundances—0.01 to 0.10—is only the *baryonic* omega. The true omega could be greater. And just as astronomers were nailing the primordial lithium abundance, a new theory was arising that predicted enormous quantities of dark matter, all of it exotic.

# EIGHT

# UNIVERSE IN HYPERDRIVE

*Had I been present at the Creation, I would have given some useful hints for the better ordering of the universe.*

—ALFONSO THE WISE

LATE one night in 1979, a peripatetic postdoc pulled out a piece of paper and forever altered cosmology—even though he disdained the field. "I considered cosmology to be a very interesting subject," said Alan Guth, "but a subject about which too little was known for it to be a serious profession. Because the observations that could be made were so limited, I felt that it was hard to go beyond a fairly broad-brush description of how the universe behaves."

Yet Guth's own theory, which he dubbed inflation, described the universe within a sliver of its creation, even before the big bang inferno coughed out its hydrogen, helium, and lithium, at a time when the cosmos was a mere 0.00000000000000000000000000000000001 second old. At that modest age, Guth said, the infant universe began to scream. According to the standard big bang theory, by the time the universe was a hundred times older—a still modest 0.000000000000000000000000000000001 second old—it had grown only ten times larger. But no, Guth said, during that brief interval the universe actually bloated up by a factor of at least 10,000,000,000,000,000,000,000,000,000, so that the presently observable universe swelled from a trillionth the size of a proton to ten times the size of a beach ball. Microscopic ripples got stretched and gave rise to density variations that would spawn the galaxies. If Guth's theory was right, it resolved two cosmological riddles, but it also erected a barrier, veiling our view of the big bang itself.

## THE ACCIDENTAL COSMOLOGIST

Since childhood, Guth had been interested in mathematics and science. "I was always fascinated by the relationship between math and the laws of nature," said Guth. "In high school, I guess, I realized that what I was most interested in was trying to understand nature at the most fundamental level—and physics was the branch of science that seemed to be most firmly centered on that goal."

Fundamental to physics itself were the forces that ruled nature, such as those through which the Moon stirs the seas and electrons illuminate a lightbulb. Physicists recognized four basic forces, and so-called grand unified theories tried to join three: electromagnetism, which makes that lightbulb glow; the strong force, which binds the particles in atomic nuclei; and the weak force, which governs such radioactive decays as the splitting of a free neutron. Thus, grand unified theories didn't quite live up to their name, for they omitted gravity. Nevertheless, they claimed that at the torrid temperatures which seared the nascent universe, the three forces united into one.

Guth hadn't worked on grand unified theories, but they would lead him to cosmology. "It all really started due to the efforts of a friend of mine, Henry Tye," said Guth. "We were both postdocs at Cornell, and he came to me one day and asked me whether or not these new theories called grand unified theories would imply that magnetic monopoles should exist."

Guth *had* worked on the theory of magnetic monopoles, but scientists had never seen one, even when they'd gone as far as searching lunar rocks and cosmic rays. Still, if magnetic monopoles existed, they were stripped-down magnets. Every known magnet bore two equal and opposite poles, a north and a south, but the elusive magnetic monopole carried one pole without the other.

Guth soon had an answer for Tye: if grand unified theories held, magnetic monopoles must exist. "I told Henry that he should just forget about it," said Guth, "because we'd never see those particles. When particle physicists try to find new particles, the usual method is to try to produce them in accelerators by colliding things together; and if the mass required is far more than the energy of your accelerator, you just can't produce them." No terrestrial accelerator could achieve enough energy to coax forth even a single magnetic monopole, which Guth calculated should weigh a whopping 10,000 trillion times more than a proton.

Nevertheless, Tye persisted. How many magnetic monopoles, he asked, had shot out of the greatest particle accelerator of all, the big bang? "This actually sounded pretty crazy to me," said Guth, "because I didn't think we knew very much about the big bang, and I certainly didn't know very much about grand unified theories." So for months, Guth put it off. Then, in the spring of 1979, physicist Steven Weinberg spoke at Cornell. Later that year, Weinberg would receive the Nobel prize for his role in unifying two of the four forces, the electromagnetic and the weak. During his Cornell talks, he discussed how grand unified theories could explain why the early universe preferred matter to its mirror twin, antimatter. "I had an enormous amount of respect for him," said Guth, "and the fact that he would work on this crazy topic of thinking about the universe at $10^{-35}$ seconds convinced me that maybe it wasn't so crazy after all."

The day after Weinberg left, Guth began to ponder how many magnetic monopoles the big bang had unleashed. The troubling answer: lots. Magnetic monopoles should outnumber protons. Since magnetic monopoles also far outweigh protons, the universe's matter density—omega—would surpass a trillion, crushing the cosmos a thousand years after its birth, long before it could foster life. Unbeknownst to Guth and Tye, a graduate student at Harvard named John Preskill had already scooped them, pointing out the same problem.

"We were of course disappointed," said Guth, who at the time had no permanent job. "So Henry and I then realized that we should shift our emphasis towards trying to figure out if there's some way of modifying the standard cosmology that might enable it to be compatible with grand unified theories."

According to those theories, the various forces broke off from one another as the universe expanded and cooled. Gravity broke off first, in a way that grand unified theories didn't describe. Then the strong force branched off, $10^{-34}$ second after the big bang, when the temperature had dropped to a billion billion billion degrees. During that phase transition, the theories said, defects arose—magnetic monopoles. Guth and Tye realized that if they could delay that transition, the expanding universe would smooth out the defects, like a painter who rubs out blemishes before they dry, and few if any magnetic monopoles would mar the observable universe.

"In our paper, we just assumed that the universe would go on expanding as it would have otherwise," said Guth. "But I'm quite sure it was Henry who suggested that we should look at whether that's true or not."

## FLAT-OUT INFLATION

On the night of December 6, 1979, Guth went home to work out the consequences of stalling the phase transition. "It had a tremendous effect, a very bizarre effect, on the expansion of the universe," he said. "It actually created an antigravity effect, causing the universe to go into a period of enormously rapid expansion."

In fact, the universe exploded on him, for it summoned that most dangerous, most unruly of mathematical functions—the exponential, which feeds on its own insatiable power, blowing up any number you give it, then blowing even *that* up. The exponential swells a mere dollar, invested for 300 years at 10 percent annually, into over a trillion dollars; because not only does the original money generate interest, but so does the interest itself. In Guth's case, space was like money, surging at a rate that depended on how much space existed. As that space tore open, so did the rate at which yet more space opened up. From $10^{-34}$ to $10^{-32}$ second after the big bang, Guth estimated, the universe puffed up by at least $10^{28}$. Stretch a tape measure from one side of the presently observable universe to the other, across some 30 billion light-years of space, and count the inches: there will be $10^{28}$ of them. Thus, in a split second, an inch-sized patch of space bloated into a region as large as the *presently* observable universe.

By the end of December, Guth named this growth spurt "inflation." "It's a curious thing," he said. "Although a lot of the history I remember quite well—I was keeping a diary at the time, so a lot of it I even have documented—I was never able to reconstruct when or exactly why the name 'inflation' came about." During the 1970s the United States faced great economic inflation. "So the word was in the air," he said, "and that certainly had something to do with my choosing it. It's even conceivable that somebody who I described it to suggested the word *inflation*. But nobody has ever told me that he was the person who suggested the name, so as far as I know, I'm the one who came up with it."

The morning after discovering inflation, Guth wrote the words SPECTACULAR REALIZATION. "I was very excited about it," he said, "because I realized immediately that it would solve the flatness problem—the problem of understanding how the mass density of the universe started out so close to the critical density."

The flatness problem centered on omega, the cosmological parameter that denotes the universe's density. Guth had heard about the problem during a lecture given by Princeton's Robert Dicke, who had first mentioned the problem a decade earlier. Although omega is a fundamental cosmological parameter, it does not necessarily stay constant. To see why, picture omega as the ratio of two enemies: gravity, which tries to pull galaxies back together; and the universe's expansion, which tries to push them away. If gravity wins, the universe will someday collapse and omega exceeds 1. Furthermore, as time goes on, this omega rises further above 1. When gravity halts the expansion, omega will be infinite, because it is the ratio of gravity to the expansion rate—and the latter will be zero. Conversely, if gravity is too weak to stop the expansion, the universe will expand forever, and omega is less than 1. As time goes on, though, this omega drops toward 0: galaxies fly away, so their gravitational tugs on one another weaken, while the universe's expansion proceeds nearly unbridled. Thus, whether the universe is open or closed, omega strays further and further from 1. Only if the universe is flat, and omega equals 1, does omega stay the same.

As Dicke pointed out during his lecture, omega *is* close to 1 today—it probably lies between 0.1 and 10—even though the big bang occurred some 15 billion years ago. In order for today's omega to be between 0.1 and 10, just a second after the big bang it had to be 1.00000000000000. Why, Dicke asked, did omega start with such a precise value? Why was the universe so flat?

Thought Guth: inflation! "That's what initially got me so intrigued by all this—that it seemed to provide a very simple explanation of this problem, which I had come to think was very, very important." No matter what numerical value omega happened to have *before* inflation, the inflationary epoch would flatten the universe and drive omega to 1. In the same way, if you blew up a beach ball to the size of the Earth, it would look as flat as the Earth's surface.

## NEW HORIZONS

A few weeks later, Guth heard of another cosmological problem. "I learned about the flatness problem by accident," said Guth. "And I also learned about the horizon problem by accident. One day some of the people at my lunch table were talking about it, and I asked them what this horizon problem was. Then I went home and thought a little bit and realized that inflation would solve the horizon problem as well, and that made it all the more exciting."

The horizon problem, which Charles Misner of the University of

Maryland first stated, showed up most dramatically in the cosmic microwave background, the afterglow from the big bang that dates back to about 300,000 years after the universe's birth. "In the context of the conventional big bang theory," said Guth, "there was not nearly enough time for the universe to come to a uniform temperature. So if different regions started out with nonuniform temperatures, they would have still had nonuniform temperatures 300,000 years after the big bang. There's no way it could have evened out." Yet wherever astronomers look, the cosmic microwave background has the same temperature, 2.7 Kelvin.

This radiation arose when light in the young universe escaped the clutches of matter—when the universe had cooled enough to allow free electrons, which scatter light, to join protons. Because this so-called recombination happened only 300,000 years after the big bang, and because space has since been nearly transparent, the light from the cosmic microwave background has traveled unhindered for almost as long as the universe is old. Thus, it originated near our horizon. Recall that the horizon exists because astronomers can't see beyond a distance farther than the universe is old. For example, if the universe is 15 billion years old, they can't see farther than a lookback distance of 15 billion light-years—light from more distant regions hasn't had time to reach the Earth. Thus, that light hasn't had time to *influence* the Earth.

Trouble is, light beams from opposite directions—for example, 15 billion light-years above the North Pole and 15 billion light-years below the South Pole—lie beyond each other's horizon and thus could not have exchanged heat, yet both indicate the same temperature, 2.7 Kelvin. In fact, according to the equations governing the universe's expansion, things are even worse: 300,000 years after the big bang, those two points were already separated by some 80 million light-years of space, a distance that light and heat could not have bridged in only 300,000 years.

Guth realized that his theory could solve this horizon problem. "With inflation," he said, "you start out with this very, very small region, which has plenty of time to come to a uniform temperature. Then inflation takes over and magnifies this tiny region to become large enough to include everything that we see." Thus, the cosmic microwave background that forms a backdrop to the stars, galaxies, and quasars started out as a microscopic speck of uniform temperature, and inflation blew it up so that today its temperature is uniform, no matter which way observers look. Guth's theory meant that the observable universe—all space within the horizon—was minuscule compared with the total universe: vast reaches of uncharted space dance beyond the limits of our vision.

Guth began giving talks about inflation, and job offers flooded in from universities. "I knew from the first night that this was exciting," said Guth, "but in fact I felt very trepidatious, because it was clearly a very dramatic change to what people had been thinking previously. Whenever one wants to propose a dramatic change, there's always the possibility that what you're proposing is dramatically wrong; and there's always the nagging worry that if this is right, why didn't someone else think of it first? So for a good period of time, I was worried that if we kept looking at this idea, we would find something about it that was drastically wrong."

In fact, that happened. Like a novelist who introduces colorful characters into an original plot but fails to bring the story to a satisfying close, Guth couldn't tame the inflationary beast he'd unleashed—instead, it tore huge irregularities into the universe. Rather than produce a universe that looked the same in all directions, inflation seemed to generate the very inhomogeneities it was supposed to get rid of.

## NEW INFLATION

The first to successfully tie up the inflationary plot thread was Russian physicist Andrei Linde. As a child, though, Linde had wanted to study the Earth, not the universe. "I liked to collect minerals and dream about going on expeditions, discovering new places," said Linde. "Then once I was traveling with my parents in our car to the south of our country, to the Black Sea, and they gave me a book to read, about astronomy. By the end of my journey, I was a completely changed person."

In graduate school, Linde studied particle physics, just as Guth had, which brought him into contact with cosmology. "When you start doing cosmology," said Linde, "you cannot stop. Cosmology is as close to the most important questions which you may ask in science. If you study how particles collide, it's interesting; but if you study how a universe was created, it's *magnificent*."

Now in the United States, Linde originated his new inflationary scenario while in Moscow. "All my best work, I did in Russia," he said. "There was complete scientific freedom, because I was not obliged to teach. I did what I believed was interesting." However, communications abroad—even scientific papers—had to be vetted by the communist authorities, delaying publication. Furthermore, Linde said that Western scientists often ignored Russian work, as he saw for himself one day when he visited a library in Switzerland. "There were some Russian journals," he said, "and they were all white and never opened. It's like you go to a cemetery and you see these pieces of knowledge, all your old friends—this is what they said, and nobody cared."

Linde knew all about the trouble Guth was having with inflation, for he had flirted with the same idea a year before—and abandoned it, because of its failure to end gracefully, and because he hadn't heard of the flatness and horizon problems, which had first been enunciated in America. Once he learned of those problems, though, he saw the inflationary theory's strengths. "I just had the feeling that it was impossible for God not to use such a good possibility to simplify his work," he once told MIT physicist Alan Lightman. This impelled him to construct a so-called new inflation model, in which the growth spurt ended gradually, so that it didn't kick up irregularities. Shortly afterward, Andreas Albrecht and Paul Steinhardt at the University of Pennsylvania independently discovered the same solution.

In 1982 advocates of inflation began to investigate the radical idea that the grand and glorious galaxies which span hundreds of thousands of light-years might have been sparked by the laws of quantum mechanics, which normally rule only the subatomic world. According to the quantum uncertainty principle, tiny ripples appeared; inflation stretched them, giving rise to variations in density. Dense regions then robbed their sparser neighbors, amplifying their own gravitational power and growing into the building blocks of galaxies.

However, the original version of new inflationary theory predicted that these density perturbations were too large, thus dooming this model to the same death that old inflation had suffered. In 1983, Linde proposed another alternative, which he called chaotic inflation. In this version of inflationary theory, some regions of the universe experienced inflation while others didn't. Because those regions which experienced inflation grew so huge, they soon dominated the universe's total volume.

If the inflation theory is right, the cosmic stampede just after the big bang erased all that came before, during the big bang itself. As Guth wrote in his book *The Inflationary Universe*, "The beauty of inflation—

that it can make predictions independent of the details of the initial conditions—becomes the cosmologist's ultimate barrier. Because the properties of the observed universe are determined by the physics of inflation, they tell us nothing about what existed before inflation." For example, because inflation flattened the universe, astronomers can't tell what the numerical value of omega was before inflation occurred.

"I don't know whether emotions here have any value," said Linde—"whether it's bad or not that we do not know something about initial conditions because everything is erased. Well, okay, maybe if you want to know how the universe was created, it would be more pleasant for you to look through the telescope and see here is the place where it was created. But even without inflation, you would have a problem, because the universe is transparent to light only starting 300,000 years after the big bang.

"But you know, you never get good presents for free. You wanted to explain the properties of the universe, and you got it—you explained the mystery—but you lost some of the abilities of observing clearly where this creation occurred."

Strangely, the shroud that the inflationary epoch threw over prior times recalls the words that Georges Lemaître wrote and then deleted from his 1931 paper proposing the big bang: "I think that everyone who believes in a supreme being supporting every being and every acting, believes also that God is essentially hidden and may be glad to see how present physics provides a veil hiding the creation."

## THE ACCIDENTAL COSMOS?

Many cosmologists, including Guth and Linde, believe the universe arose from nothing—though it's a "nothing" with a crucial qualification. According to the quantum uncertainty principle, quantum fluctuations cause tiny particles to pop into existence out of nothing. The larger the particle, the less likely this quantum creation is, which explains why we don't see rocks or mountains coming out of the void. In 1973, however, physicist Edward Tryon suggested that a quantum fluctuation had given birth to the entire universe. Since the universe is anything but small, the chance that it could pop into existence might seem slim. If the inflation theory is correct, though, then a small universe—perhaps containing just a few particles—could spring forth, and inflation would then take over and enlarge it to its presently monstrous size.

On repeated occasions, Guth has called such a something-from-nothing universe "the ultimate free lunch."

However, closer examination reveals a sleight of hand, for even when there's "nothing," there is still, inexplicably, something: the laws of physics that allow such quantum fluctuations. "One doesn't have any matter or any space or any time," said Guth, "but one does have the laws of physics, and in the vague use of the word *something*, that is something, and I completely agree that it's a very significant something. There's an important part of the question here that one could ask and that we're not offering any answer to whatever: where did these laws of physics come from?"

If the universe is indeed a quantum fluke, it may simply be an accident, devoid of any purpose. "In answer to the question of why it happened," Tryon wrote, "I offer the modest proposal that our Universe is simply one of those things which happen from time to time." Then again, perhaps the purpose of the universe is to create beings who think the universe has no purpose.

Guth pointed out that such questions lie beyond physics. "It's okay to ask those questions," he said, "but one should not expect to get a wiser answer from a physicist. My own emotional feeling is that life has a purpose—ultimately, I'd guess that the purpose it has is the purpose that we've given it and not a purpose that came out of any cosmic design. But I think there's no real way of answering questions like this, and it's the kind of question which each individual will be seeking his or her own answer to."

"For me," said Linde, "consciousness is the biggest miracle. I study the universe, but eventually it would be my desire to understand more about *us*. Some believe that when they study matter, and they finish this study, that will be it—the full picture. But I think that there will be some parts missing, and these parts are extremely important, and one of them is this study of consciousness, which I take very seriously."

Our universe may not even be unique. Instead, other universes may have sprouted off from it, possibly an infinite number of them, via quantum fluctuations amplified by inflation. Indeed, our own universe may have sprung from some other one. Although all stars within our universe will eventually die, new universes could emerge so that the entire ensemble of universes—the mega-universe—lives forever.

In 1986, Linde found that this unusual possibility actually occurred in most versions of chaotic inflation. He called such a mega-universe "the eternally existing, self-reproducing inflationary universe." "It's like when a river flows," he said. "It becomes diverted into many smaller rivers, and each of these small rivers looks like a river. It does not seem to change. But nevertheless, there is change. Our part of the universe *is* expanding; our part of the universe *is* changing; but in a self-reproducing universe, if you look at the whole tree of universes producing quasi-universes, the whole tree is an eternally growing tree which does not seem to change."

In a way, such a mega-universe resembles the big bang theory's arch rival, the steady state theory. "If one assumes that the steady state theory was motivated more by philosophy than physics," said Guth, "this I think achieves the same philosophical goal: a universe which goes on forever and would not change with time, even though within any 'pocket' universe, one would see definite evolution, which we identify with the big bang theory."

Even if ours is the only universe, inflationary theory proclaims that it could be far more enormous than the observable universe, possibly spanning trillions, quadrillions, or quintillions of light-years—and conceivably even more. "One can consider it depressing that we can see such a small fraction of it," said Guth, "but alternatively, depending on one's psychology, one can consider it exhilarating to know that there's so much out there. Although we don't have any way of learning about it at this point, we can at least speculate about it, rather than just to believe that it's not there at all." Since distant light hasn't had time to reach the Earth, the standard view is that astronomers can't see these far-off reaches of the cosmos. Nevertheless, this pessimistic view might prove as wrong as nineteenth-century French philosopher Auguste Comte's pronouncement that scientists would never know what the stars are made of.

## TESTING INFLATION

Other universes can get intoxicating: you can say anything you want about them and never be proved wrong, as long as astronomers can't see them. Nevertheless, the inflation theory itself predicted that the space of the observable universe should be flat. If there is no cosmological constant, that means omega must be precisely 1.0. However, the primordial abundances of deuterium, helium-4, and lithium-7 imply that baryonic matter—ordinary material, built of protons and neutrons—

gives an omega of only about 0.01 to 0.10 (see the discussion at the end of Chapter 7), far short of inflation's prediction. Thus, if omega is really 1.0, then the universe must harbor vast quantities of some mysterious, *non*baryonic matter that constitutes 90 to 99 percent of the universe's mass.

But many astronomers never bought the idea that omega was 1. "It was theoretical physicists who argued that way, and it never struck me as a very powerful argument," said Jeremiah Ostriker, the Princeton astronomer who recognized that dark halos stabilized galactic disks. "It was more based on desire. It was like the arguments for steady state, from Hoyle. It was sort of on the level of, 'Wouldn't it be nice if?' Without any observational evidence for it *ever*."

That didn't bother Guth. "I was still pretty confident that inflation was right, in spite of that observational evidence. I always took these fundamental problems—the flatness and horizon problems—as being observational evidence for inflation. It just happened to be a measurement that was made before inflation was invented rather than a measurement that was made after. I considered that a more important observational triumph than the observational discrepancy over the value of omega, which everybody agreed was pretty uncertain."

After the invention of the inflation theory, astronomers were even more motivated to measure omega's numerical value. Meanwhile, as the theory was taking hold among many theorists, convincing them how gargantuan the cosmos might be, observers were discovering how lumpy it was, crisscrossed by vast chains of galaxies hundreds of millions of light-years long, with one mighty conglomeration yanking the entire Local Group of galaxies.

# NINE

# THE ARCHITECTURE OF THE COSMOS

ALBERT EINSTEIN assumed it, Edwin Hubble "proved" it: the universe was uniform, homogeneous, isotropic, *boring*. Galaxies sprinkled space like flowers adorning spring. Sure, they clustered in occasional bouquets—in Virgo, in Coma Berenices—but elsewhere these cosmic blossoms dotted the universe at random, wildflowers strewn across a vast field. Indeed, astronomers called galaxies lying outside obvious clusters "field" galaxies, and they enshrined the universe's dullness in a celestial statute known as the cosmological principle.

"There were voices in the background saying, 'Wait a minute; wait a minute,'" said Steve Gregory of the University of New Mexico, one of the first to attack the cosmological principle. "But there was one voice that beat out all others, and that was Edwin Hubble himself. Because of the weight of his name, his argument carried the day."

After discovering the universe's expansion, Hubble attempted to probe the galactic distribution by taking more than a thousand photographic plates in different directions. On these plates he counted a total of 44,000 galaxies. Hubble wrote, "The counts . . . conform rather closely with the theory of sampling for a homogeneous population. Statistically uniform distribution of nebulae appears to be a general characteristic of the observable region as a whole."

Thereby apparently confirmed, the cosmological principle impeded further progress in what might be called cosmography. Said Gregory,

"Even after my colleague Laird Thompson and I had made the first set of discoveries, and we tried to get money from the National Science Foundation, I remember one referee saying, 'This is complete rubbish. We all *know* the universe is homogeneous and isotropic. These superclusters and voids cannot exist.'"

## SUPERCLUSTERS AND VOIDS

Even during Hubble's heyday, hints had emerged that the galaxies he studied had never heard of the cosmological principle. For example, in 1932 an obscure Austrian astronomer named Walter Bernheimer reported that he had traced an enormous filament of galaxy groups and clusters across much of the autumn sky. "The density of nebulae in this part of the sky is remarkable," Bernheimer wrote. "Thus we count 117 N.G.C. objects in the central region, but only 8 and 3 respectively in equal large areas, situated north and south of the said region." This cluster of clusters, or *supercluster*—Bernheimer's word—started in the square of the constellation Pegasus, draped into the northern part of the faint zodiacal constellation Pisces, swept through the constellations Andromeda and Triangulum, then disappeared behind the Milky Way in the constellation Perseus. Today astronomers know that the Pisces-Perseus supercluster is one of the nearest, 270 million light-years from Earth. It stands out in part because few galaxies intervene between it and us.

In Pisces this supercluster approached the zodiac, the zone harboring the Sun's planets. Two years earlier, examination of the zodiac had yielded far-off Pluto. In 1936, as Pluto discoverer Clyde Tombaugh hunted for a farther planet, he rediscovered the Pisces-Perseus supercluster and later tried to persuade Hubble himself of the choppy galaxy distribution he was seeing. "I knew that the distribution was extremely patchy, but I couldn't convince him," Tombaugh told his biographer, David Levy. "I began, 'Dr. Hubble, on my plates, I have made counts. I have seen the area, and I don't agree with the conclusion you have drawn in the distribution. I find marked irregularities. I see the voids and the concentrations, even at high Galactic latitudes. What I have seen on my plates does not agree with what you have said.' He seemed a bit shocked about it. But I couldn't convince him." Two of Hubble's nemeses—Harlow Shapley at Harvard and Fritz Zwicky at Caltech—also noted the irregular distribution of galaxies, Zwicky going so far as to claim that the Milky Way and its neighbors belonged to the Virgo cluster.

During the 1950s, pugnacious French astronomer Gérard de Vaucouleurs challenged the establishment by proposing that the Milky Way resided at the periphery of an enormous disk of galaxies centered on the cluster in Virgo. Other astronomers refused to believe him; de Vaucouleurs didn't care. He once told MIT physicist Alan Lightman, "If it's there, damn it, I'm going to say it's there."

De Vaucouleurs called this disk of galaxies the Local Supercluster. What stars were to the Milky Way, entire galaxies were to the Local Supercluster. Thus, the Milky Way was one of its "stars," shining near its edge. Because the Local Supercluster's heart lay in Virgo, most bright galaxies populated the celestial hemisphere centered there. Astronomers had long known of this galactic imbalance. For example, Virgo boasts ten galaxies bright enough for eighteenth-century astronomer Charles Messier to have catalogued, whereas the constellation in the opposite direction—Cetus, which is nearly as large—hosts but one. In Cetus, observers gaze toward the Local Supercluster's edge, into the vast gulf of "intersuperclusteral" space beyond.

Because the Sun nestles within the Milky Way's disk, a ribbon of starglow—the combined radiance of disk stars, often called "the Milky Way"—rings the sky. In like fashion, because the Milky Way itself nestles within the disk of the Local Supercluster, de Vaucouleurs could trace its member galaxies in a narrow ribbon around the sky: from the northern constellations Cassiopeia and Camelopardalis (home of the spiral galaxy NGC 2403, a member of the M81 group) to Ursa Major (home of the spiral galaxy M81 and its companion M82, as well as the majestic spiral M101) to Canes Venatici (home of the beautiful Whirlpool Galaxy, M51, and the Sunflower Galaxy, M63) to Coma Berenices and Virgo (both home to the Virgo cluster), on south through Hydra (home of the exquisite spiral M83) and Centaurus (home of the peculiar elliptical galaxy Centaurus A) to the far southern constellations Circinus, Pavo, and Tucana, then back up north to Grus, Phoenix, and Sculptor (home of the Sculptor galaxy group) and on to Cetus and Pisces (home of the spiral galaxy M74), and then to Andromeda (home of the great galaxy there) and back to Cassiopeia. Although this ribbon covered a mere tenth of the sky, de Vaucouleurs said it contributed two thirds of all bright galaxies that did not belong to another supercluster he identified, which snaked through Cetus, Fornax, Eridanus, Horologium, and Dorado.

Astronomers call the plane containing the Milky Way's disk the

Galactic plane, so de Vaucouleurs called the plane containing the Local Supercluster's disk the *super*galactic plane. Fortunately, the supergalactic plane cuts at nearly right angles to the dust-clogged Galactic plane, allowing astronomers on Earth, and throughout the Milky Way, a clear view of most of its members. To picture this situation, lay a plate on a table, then touch a coin's face to the plate's edge. The plate is the Local Supercluster, the coin the Milky Way's disk. However, in order to illustrate the relative sizes, the plate would have to be dozens of feet across.

Unfortunately, de Vaucouleurs' idea was ahead of the technology needed to prove it. He could see in which directions the galaxies lay but not how far most were from the Milky Way, preventing him from mapping the Local Supercluster's three-dimensional structure in a way that would have convinced the skeptics. Galaxy distances, of course, could be derived from redshifts—the greater a galaxy's redshift, the greater its distance—but in those days measuring a galaxy's redshift was so arduous that known redshifts existed for only a hundred-odd galaxies. De Vaucouleurs was like an explorer in St. Louis who knew that Boston, New York, Philadelphia, Baltimore, and Washington, D.C., lay in roughly the same direction but did not know their distances and thus could not prove that they formed a metropolitan corridor along the East Coast.

The same obstacle confronted California astronomer George Abell, who in 1958 published a catalogue of 2,712 galaxy clusters. Abell thought that the clusters themselves clustered, forming superclusters, but without redshifts he couldn't prove it. Nor could two other California astronomers, Donald Shane and Carl Wirtanen. Shane and Wirtanen spent over a decade counting more than a million galaxies across the sky, finding filaments and holes that critics dismissed as tricks of the eye.

Finally, in the 1970s, sensitive new detectors began to yield redshifts of fainter galaxies. One of the first astronomers to exploit the new technology was Steve Gregory, who traced his interest back to childhood. "When I was seven or eight years old, my parents bought a set of encyclopedias for my brother and me," Gregory said. "I sat down with volume A, wanting to go through all twenty-six volumes. And I never finished volume A—I got to this part on astronomy, and the pictures just knocked me out." When he entered college, however, he chose to major not in astronomy but in math. "That's because I had one fatal flaw: I was afraid of the dark. I knew that might cause me problems as an astronomer!" After one semester, though, he switched to astronomy, but he didn't conquer his fear of the dark until years later.

For his doctoral work, at the University of Arizona, Gregory studied the rich cluster of galaxies in Coma Berenices, a dim constellation south of the Big Dipper and east of Leo. The Coma galaxy cluster had an illustrious history—here Fritz Zwicky had first detected dark matter.

"The Coma cluster is an ideal place for an observer," said Gregory. "It's perfectly situated at the north Galactic pole, so there's no obscuration from our own Galaxy, and it's up in the springtime, when typically the weather is best." The Coma cluster is in nearly the same direction as the Virgo cluster but five to six times farther away, some 350 million light-years from the Milky Way. At first, Gregory planned to study the Coma cluster theoretically, but few of its galaxies had known redshifts. He therefore wanted to rectify the situation by measuring the redshifts himself, a project that struck his advisors as pointless. At that time, however, Wallace Sargent—the same Caltech astronomer who had helped discover the primordial helium abundance in two small galaxies—came through town. "You know," Sargent said to one of Gregory's superiors, "I think that's a really good idea."

Gregory got the go-ahead. He measured the redshifts of galaxies in the direction of Coma. Those galaxies which belonged to the Coma cluster would share its redshift, 7,000 kilometers per second. Those galaxies which lay in the foreground, between the Milky Way and the Coma cluster, would have smaller redshifts, and those galaxies which lay beyond the Coma cluster would have greater redshifts.

The universe, however, held a secret. Said Gregory, "Probably in 1973, I was showing my work to Bill Tifft—he was my main dissertation advisor—and he made the comment, 'Gee, where's the foreground of the Coma cluster?'" Nearly all the galaxies Gregory had observed matched Coma's redshift; hardly any had smaller redshifts. "That really was the discovery of voids," said Gregory, "so Bill Tifft should be given a lot of credit for that. Together we developed an understanding that there was a hole in the foreground, where there were no galaxies." Good-bye to the uniform "field" of galaxies strewn across the cosmos like wildflowers.

After completing his dissertation, Gregory teamed up with another astronomer who had just earned his doctorate at the University of Arizona, Laird Thompson. Thompson had once taken a course on extragalactic astronomy from galaxy cluster expert George Abell. "Although the current view was that things were homogeneous," said Thompson, "Abell insisted that the clusters of galaxies themselves were not uniformly distributed. I took that to heart. Yet without actually measuring the redshifts, no one would believe it."

Gregory and Thompson were about to demonstrate it themselves. While studying the Coma cluster, Gregory had noticed an extension of galaxies that jutted westward, toward the tail of Leo, toward *another* galaxy cluster, named Abell 1367. "Laird and I were both thinking the same thing," said Gregory: "it looks like the Coma cluster and Abell 1367 are at the same redshift. They're close together on the sky. We need to go out and see if we can find a bridge between the two and prove there's a supercluster."

Actually, the two clusters weren't *that* close together. They were separated by 19 degrees—37 times the apparent diameter of the full Moon. At Coma's distance, this angular separation corresponded to an expanse of 120 million light-years of space. But did the galaxies that appeared between the two clusters truly share their distance and redshift? To find out, Gregory and Thompson began to observe the galaxies' redshifts.

At home one day, Gregory, who had moved to a small college in New York State, examined the positions and distances of the galaxies. "It was a moment of epiphany," he said. "It was a winter's day in upstate New York, and I was working on the redshifts from data that Laird and I had taken in 1975 and 1976. I plotted the galaxies up and found that they linked together. It was a moment of sitting there at my dining room table, staring off into a snowstorm, and saying, 'I know something that no other human being knows.'"

He said, "If you're in a forest and you don't know what's out beyond it—you can't see past a few hundred yards, because there're trees—and someone says, 'Oh, we're in a forest,' you say, 'What importance is that?' Laird Thompson and I sort of skinnied up the tallest tree and looked out, and what we found was, beyond the forest we were in, there was a meadow—the equivalent of a void—and then there was another forest—the equivalent of a supercluster."

Gregory and Thompson published their discovery in a landmark 1978 paper. In the direction of the Coma supercluster, they found a few galaxies at low redshifts, nearby galaxies that belonged to the Local Supercluster, the equivalent of trees in our own forest. At greater distances, a few small groups of galaxies scattered in front of a void that yawned across 90 million light-years of space—a large "meadow" with no galaxies at all. Then, at a distance of 350 million light-years, came the monstrous Coma supercluster itself, looking like a bank of distant trees that swept from Coma Berenices to Leo, across 120 million light-years of space. Galaxies speckled the entire structure.

Especially intriguing, though, were the vast expanses of space lacking galaxies, regions that Gregory and Thompson called voids. As Thompson wrote in their paper, "It is an important challenge for any cosmological model to explain the origin of these vast, apparently empty regions of space. There are two possibilities: (1) the regions are truly empty, or (2) the mass in these regions is in some form other than bright galaxies."

Some astronomers, however, greeted the discovery less enthusiastically than Gregory and Thompson had hoped. "We saw it as being quite obvious, just based on the data," said Thompson, "yet it wasn't as though there was some transformation of the astronomy community. There were all these people who were very reluctant to either change their mind or even look at it. The most astounding reaction was that people whose names I won't mention said that these patterns were just a figment of your imagination—that your eye tricks you into thinking that there's some structure there. That was most disconcerting."

Indeed, around the same time, a team including Princeton theorist James Peebles transformed the galaxy counts from Donald Shane and Carl Wirtanen into a dramatic black-and-white sky map. Areas with lots of galaxies were colored white, those with few or none were colored gray or black. Bright, interlocking filaments of galaxies crisscrossed the map, separated by gaping holes—yet the Princeton scientists argued *against* the reality of these features. "There is a strong temptation to conclude that the galaxies are arranged in a remarkable filamentary pattern on scales of ~5° to 15°," they wrote, "but we caution that this visual impression may be misleading because the eye tends to pick out linear patterns even in random noise."

Such statements didn't bother Gregory. "I never put much stock in theorists," he said. "I remember vividly one meeting where a theorist gave a paper on how much you can learn from redshifts in a cluster of galaxies. At the lunch break I walked up to him and said, 'Hey, this probably will interest you: I've got data—hundreds of redshifts in the Coma cluster.' And he sort of sniffed at me: 'Well, I don't really want to look at data.'"

In contrast, many observers recognized the importance of Gregory and Thompson's discovery. Some were already doing similar redshift surveys. For example, Guido Chincarini, then at the University of Oklahoma, and Herbert Rood, then at Michigan State University, had also been mapping the Coma cluster, finding it larger than had been thought and the galaxy distribution outside the cluster patchy. A team in Estonia—Mihkel Jôeveer, Jaan Einasto, and Erik Tago—reported superclusters and voids, too. Within just a few years, redshift surveys confirmed the reality of the Pisces-Perseus supercluster, 270 million light-years distant, as well as other superclusters, such as one in Hercules, 550 million light-years distant, that both Harlow Shapley and George Abell had long suspected. Voids lay in front of both superclusters. The one in front of the Hercules supercluster was 200 million light-years deep.

In 1981, Robert Kirshner, then at the University of Michigan, and his colleagues made headlines by discovering a huge void in Boötes, a constellation near the Big Dipper that features the bright orange star Arcturus. The Boötes void was 300 million light-years deep. Meanwhile, Marc Davis, then at Harvard, and his colleagues measured the redshifts of some 2,400 galaxies, finding their distribution "frothy," marked by large superclusters with holes in between. "Our results present a severe challenge to all theories of galaxy and cluster formation," wrote Davis and his colleagues. In 1982 radio astronomers Riccardo Giovanelli and Martha Haynes measured the redshifts of radio waves from galaxies and found a supercluster linking the constellations Lynx and Ursa Major.

Any remaining doubt that superclusters and voids ruled the cosmos evaporated in 1986, when astronomers at the Harvard-Smithsonian Center for Astrophysics unfurled a redshift map that swept all the way from Hercules, through the Coma cluster, and on to Cancer. The astronomers had expected to disprove the existence of large features; instead, they found long chains of galaxies stretching for hundreds of millions of light-years, separated by voids that typically spanned 125 million light-years. A press release appeared, and the work made the front page of the Sunday *New York Times*. One of the astronomers—Margaret Geller—appeared on *Good Morning America* and in *Time*. By contrast, when Gregory and Thompson made their discovery, they issued no press release, and not even astronomy magazines mentioned their work. "We were naive," said Gregory.

Gregory and Thompson weren't happy with the consequences. The media blitz led many to believe that Geller's team had been the first to discover the superclusters and voids. Furthermore, said Gregory and

Thompson, Geller refused to acknowledge earlier work. Geller denied the charge, saying she *has* referenced their work. Nevertheless, the 1986 paper swept over part of the same region of the sky that Gregory and Thompson had mapped and confirmed their results there, but nowhere did the paper state this. Allan Sandage decried this "rewrite of the history" as "a travesty of justice," comparing Steve Gregory and Laird Thompson to Ralph Alpher and Robert Herman—the two scientists who predicted the cosmic microwave background, only to find others claiming credit when it was confirmed. Geller countered that the 1986 work was distinctive because it swept over a wide swath of sky, included far more galaxies—1,099 versus Gregory and Thompson's 238—and studied regions not previously known to have unusual structures, thereby better sampling the universe at large.

Thompson, though, didn't see any qualitative difference between the results of the two studies. "I don't begrudge anybody getting publicity," he said, "but to characterize their work as a great discovery is a distortion. To have called it an extremely beautiful and complete catalogue based on someone else's discovery would have been the fair thing to do. But you can look through all that publicity you want, and you never see the names of Steve Gregory and myself."

Gregory said, "I remember seeing Margaret Geller on *Good Morning America*, describing how she discovered voids. It verges on criminal." Ironically, just two years earlier Geller had questioned the reality of the filaments and voids in the Shane-Wirtanen galaxy map. "But I don't want to go through life being bitter about this," said Gregory. "And for the most part, 99 percent of the time, I'm happy—I did it, I know I did it, people who I care about know I did it; and 1 percent of the time I say, 'Argh! What's-her-face at Harvard gets the credit.'"

## THE COSMIC ARCHITECT

Even before redshift surveys confirmed the galaxy superclusters and discovered the voids between them, astronomers had contemplated how galaxies arise and organize. Perhaps not coincidentally, the two main theories reflected the political climates in which they were championed. One theory put the individual galaxy first and blossomed in that bastion of individualism, the United States. The competing theory put the group first and originated in a dictatorship that condemned individualism, the Soviet Union.

The first theory—the so-called bottom-up theory—said that individual galaxies formed first. Then, over time, these galaxies attracted others through gravity, gathering into clusters and finally superclusters. Somewhat analogously, in an unswept house, dust gathers in corners; over time, more and more dust gathers, forming larger and larger dust balls. Big bang formulator Georges Lemaître had suggested the idea, and in 1965 Princeton's James Peebles developed it further.

The second theory—the top-down theory—arose from Soviet cosmologist Yakov Zel'dovich and his colleagues in the 1970s. They said that the first structures were "pancakes," colossal entities that resembled superclusters. Then these pancakes fragmented into individual galaxies. Analogously, drop a windowpane so that it shatters; the resulting bits of glass belong to a pattern that reflects the larger structure.

Both theories, however, ran into problems with the cosmic microwave background. In the early universe, said both theories, some regions had been denser than others. These dense regions then grew into either galaxies, in the American bottom-up model, or superclusters, in the Soviet top-down model. Trouble was, such density variations should have imprinted themselves on the cosmic microwave background, but measurements made during the 1970s and 1980s indicated that it was smooth. Because the cosmic microwave background is the light set free by ordinary, or baryonic, matter 300,000 years after the universe's birth, the lack of observed fluctuations suggested that baryonic matter alone couldn't do the job.

The solution: *non*baryonic matter. This interacted little with either light or ordinary matter and so could begin clumping long before the universe celebrated its 300,000th birthday, *without* distorting the cosmic microwave background. Afterward, the nonbaryonic matter could attract the baryonic matter needed to assemble the galaxies or superclusters, whichever came first. Other work also argued for the existence of nonbaryonic matter. For example, the primordial abundances of the light elements—hydrogen, helium, and lithium—suggested that the amount of baryonic matter wasn't enough to explain the total dark matter abundance in galaxy clusters. Furthermore, those who believed in the inflation theory—that the universe had hyperexpanded a split second after birth—thought that omega was 1, which demanded even greater quantities of nonbaryonic matter.

For nonbaryonic matter, the American, bottom-up theory invoked *cold* dark matter, whereas the Soviet, top-down theory invoked *hot* dark matter. The terms indicated how slowly or swiftly the nonbaryonic particles moved when the universe was young. In the same way, air molecules move more slowly in a cold room than in a hot room. Cold dark matter, then, consisted of heavy, exotic subatomic particles that moved slowly when the universe was young. No such particles had ever been discovered, but they fit the bottom-up theory perfectly. That's because such lethargic particles would gather into even small density enhancements, which would build individual galaxies.

In contrast, the Soviet top-down theory—the one that formed the superclusters first—needed *hot* dark matter. Its particles were lightweight, flitting about so fast that they ignored small enhancements in density and settled only into the largest ones, those which would go on to form superclusters. Indeed, the superclusters and voids that astronomers were discovering seemed to support the Soviet top-down theory. Moreover, whereas no one had seen a single particle of *cold* dark matter, *hot* dark matter had an obvious candidate: the neutrino.

In the next second, trillions of neutrinos will shoot through your body. Don't look for them now, though—they've already made it past the Moon. Neutrinos penetrate the Sun and Earth just as easily. Nuclear reactions at the Sun's center spew out both photons and neutrinos, but whereas the photons take tens of thousands of years to escape the solar furnace, the neutrinos slip out in just two seconds. During the day, solar photons and neutrinos both rain down on Earth; during the night, the ground blocks the photons, but the neutrinos keep on coming. If human eyes saw neutrinos the way they do photons, they'd see the Sun day and night, the difference being that at night the Sun would shine through the ground. Despite their astonishing properties, neutrinos vastly outnumber particles of ordinary matter. The Sun makes them, other stars make them, and so did the big bang itself.

Numerous though neutrinos are, as potential supercluster sculptors they still had a problem: physicists had traditionally assumed their mass to be zero. No mass, no gravity, no superclusters—no luck. In 1980, though, physicists "discovered" that the neutrino had mass. It wasn't much—about 1/10,000 of the electron's mass—but with all the neutrinos darting around, it added up, giving an omega around 1, as the inflation theory demanded. Thus the top-down, hot dark matter model could erect superclusters and voids, account for the dark matter that pervaded the cosmos, and agree with the key prediction of inflation theory, which itself solved the flatness and horizon problems.

Alas, trouble emerged. First, no one could confirm the "discovery" of the neutrino's mass. Second, although the top-down model formed superclusters, like many group plans it neglected the individual, failing to confer enough dark matter onto each galaxy. Third, because the theory predicted that galaxies formed *after* superclusters, it also predicted that astronomers should see no distant galaxies—for such galaxies would have emitted their light toward Earth long ago, before the top-down theory said they had formed. Yet astronomers saw lots of quasars, which were probably the luminous cores of newborn galaxies, at large distances. And fourth, the theory tripped up in a tiny galaxy gracing the northern constellation Draco the Dragon.

## THE DRAGON'S ROAR

Ironically, the top-down, hot dark matter model—designed to construct the grandest structures in the cosmos, superclusters that spanned hundreds of millions of light-years—stumbled much closer to home, in the dimmest galaxy ever seen, a smudge of stars orbiting the Milky Way. The Draco Galaxy was a dwarf spheroidal galaxy, its few million widely separated stars emitting less light than the single brightest star in the Milky Way. Thus it went undetected until 1954, though it lay a mere quarter million light-years away, a bit farther than the Magellanic Clouds.

Yet the little galaxy held a big secret, one that Marc Aaronson, a young astronomer at the University of Arizona, would uncover. "The best word to describe Marc was *passion*," said Edward Olszewski, who began working with him around the time of the discovery. "He *loved* observing, and he was really, really good at it. He could be really annoying, because he was so driven that he sometimes would stomp over you. He was so sure of his scientific abilities that he took on established dogma." In the late 1970s, before turning thirty, Aaronson tangled with Allan Sandage over the Hubble constant. He also challenged conventional thinking about globular clusters in the Magellanic Clouds, proclaiming that some of the clusters which had been thought ancient weren't as old as their counterparts in the Milky Way.

In 1982, Aaronson began to measure the velocities of stars in the Draco Galaxy. "At the time," said Olszewski, "everybody thought that dwarf spheroidal galaxies were analogues to globular clusters. They were bigger in size, lower in density, had approximately the same number of stars, and should have had the same stellar population. And everybody knew that globular clusters had mass-to-light ratios of 1 to 2." That meant globular clusters had no dark matter. If the Draco Galaxy was just a puffed-up globular cluster—same material, spread over more space—it, too, should lack dark matter. Without dark matter, Draco had little gravity with which to stir up its stars, so they should share the same velocity. In astronomical parlance, Draco's velocity dispersion should be low, only about 1 kilometer per second.

Aaronson observed the Doppler shifts of three stars in Draco and found just the opposite. Their velocity dispersion was about 10 kilometers per second, 10 times higher than expected, implying a dark matter content 100 times greater—because the mass goes as the square of the velocity dispersion. Furthermore, Draco's dark matter couldn't be neutrinos, the darling of the top-down, hot dark matter model for galaxy formation. They would have zipped right out of tiny Draco.

Based on the high velocity dispersion of the three Draco stars, Aaronson submitted a paper to *The Astrophysical Journal Letters*. "The referee rejected the paper," said Olszewski. "Basically she said two very sensible things. One was, are the stars velocity variables?" A star's velocity varies if it orbits a companion; thus, binary stars might inflate the galaxy's velocity dispersion. "And two was, get some more stars and come back later." Trying to determine a velocity dispersion from just three stars struck most astronomers as absurd. Aaronson countered that if Draco's true velocity dispersion were *low*, the chance that he could observe such divergent velocities among three stars was nearly nil.

"The referee, of course, was right," said Olszewski. "I mean, Marc should have waited till he got two or three more stars; but Marc was right because he was right. Marc's genius was that it didn't matter what the party line was. He was confident enough in his observing ability and his analysis ability that he said, 'Here's something, and it's time to explain it.'" Aaronson therefore rejected the rejection. "One thing Marc could be very good at was whining. When he didn't get his way, he would whine until you gave in. He just didn't take no for an answer. So he wrote a reply to *The Astrophysical Journal Letters*, saying, 'Everybody in the world is already talking about this result; publish it anyway.' And so they published it anyway." Two days after the paper was accepted, Aaronson observed a fourth star in the Draco Galaxy. Its velocity differed from the other three, confirming Draco's high velocity dispersion.

"At the time that paper came out," said Olszewski, "the word on the street was that it was discussed at every graduate school in the United States, and everybody basically said this was hogwash." Olszewski himself thought the same. "But I thought that some of the Magellanic Cloud stuff that Marc was doing was pretty silly, too, and time proved him right and me just to be stupid."

Aaronson and Olszewski observed additional stars in Draco that further confirmed the small galaxy's high velocity dispersion. They also observed another dwarf spheroidal galaxy near Draco named Ursa Minor. It, too, had a high velocity dispersion and therefore lots of dark matter. Whatever dark matter was, it wasn't neutrinos.

However, on the night of April 30, 1987, Aaronson was killed at the telescope. "That afternoon, Marc and I had sat down and talked about the various projects that we were doing," said Olszewski. "Marc was also telling me that he had this cool new way of focusing at the four-meter telescope." That was the telescope atop Kitt Peak which Aaronson would use that night. Olszewski would be using another telescope, the 90-inch. "That night," said Olszewski, "the weather was really crummy, and nobody opened at sunset. Finally, the weather got good enough that they would let us all open. And then the phone rang." Olszewski's colleague answered; it was from the four-meter telescope, for Olszewski. "Marc and I had talked about how I ought to come up to the four-meter telescope and he'd show me this cool new way to focus, and I'm thinking to myself, 'Marc, wait till I get started observing.' So I said something like, 'Tell them to go away.'" But Olszewski's colleague said, "No, no—you've got to go up there. Something's wrong." "So I started running up to the four-meter," said Olszewski. "I got around the corner, and I saw lights everywhere. Of course, no one would tell me what's going on—'There's been an emergency' was the phrase. I got into the elevator with a guy carrying an empty stretcher. Whatever floor we got off on, suddenly I saw Marc—lying there, dead." Aaronson had opened a door to the telescope's catwalk, to check the weather, but the dome had been rotating at full speed; it slammed the door shut, killing him. He was only thirty-seven years old.

## THE GREAT ATTRACTOR

With the demise of the hot dark matter model for the formation of galaxies, its competitor—the cold dark matter model, which formed galaxies from the bottom up—began its reign. However, the cold dark matter model had problems itself. Whereas it could make galaxies large and small, it had trouble herding them into superclusters. In addition, most theorists assumed that omega was 1, as the inflation theory demanded, but the motions of galaxies within clusters implied a much lower omega. Thus, if the true omega was as high as 1, large quantities of dark matter must exist outside the clusters and superclusters, in the voids; but for some reason it had failed to foster any galaxies.

The cold dark matter model suffered another blow in 1986, when a team of seven astronomers discovered that the Milky Way, the Local Group, the Local Supercluster, and ever farther-flung superclusters were being yanked by a huge, uncharted system of galaxies. Its heart hid behind the Milky Way's dusty veil around Centaurus, and its discovery was an accident.

So was its name. "I didn't plan it," said Alan Dressler of the Carnegie Observatories in Pasadena, California. "We had a news conference in Washington, D.C., and I had a roomful of reporters there. I remember stretching my hands as large as possible, just trying to think of a way to explain it, and somehow 'Great Attractor' came out. The reporters loved it. Some of my colleagues hated it."

The Great Attractor revealed itself as Dressler and his six colleagues—Sandra Faber, David Burstein, Roger Davies, Donald Lynden-Bell, Roberto Terlevich, and Gary Wegner—were investigating elliptical galaxies, the round or oval balls of stars that dominate many galaxy clusters. In particular, the astronomers wanted to see how such basic galactic properties as size, luminosity, and velocity dispersion depended on one another.

Unlike the photogenic spirals, elliptical galaxies look boring—exactly why Dressler's team chose them. Said Dressler, "If you went to a spiral galaxy, where there's ongoing star formation, it would be more complicated. In some sense, spiral galaxies are still in the act of being formed, whereas ellipticals are finished products."

The seven scientists observed about four hundred elliptical galaxies across the sky, wherever the Milky Way didn't block the view. To estimate each galaxy's distance from Earth, they derived a relation that linked an elliptical galaxy's diameter, surface brightness, and velocity dispersion to its true, or intrinsic, brightness. The greater the gulf between the galaxy's intrinsic brightness and its apparent brightness, the farther the galaxy must be from Earth.

The seven astronomers then compared each galaxy's distance with its redshift. As Hubble had found, the greater the one, the greater the other. However, other astronomers had discovered that the gravity of the Virgo cluster, heart of the Local Supercluster, tries to pull the Local Group of galaxies toward it. So far it has failed—the Local Group is still moving away from it—but Virgo's redshift is not as great as its distance says it should be. Dressler's team put the shortfall at 250 kilometers per second.

More than just the Virgo cluster was trying to reel the Local Group in. Sixty-five degrees to its south, a large supercluster in Centaurus seemed to be exerting its gravitational strength. The seven scientists put our speed "toward" Centaurus at 570 kilometers per second—although, as with Virgo, we are actually heading away from the Centaurus galaxies; we're just doing so more slowly than we "should," by 570 kilometers per second. The two motions—250 kilometers per second toward Virgo, 570 kilometers per second toward Centaurus—explained most of the Local Group's motion through the universe at large. In the late 1960s and 1970s, scientists had determined this motion by observing the cosmic microwave background, which was blueshifted in the direction we're approaching.

"That would have probably been quite believable," said Dressler. "We would have said, 'Okay—finished. We've accounted for why our Galaxy is moving.' The big surprise, the thing that made it unbelievable and incredible, was that the Centaurus supercluster itself was moving—*even faster* than we were." Thus, something beyond the Centaurus supercluster was pulling it, and the Milky Way, toward it. The center of the Great Attractor, estimated the astronomers, lay about 220 million light-years from the Milky Way.

Then the scientists asked another astronomer, Ofer Lahav at the University of Cambridge, to plot the positions of some 15,000 galaxies and to center the map on Centaurus. "Nobody had ever really made a map like that, showing this huge structure in which each galaxy is like a star," said Dressler. "You see this swath of light that goes behind our Galaxy and comes out the other side. There was this really striking thickening in the supergalactic plane that said, 'Gosh, there's more galaxies there than anything we've been talking about.' Not the Local Supercluster, not the Centaurus supercluster—this is much bigger than anything like that, maybe ten times as big as those two things."

On the other side of the Milky Way from the Great Attractor's center is the Pisces-Perseus supercluster, the one reported back in the 1930s. The Pisces-Perseus supercluster also tries to pull the Galaxy its way, but Dressler cited three reasons it loses the cosmic tug-of-war. First, the Great Attractor has roughly twice the mass, giving it twice the gravitational strength. Second, the Great Attractor is a bit closer to the Milky Way. And third, a large void—discovered by Gregory, Thompson, and Tifft, and confirmed by Haynes and Giovanelli—sits between the Milky Way and the Pisces-Perseus supercluster. "If that void were populated with galaxies," said Dressler, "it would be more of a fight. We would still be going toward the Great Attractor, but we would be pulled partly back the other way."

Whereas the Pisces-Perseus supercluster was first spotted long ago, the more dominant Great Attractor, with its heart near the dust-clogged Galactic plane, eluded detection until the 1980s. "That region was not very well studied," said Dressler, "because why muck around near the Galactic plane? There was so much else to do in those days. People did not go looking for difficult regions in which to work, and they didn't know there was anything special about that area. In addition, it was in the southern hemisphere; most telescopes are in the north."

The theorists didn't like the Great Attractor. It further threatened the cold dark matter model for galaxy formation. If, as that model said, galaxies formed first, then gathered into clusters and superclusters, not enough time had elapsed to build an entity as huge as the Great Attractor.

"A lot of people would come up and say we were wrong," said Dressler. "They didn't know why we were wrong or how we were wrong, but they were sure we were wrong." One theorist—Amos Yahil—gave the seven scientists their name. "He was one of those people who was exasperated by the whole thing," said Dressler. "We'd thrown this monkey wrench into the works when everything was going so well. He named us the Seven Samurai, because we were these troublemakers, brandishing our saber swords. I thought it was charming, actually. There had been much less polite epithets used, which I'm not going to mention, because I don't want them to wind up in print!"

In 1989, two of the Seven Samurai—Dressler and Faber—confirmed the Great Attractor's existence by observing elliptical galaxies in Centaurus that lay *beyond* the Great Attractor's center. Whereas the galaxies in the foreground, between the Milky Way and the Great Attractor, were being yanked away from us and thus had higher-than-expected redshifts, the galaxies beyond the Great Attractor's center had *lower-*

than-expected redshifts, because the Great Attractor's center was yanking them toward the Milky Way. Later work, employing more accurate distances to elliptical galaxies, confirmed the general direction of the Local Group's motion toward the Great Attractor but cut the speed somewhat. Also, astronomers led by Renée Kraan-Korteweg identified a massive cluster of galaxies as the Great Attractor's heart. The cluster lay behind the Milky Way's obscuring gas and dust in the constellation Norma, east of Centaurus; its mass rivaled that of the Coma cluster.

Larger redshift surveys were turning up larger structures, too. For example, in 1989 Harvard-Smithsonian astronomers Margaret Geller and John Huchra reported the discovery of what they called a "Great Wall"—a chain of galaxies stretching across 850 million light-years of space. Even as astronomers were uncovering such architectural monuments, NASA was about to launch a satellite that would discern tiny irregularities in the big bang's afterglow, irregularities which had spawned the superclusters and voids that decorated the cosmos.

# TEN

# RIPPLES IN SPACE

IT IS the farthest that telescopic eyes have ever peered, a glowing backdrop to the galaxies, clusters, superclusters, and voids: a wall of primordial light that surrounds us on all sides and arose a mere 300,000 years after creation. Before that time, hot fog pervaded the universe, scattering light every which way; but as the universe expanded and cooled, the fog lifted, and the light tore free. That light has sped toward us ever since, its wavelength getting stretched by the universe's expansion so that it now shines prominently at microwave frequencies. Otherwise, though, this ancient light has streamed toward us nearly unaltered, still carrying clues to the universe's origin.

"The cosmic microwave background is the primary remnant from the big bang that we can look at directly," said John Mather of NASA's Goddard Space Flight Center in Greenbelt, Maryland. Mather started studying the cosmic microwave background as a graduate student, and soon after, he proposed that NASA launch a satellite to scrutinize this afterglow of creation. In 1992, that satellite garnered front-page headlines by discerning ripples in the cosmic microwave background that presaged the formation of galaxies.

## FOG AT COSMIC DAWN

The universe was born hot and therefore bright—so bright that light ruled the matter. Because of the heat, photons sizzled with energy. That energy, via Einstein's equation $E = mc^2$, counted as mass, so during this so-called radiation-dominated era, the photons outweighed the matter.

The radiation-dominated era survived just a few tens of thousands of years, the exact figure depending on the numerical values of the Hubble constant and omega. The era ended because the universe expanded, diluting both the photons and the matter. But the photons suffered worse: the expansion stretched their wavelengths, sapping their strength. When the mass density of the photons slipped below that of the matter, the matter-dominated era began.

During this era, protons sought the electrons that were whizzing around, but the photons ripped them away again. Free electrons scatter light ferociously. As a result, even though photons outnumbered electrons roughly a billion to one, light couldn't penetrate the early universe. Instead, it ricocheted from free electron to free electron, like a pinball in a pinball machine, never getting anywhere. A heat-resistant observer would have seen a bright fog in all directions. The light pressed on the ordinary, or baryonic, matter, preventing it from clumping together and thwarting the formation of future galaxies.

Fortunately, the photons couldn't stop the ghostly *non*baryonic matter from getting together. Nonbaryonic particles sailed past the light, so as soon as the matter-dominated era began, they could attract other nonbaryonic particles through their gravity, like unseen stagehands arranging props for a play, awaiting the actors' arrival.

The universe continued to expand and cool. Roughly 300,000 years after the big bang—again, the exact figure depends on the Hubble constant and omega—the temperature dropped to 3,000 Kelvin, and the photons grew too weak to dislodge the electrons that the protons were capturing. As a result, the protons mopped up nearly all of the unruly electrons that had been harassing the light, a momentous event that cosmologists call recombination. With the electrons tamed, the light exerted almost no pressure on the baryonic matter, which now started to fall toward the clumps of nonbaryonic matter that had earlier set the celestial stage. Furthermore, because the light streamed free of the clutch of matter, astronomers today can still see it, redshifted by a factor of about 1,100.

Thanks to the clumps of nonbaryonic matter, the early universe was not perfectly smooth. "By looking at the cosmic microwave background," said Joseph Silk, who had explored how galaxy formation might alter the ancient light, "one can see the imprint of these density fluctuations on the universe. They left their trace as temperature fluctuations." Primordial photons that streamed away from a dense region lost energy as they fought gravity, so they should appear slightly cooler, or redder, than other photons in the cosmic microwave background. Likewise, light from sparser areas should look slightly warmer, or bluer. Therefore, by measuring temperature fluctuations in the cosmic microwave background, astronomers had the chance to observe the density fluctuations that existed a mere 300,000 years after the big bang, density fluctuations that had triggered the development of the galaxies, clusters, superclusters, and voids.

In short, the cosmic microwave background is a snapshot of the cosmos a mere 300,000 years after its birth—to an eighty-year-old man, the equivalent of a photograph at the age of just fifteen hours. Thus the cosmic microwave background probes the deep past, and the deep distance. For example, if the universe is exactly 14 billion years old, the light from the cosmic microwave background has a lookback distance of 13,999,700,000 light-years—99.998 percent of the distance to the edge of the observable universe. Seeing the cosmic microwave background is like standing on the California coastline and seeing all the way across the United States to within just 300 feet of the *Atlantic* coastline. No telescope that detects any form of light, visible or invisible, will see farther. Only a telescope sensitive to particles that pierced the primordial fog, such as neutrinos, could peer past the cosmic microwave background.

Since the cosmic microwave background represents light from only 300,000 years after the big bang, why didn't it pass us by long ago? Actually, light from that era has passed us by continuously; but every year, the light comes from farther away. To picture this, imagine long-lived observers who forever occupied the Milky Way's position in space. When light first broke free of matter during the recombination era, it filled the whole universe, both within and beyond the observers' horizon. At first, however, the observers saw only light nearby, because distant light hadn't yet reached them. As time went on, they saw more and more distant light. A million years after the light broke free, they saw light from a million light-years away; a billion years after the light broke free, they saw light from a billion light-years away; and so on. Meanwhile, the light that had earlier passed the observers became the background radiation for observers in other galaxies. Today the light that originated at the Milky Way's position during the recombination era forms part of the cosmic microwave background for those who are some 14 billion light-years from Earth.

## THE COBE MISSION

By 1965, when Arno Penzias and Robert Wilson discovered the cosmic microwave background, the universe's expansion had chilled it to a mere 2.7 Kelvin. Nevertheless, its photons harbor far more energy than all the universe's starlight. Because some of this radiation can't penetrate the Earth's atmosphere, in the 1970s NASA approved a satellite mission named COBE ("coby"), for COsmic Background Explorer. The COBE scientists planned to launch the satellite on an ordinary rocket. However, such rockets competed with NASA's soon-to-be-flown space shuttle, so NASA began to phase them out, ordering the scientists to launch COBE on the shuttle.

"I figured, I'm just an underling in the trenches here, and I have to do what they say," said John Mather, "but I thought it was a bad idea. The shuttle is not the natural way to get things up there, and we particularly needed something special from it: a launch from the West Coast, and as time went along, it became clear they were never going to launch any shuttles from the West Coast."

A West Coast launch could put COBE into a so-called polar orbit, one that passed over the North and South Poles, thereby sheltering the satellite's instruments from sunlight. In principle, an East Coast launch could also propel COBE into a polar orbit, but if the shuttle launched northward from Florida, it would pass over populated areas in the United States, and if it launched southward, it would pass over Cuba, hardly America's best friend.

Mather never liked watching the shuttle take off. "I had always been afraid to watch the launches," he said, "because I was afraid they were going to blow up, as so many other rockets have—not carrying people, but carrying other kinds of payloads."

On the morning of January 28, 1986, Florida was so cold that several engineers at the Morton Thiokol company, manufacturer of the shuttle's solid rocket booster motors, had recommended against launching the shuttle. NASA officials pressured them—"My God, Thiokol, when do you want me to launch, next April?" Thiokol's managers, fearful of losing their lucrative NASA contract, overruled their own engineers; and seventy-three seconds after liftoff, the shuttle exploded, killing all seven astronauts aboard.

"It was really devastating," said Charles Bennett of NASA's Goddard Space Flight Center and another member of the COBE team. "The first thought that was on everybody's mind was the astronauts that were killed. In fact, I didn't even start thinking about the effects on COBE until a little later."

COBE wasn't aboard that shuttle, but NASA grounded the shuttle for years, blocking the very route NASA had forced COBE to take. Furthermore, because of NASA's plan to eliminate competition with the shuttle, few rockets remained to launch the satellite. Thus, the COBE team scrambled for alternate launch vehicles, even exploring possibilities in France, Russia, and China.

Finally, COBE managers persuaded NASA to let them launch the satellite on one of the few rockets NASA had left. Said Bennett, "NASA headquarters realized that COBE was a very important science project—that it would resonate with people, that it would be good for the agency and good for science, and that would be a good step on the road back for NASA. So headquarters said, 'Okay, we'd like to launch you; but for it to do us any real good, it would have to be soon.'" Of course, the scientists had originally wanted to launch COBE on an ordinary rocket. Trouble was, COBE wouldn't fit. Because NASA had forced it onto the shuttle, it now weighed too much, and it was too big. Stripping down COBE and testing the redesigned satellite took two and a half intense years.

COBE lifted off from California on November 18, 1989. Present at the launch were Ralph Alpher and Robert Herman, who had first predicted the existence of the radiation COBE would study. NASA had invited them to speak, helping heal the wounds they had suffered after failing to receive credit following the cosmic microwave background's discovery. Unfortunately, during what should have been COBE's most triumphant moment, exactly the same problem would tear the COBE team asunder.

## THE UNIVERSE'S BIRTH CERTIFICATE

COBE delivered its first result fast. For decades, astronomers had been struggling to measure the cosmic microwave background's full spectrum, but they had never succeeded. COBE did it in nine minutes.

According to the big bang theory, that spectrum should have been what physicists call a blackbody—an unfortunate and misleading term. First, blackbodies have nothing to do with black holes. Second, blackbodies aren't dark. The Sun—yes, the Sun—is approximately a blackbody. A perfect blackbody absorbs all radiation falling on it and gives off radiation whose spectrum has a characteristic shape that depends solely on its temperature. The blackbody shines most intensely at one particular wavelength, less intensely at longer wavelengths, and much less intensely at shorter wavelengths. The hotter the blackbody, the shorter is the wavelength at which the blackbody glistens most. For example, a cool star, like Orion's red Betelgeuse, shines brightest at infrared wavelengths, longer than those visible to the eye; a warm star, like the yellow Sun, shines brightest at visible wavelengths; and a hot star, like Orion's blue Rigel, shines brightest at ultraviolet wavelengths, shorter than the eye can see.

In the early universe, photons achieved what physicists call thermodynamic equilibrium, which should have led to a blackbody spectrum. Furthermore, as the universe's expansion stretched this radiation, the spectrum should have retained its blackbody shape. Nevertheless, two years before COBE's launch, scientists in Japan and Berkeley sent up a rocket that glimpsed large deviations from a blackbody spectrum, threatening the big bang theory.

In January 1990, Mather prepared to present COBE's spectrum. Said Bennett, "A lot of people had stopped me in the hall ahead of time: 'What's he going to say? What's he going to say?' There were rumors floating around that he saw a big nonblackbody, or he saw a blackbody—every rumor you could imagine. Of course, sometimes people just make up the rumors, to see how you'll react to them. I just said, 'Well, you'll have to hear the talk and see what he has to say.'"

Mather spoke before an audience of over a thousand astronomers. He described the COBE project, then matter-of-factly presented a graph of the cosmic microwave background's spectrum: a perfect blackbody, having a temperature of 2.73 Kelvin, or −454.76 degrees Fahrenheit, some 60 degrees Fahrenheit colder than Pluto.

The audience greeted the announcement with a standing ovation. "I was not expecting it at all," said Mather. "We had all just been working so hard to get the data ready that we hadn't been fully thinking about how much it meant to other people; so I was amazed." The blackbody spectrum vindicated the big bang theory.

Ironically, Mather's talk was chaired by Geoffrey Burbidge, an astronomer at the University of California at San Diego and a longtime foe of the big bang theory. He reportedly said, "They've swallowed it hook, line, and sinker."

## GALACTIC GENESIS

Even amid all the cheers over COBE's spectrum, the satellite was still hunting for the tiny ripples that should have lurked in the cosmic microwave background—the ripples which represented the density fluctuations that had spawned the galaxies. During the 1990 meeting, George Smoot, of the Lawrence Berkeley Laboratory in California and the principal investigator for the COBE instrument that sought the ripples, reported that the satellite had seen none. "Using the forces we know now," Smoot told *The New York Times*, "you can't make the universe we now know." As a result, despite the blackbody spectrum, media reports suggested that the big bang theory was in trouble, because the early universe had been too smooth to give rise to galaxies.

Actually, the failure to find the ripples meant that the temperature fluctuations had to be tiny, challenging even COBE. Over the following months, COBE continued to observe the cosmic microwave background, and COBE team members began to analyze the data. In the summer of 1991, COBE team member Ned Wright, of the University of California at Los Angeles, thought he saw what they were looking for.

"I don't think anybody was convinced when they first heard," said Bennett. "Everyone was tantalized, but I think everybody, including Ned Wright, realized that there was still a lot of work to do—although I'm sure if you were taking bets, he would have bet that the fluctuations were real. I don't think he would have bet his house. By the time we were finished, I think we *all* would have bet our houses!" COBE team members checked and rechecked the data, making sure that the Sun, the Earth, the solar system, and the Galaxy weren't contaminating the data. In addition, COBE team member Alan Kogut investigated whether COBE's instruments and software might introduce spurious ripples. All the while, the scientists kept the discovery secret. They had agreed to release the news only when everyone on the team was convinced the find was real. Furthermore, no one would disclose the discovery separate from the others.

Finally, in April 1992, the scientists were ready. They selected Smoot to announce the momentous discovery. Beforehand, NASA headquarters pressured Smoot to make a dramatic presentation. Since the shuttle explosion, NASA's problems had worsened, because just two years earlier it had committed the biggest astronomical blunder of the twentieth century: it had launched the high-priced Hubble Space Telescope but failed to properly check its optics—and now the telescope couldn't see straight. NASA therefore wanted the COBE discovery emblazoned on the front page of every newspaper in the country. One NASA official told Smoot, "Don't screw up"—though he used a much saltier verb than *screw*.

Smoot succeeded beyond NASA's wildest dreams. During the press conference, he showed a map speckled with pink and blue blobs, regions of the cosmic microwave background 0.00003 Kelvin warmer and cooler than the average. "We have observed the oldest and largest structures ever seen in the early universe," Smoot told the reporters. "These were the primordial seeds of modern-day structures such as galaxies, clusters of galaxies, and so on. Not only that, but they represented huge ripples in the fabric of space-time left from the creation period." When asked about the discovery's importance, Smoot uttered the remark that got quoted more than any other: "If you're religious, it's like seeing God."

The next day, the front page of *The New York Times* read SCIENTISTS REPORT PROFOUND INSIGHT ON HOW TIME BEGAN. *The Washington Post*'s front page said NEW FINDINGS SUPPORT THEORY OF 'BIG BANG,' and the front page of *The Los Angeles Times* carried the COBE photograph beneath the headline RELICS OF 'BIG BANG' SEEN FOR 1ST TIME. British physicist Stephen Hawking said, "It is the scientific discovery of the century, if not of all time." The press picked up this hyperbolic quote but rarely asked how the discovery of ripples in the cosmic microwave background could outrank the discovery of the background itself. It was like saying the discovery of the Grand Canyon outdid the discovery of America.

The British magazine *New Scientist* took a more jaded view. "It is only thirty-millionths of a degree Celsius, and yet it has raised the temperature of the media—and the public—to unprecedented levels. Bars and restaurants around the world are said to have echoed last week to gossip about the big bang, cosmology and 'The Meaning of It All.'" The magazine asked, "Was the measurement of 'ripples' in the cosmic background radiation the discovery of the decade, or the century—or of all time? Well, you can argue that it rates as the scientific advance of the year so far."

Smoot's remark about seeing God unsettled some members of the COBE team. "When I first heard him do it," said Mather, "I thought, 'Okay, George, you're getting in trouble now,' because a lot of people will remember that and only that. We got an awful lot of religious-oriented people asking us whether our observations supported their theory of the universe. The problem is that scientists have been fighting religionists for too many centuries now for it to be an easy relationship."

COBE's discovery was good news for inflation, the theory that said the universe had blown up a split second after its birth. The pattern of the fluctuations matched the theory's predictions. On the other hand, the standard cold dark matter model for galaxy formation—in which omega equaled 1—didn't fare as well. The theory had already run into trouble from the discovery of superclusters, voids, and the Great Attractor. With the COBE result, theorists finally had an anchor in the early universe. Extrapolating those ancient fluctuations to the present universe led to the expectation that two galaxies close together—for example, the Milky Way and Andromeda—should move much faster relative to one another than they typically do. Princeton astronomer Jeremiah Ostriker wrote, "*Thus, the standard CDM [cold dark matter] model, when normalized to the COBE-determined amplitude is in serious conflict with the observed pairwise velocity dispersion of galaxies.*"

Mather said that his experience with the media was mostly positive. "I enjoyed talking with the press," he said. "A lot of people were very interested in this result—they wanted to know what it meant; they wanted to be able to explain it to their readers and their viewers. And they were patient with us scientists, who are not so practiced in the ways of the press. The main problem is that when you tell people there was a big bang, they don't believe you. So whatever you say after that doesn't matter!"

## THE DARK SIDE OF THE AFTERGLOW

What the public didn't then know was that behind the scenes, members of the COBE team were seething—not at Smoot's remark about God, but about what they saw as his attempt to steal credit for himself. "The odd thing is," said Mather, "he deserves a lot of credit. It's just that when you take credit without giving it to other people, it diminishes you."

Mather found out something was awry two days before the announcement. His phone rang. A reporter from the Associated Press had questions about COBE's great discovery. Mather was stunned: no one outside NASA and the COBE team was supposed to know of the discovery. But the reporter had received a press release from Smoot's institution in Berkeley, a press release that violated the COBE publication policy. Worse, as Mather would later learn, the press release glorified Smoot and Berkeley, barely mentioning the rest of the COBE team.

Bennett didn't see anything wrong until the press conference itself.

"I was surprised that so many of the questions from the reporters were a little off-base," he said, "and it wasn't until later that I realized that was because this off-base press release had come out. Things like, 'When did you first know that the Berkeley scientists had found this result?' 'How well do you know Mr. X at Berkeley who helped Dr. Smoot with this result?'" In fact, of the nineteen members on the COBE science team, the only one at Berkeley was Smoot himself. However, both Mather and Bennett acknowledged that the surreptitious Berkeley press release helped attract the media.

Smoot denied trying to steal credit for the discovery. "At that time," he said, "I was a fairly shy person, not in terms of running a project but in terms of going out in public; so to think that I wanted to hog the spotlight is kind of funny. I was naturally shy and a private person." Nevertheless, the media continued to focus on Smoot, neglecting the other scientists. *Scientific American* profiled Smoot in an article titled "COBE's Cosmic Cartographer" and described what it called "his" discovery. Even *People* magazine featured Smoot, calling up Bennett to ask the vital cosmological question: how often do you play basketball with Dr. Smoot?

"Many of us had hoped that it would be reported as the COBE team's result," said Bennett. "It was Ned Wright who first noticed the stuff; it was Al Kogut who did an enormous amount of work going through the systematic errors; I had done this work on the Galaxy. I would really have described it as a wonderful *team* effort, so it was definitely inaccurate to portray it as a George Smoot result, when in truth he didn't find it in the first place and he was out of town in Antarctica for much of the work. It was very sad that the COBE team, which had worked so well together for so long, broke apart like that. In some sense, it was snatching defeat from the jaws of victory."

Meanwhile, the book world was smiling on Smoot. A literary agent wanted him to write a book with science writer Keay Davidson. It was a smart time to strike. Four years earlier, for better or for worse, publishers had brought out Stephen Hawking's much-bought but little-read bestseller *A Brief History of Time*, and publishers, like lemmings, raced to throw large sums of money at imitations. The worldwide advance for Smoot's book reportedly reached $2 million, but Smoot denied the figure, and there's good reason to believe him, since reported advances often get inflated. Still, Smoot and his coauthor were paid well—too well, according to the publisher. "We paid lots," moaned one executive to

*The New Republic*, "and we lost lots." Smoot's photograph appeared on the book's cover, with Hawking's hyperbolic quote beneath the title, but the book failed to take off. As a publishing phenomenon, Smoot was no Hawking.

Mather's reaction when he heard of Smoot's book: "There he goes again." Mather and many other COBE members saw it as yet another affront. Some contemplated legal action; others wanted to force Smoot off the COBE project. In the end, however, they decided a public fight would tar them all. They demanded and after months of negotiation got a letter of apology from Smoot. But the apology struck many team members as half-hearted, the sort a politician issues: he's not sorry he did it; he's sorry he got caught doing it.

Mather decided to write his own book about COBE. "It was prompted largely by that other book," said Mather. "I would have never had the time and effort to spend on it if I hadn't had a need to set the record straight. There's nothing like a little annoyance to get you going about something." Unlike Smoot's book, which said nothing of the controversy, Mather devoted most of a chapter to the matter, entitled "The Private Life of the Cosmic Background Radiation."

Mather's book, cowritten with science writer John Boslough, carried neither the author's photograph nor Hawking's quote on the cover. Whereas Smoot made a fortune on his book, Mather made no money on his. "I work as a civil servant," he said, "so I didn't get paid at all for the book. I'm well paid as a civil servant. The federal civil service is protected from conflicts of interest by making sure we work for only one: you, the taxpayer."

One thing Mather and Smoot agreed on: neither liked the other's book. Several COBE team members unaffiliated with either Mather's institution or Smoot's were asked which book they preferred. Most chose Mather's.

Smoot decried what he called the negative tone of Mather's account. "I think the legacy of COBE ought to be positive," said Smoot. "In some sense, COBE was a linchpin of modern cosmology. It really energized the cosmologists themselves—to see, first, that there was progress being made, and second, that the public really cared. In the long run, people are only going to remember the word *COBE*," not the individual scientists.

COBE's discovery of ripples in the cosmic microwave background was the first shock wave to hit cosmology in the 1990s. The next came just a year later. While COBE was gazing into the deep distance, astronomers had embarked on a hunt close to home—and discovered what seemed to be three dark stars on the outskirts of the Milky Way that promised to divine the nature of the mysterious dark matter pervading the cosmos.

# ELEVEN

# MACHOs VERSUS WIMPs

OURS is a dark galaxy. Although the Milky Way's beautiful disk speckles the night with the glow of countless stars, these stars mask the Galaxy's dark truth: as astronomers recognized decades ago, most of its mass is invisible, beyond the purview of either human or telescopic eye. An enormous dark halo surrounds the luminous, spiral-sculpted disk and stretches across hundreds of thousands of light-years of space. This dark halo harbors some 90 percent of the Galaxy's mass, but it emits little or no light. Furthermore, dark halos also engulf other giant galaxies.

"I am completely in the dark about what dark matter is made of," said Princeton astronomer Bohdan Paczyński. "I really have no clue whatsoever, and whatever people claim is just guesswork." Yet as Paczyński himself realized, our Galaxy offers unique clues to the nature of the dark matter that pervades the cosmos. In 1993, by following Paczyński's suggestion and scrutinizing the outer precincts of the Milky Way, astronomers discovered what they thought were three chunks of dark matter, floating in the dark Galactic halo.

## THE GRAVITY OF THE SITUATION

Born in Poland, where his father first introduced him to astronomy, Paczyński had long been visiting the United States. "In 1981," he said, "a state of war was declared in Poland. It was called here 'martial law.'"

Poland's communist dictator was under the gun: unless he cracked down on dissidents, the Soviet army would invade and crush them itself, just as it had in Hungary in 1956 and Czechoslovakia in 1968. "It looked pretty scary," said Paczyński, "so we decided we'd better stay in the United States another year, and then another year, and before we knew it, we were here for good."

Scientifically, Paczyński first earned notoriety not for his work on dark matter but for his speculations on splendiferously bright matter: gamma-ray bursts, sudden flashes that rock the sky about once a day. "The reaction from astronomers was uniformly negative," said Paczyński. "'He's crazy,' they thought. And for that reason, I was doing other things on the side, because if I had been working only on gamma-ray bursts, I would have been pronounced cuckoo."

Gamma-ray bursts, astronomers thought, originated in the Milky Way. In 1986, however, Paczyński boldly proclaimed that the outbursts arose billions of light-years beyond. To appear so bright from Earth, they had to be extremely powerful. Other astronomers therefore scoffed at Paczyński's idea, but observations during the 1990s proved him right. Aside from the big bang, gamma-ray bursts are the most powerful explosions known.

The same year that Paczyński sparred with his peers over gamma-ray bursts, he published a pivotal paper that spurred a hunt for dark matter. "I was lucky on two accounts," he said. "First of all, the technology was available to implement the idea, even though I personally had not realized that. And the second thing was this buzz word—'dark matter'—which caught the attention of many people."

Paczyński proposed a clever method to figure out what constituted the Milky Way's dark halo—and, by implication, the dark halos of other giant spirals throughout the universe. Astronomers already knew that the Galactic dark halo couldn't be made of main-sequence stars, which generate their light as the Sun does and which observers could see. Even the faintest such stars—red dwarfs, like Proxima Centauri and Barnard's Star—would have been detected. That left two broad possibilities for the dark halo's composition: dim astronomical objects, and exotic subatomic particles.

If you're an astronomer, you'd naturally root for the former, known in the trade as MACHOs, or MAssive Compact Halo Objects, which, though dim, are at least within your range of experience. MACHOs consist of ordinary, or baryonic matter—the same stuff that constitutes

DARK HALO MACHOs

| Object | Mass (Sun = 1) | Typical Duration of Microlensing Event |
|---|---|---|
| Massive black hole | 1 million | 200 years |
| Stellar black hole | 10 | 8 months |
| Neutron star | 1.4 | 3 months |
| White dwarf | 0.6 | 2 months |
| Brown dwarf | 0.02 | 2 weeks |
| Jupiterlike planet | 0.001 | 2 days |
| Earthlike planet | 0.000003 | 3 hours |

the Earth. From low-mass to high-mass, possible MACHOs include asteroids; planets, either small (like Earth) or large (like Jupiter); brown dwarfs, stars born with too little mass to ignite their fuel; white dwarfs, burned-out Sunlike stars; neutron stars, stellar cinders that outweigh the Sun but are only the size of a small asteroid; stellar-sized black holes, collapsed stars whose gravity is so great they swallow even light; and massive black holes, which exceed the Sun's mass thousands or millions of times over.

On the other hand, if you're a particle physicist, you might lean toward some type of exotic subatomic particle, such as WIMPs, or Weakly Interacting Massive Particles. Unlike MACHOs, WIMPs are unearthly, *non*baryonic matter. Since by definition they only "weakly interact," WIMPs have yet to be identified even on Earth. Still, they could exist in such large numbers that they would account for all of the dark halo's mass. Many astronomers didn't like WIMPs, however, since it's hard to relish the thought that most of the Galaxy—and the universe—is made of something your telescope can't detect.

Paczyński offered a way to find out which it was, MACHOs or WIMPs. If MACHOs populated the dark halo, he said, a MACHO should occasionally drift in front of a star in the Magellanic Clouds, the Milky Way's brightest satellite galaxies. In so doing, the MACHO's gravity would magnify, or "microlens," the star's light, bending what would otherwise be diverging light rays together so that, as observed from Earth, the background star would appear to brighten, then fade back to its normal brightness. Furthermore, the more massive the MACHO, the longer the microlensing event would last, giving astronomers a clue to the MACHO's mass. That's valuable information, since different types of stars have different masses. On the other hand, if subatomic particles such as WIMPs made up the dark halo, astronomers should see no microlensing events at all.

Popular and even technical articles often claim that gravitational lensing results from Einstein's general theory of relativity. This is true but misleading, because good old-fashioned Newtonian gravity also predicts the phenomenon. In fact, the first description of the gravitational bending of light appeared in 1801, long before Einstein was born.

"The formula is exactly the same," said Paczyński. "There's only a factor of two difference." Einstein's theory predicts twice the deflection of light. "The very first paper Einstein wrote about the subject has a mistake in it," said Paczyński, "and he got the same answer as in the Newtonian case. As soon as Einstein published his first result, an expedition was set up to look for the effect in the solar eclipse at the Crimean Peninsula." During a solar eclipse, with the Sun's glare blotted out, astronomers could see how much the Sun's gravity bent the light rays of stars beyond it, altering their apparent positions. "Luckily for Einstein," said Paczyński, "World War I broke out, and they couldn't do the observations. Had they gone, they would have found the effect was twice as large as he predicted." Einstein eventually realized his mistake, so that by 1919, when British astronomer Arthur Eddington and his colleagues observed another solar eclipse, they found just the deflection Einstein predicted, transforming him into an international celebrity.

## A MACHO START

In principle, Paczyński's idea of using gravitational microlensing to detect dark matter was a clean test that would separate the MACHOs from the WIMPs. But it was hardly a plan for the, uh, wimpy, because it called for a huge observational campaign. Even in the best possible case—a completely MACHO dark halo—at any particular time only one star in a million would be getting microlensed. Astronomers would therefore have to monitor millions of stars, night after night, to catch just one instance of MACHO-induced microlensing.

"Before we got started," said Paczyński, "all astronomers I know were extremely skeptical that anything like this could be done. They said we shall be swamped with data, we shall never get it, and there will be all those variable stars." The variations of such stars—Cepheids, RR

Lyraes, eclipsing binaries, and more—would far outnumber and could masquerade as the rare microlensing events the astronomers were seeking. "I was never too much deterred by all of those arguments," said Paczyński, "because once upon a time, I myself was a variable-star observer. For me, getting microlensing events would have been just wonderful, but if we could not have separated them from variable stars—just as well, we'd learn so much about variable stars."

Fortunately, microlensing events stand out from variable stars in several ways. First, microlensing doesn't discriminate: it affects all colors equally. If a MACHO magnifies a star's red light tenfold, then it should do the same to the star's yellow light and its blue light. In contrast, most variable stars vary by different amounts at different colors. Second, the brightness of a microlensed star should not vary again, because the chance that a second MACHO would pass in front of the same star is minuscule.

Enchanted by the promise and undaunted by the problems, several teams of astronomers organized to seek gravitational microlensing. The three oldest groups adopted acronyms that *Physics Today* described as "a coordinated assault on political correctness": in addition to a primarily American-Australian collaboration simply called MACHO, there arose a French group named EROS (Expérience de Recherche d'Objets Sombres) and a Polish-American group, including Paczyński, named OGLE (Optical Gravitational Lensing Experiment).

Members of the MACHO, EROS, and OGLE teams needed more than just cute acronyms. They also needed enormous persistence and patience, since they had to monitor so many stars in order to see any microlensing events. For example, the MACHO team measured the brightnesses of between 10 and 20 million stars per *night*. The experiment would have been impossible without advanced technology—in particular, charge-coupled devices (CCDs) to acquire the data, and powerful computers to analyze it.

The first genuine success came Sunday, September 12, 1993, when the MACHO team detected a microlensing event toward a star in the Large Magellanic Cloud. "It was enormously exciting," said Charles Alcock, a British-born astronomer at the University of Pennsylvania. "It was one of the high moments of my career and my life." Although the MACHO team's telescope was in Australia—the Magellanic Clouds are in the deep southern sky—the data were analyzed in California. Said

Alcock, "Members of the MACHO team spent hours every day just looking at star after star, light curve after light curve, and Will Sutherland, who was one of the young people, one of the postdocs on the project, was the first to see the event.

"That first event was extremely clean and beautiful. It occurred against a relatively bright star, so the measurements were good. It was a great way to start, because there was very little doubt amongst us that this was gravitational microlensing."

Another MACHO team member, Kim Griest, had coined the term *MACHO*. "That was a joke," said Alcock, "but it stuck. Kim Griest was being teased at lunch that the people looking for WIMPs clearly had an edge over us, because they had a good acronym. And he made up 'MACHO' on the spot, just as a riposte to this joke, not expecting it to have any consequence. People started using it in e-mail, and the name emerged as the accepted term."

Before the MACHO team announced its discovery, Alcock discovered that he had competition: the EROS team found not one but two apparent instances of microlensing toward the Large Magellanic Cloud. Soon after, the OGLE team found a microlensing event toward the center of the Milky Way. With these discoveries, a wave of euphoria swept over astronomers, for they had finally "seen" some of the mysterious dark matter that dominated the Galaxy. Many pundits proclaimed that the heroic MACHOs had won their battle with the evil WIMPs—which astronomers can never hope to detect—and that the dark halo was totally MACHO.

But such claims were misguided. "Quite a lot of people jump to conclusions before the experimental groups are willing to reach those conclusions themselves," said Alcock. "In our first paper, we published the discovery of one event, and an enormous amount of speculation was based on that one event, plus two events of lower quality from the EROS group. And it was entirely premature."

Not only were such statements wrong, but they were also based on faulty data. Although the MACHO team's one microlensing event withstood scrutiny, the two EROS events came under attack, from no less than Paczyński himself. Both of those brightenings turned out to be caused by the antics of variable stars rather than dark-halo MACHOs.

"The whole series of announcements in 1993 was really triggered by the EROS claim," said Paczyński. "The EROS people told the MACHO people that they had two microlensing events, and they wanted to go public, so the MACHO team decided to go public; and as soon as OGLE detected their first event, they went public as well. Now we know that what EROS was announcing was probably not really microlensing. It just shows that people under stress claim things which don't last."

## REVENGE OF THE WIMPs

By 1995 the MACHO-versus-WIMP showdown had become incredibly volatile. In the first half of the year the MACHO team reported that dark-halo MACHOs were scarce, falling well below the number expected if MACHOs in the mass range to which the experiment was sensitive made up all of the dark halo. This led many people to declare WIMPs the winners. But that, too, was a premature conclusion.

"In some sense," said Alcock, "we hoped the dark halo was made more of MACHOs, because frankly it is more fun to find most of the stuff than some of the stuff. But what we actually expected is that we'd see either no events or lots of events. So finding an intermediate number was very surprising."

It was also confusing. At first the MACHO team said that 20 to 25 percent of the dark halo was made of MACHOs having the mass of brown dwarfs. Brown dwarfs range from 1 to 8 percent of the Sun's mass, not enough to spark the nuclear reactions that power the Sun. Such stars glow briefly, as they convert gravitational energy into heat, then fade from view, making them ideal candidates for dark matter.

The 20 to 25 percent estimate was widely quoted, convincing many that WIMPs made up the remainder. In late 1995, however, the tables turned again, this time in favor of the MACHOs. When the MACHO team analyzed additional observations, Alcock was stunned: the brown dwarf MACHOs were gone, and in their place were more massive MACHOs. These MACHOs, said the scientists, constituted anywhere from 20 to 100 percent of the Milky Way's dark halo—which meant there might be no WIMPs at all. The MACHO team at first put the mass of the dark-halo MACHOs at a fifth that of the Sun, then raised it to a half the Sun's mass. Then, in 2000, the team reported yet more data and came to a still different conclusion: each MACHO still weighed about half the Sun's mass, but they constituted only about 20 percent of the dark halo, leaving the remainder for the WIMPs.

The mass of the putative MACHOs, half the Sun's, seemed odd, because it was so normal—for red dwarfs. Red dwarfs resemble the Sun, because they burn hydrogen into helium at their cores, but they differ by being smaller, fainter, and cooler. Even so, red dwarfs glow brightly enough that they could hardly hide in the dark halo. Thus, the MACHO team speculated that the MACHOs were another type of star with a similar mass: faded *white* dwarfs. These dense, burned-out remnants of Sunlike stars account for 5 percent of the stars in the Galaxy's disk; the nearest white dwarf orbits the bright star Sirius, only 8.6 light-years from Earth. As white dwarfs age, they cool and fade, so they could populate the dark halo while eluding terrestrial astronomers.

However, white dwarfs weren't popular. Many astronomers, Paczyński included, dismissed a white-dwarf dark halo as nonsense. "How is it possible that there are so many white dwarfs in the dark halo and nothing else stellar?" he asked. The dark halo weighs roughly a trillion times more than the Sun. If 20 percent of that consists of white dwarfs, the dark halo has 400 billion of them—more than all the stars in all other parts of the Milky Way put together.

Furthermore, that throng of white dwarfs seems to be unaccompanied by any other type of star. At present, when the Milky Way gives birth to new stars, it makes them in all masses, from small to large. The trouble is, white dwarfs evolve only from stars born into a limited mass range. Stars born with more than eight times the Sun's mass don't become white dwarfs. Instead, they explode as supernovae, showering the Galaxy with heavy elements. If the dark halo spawned lots of supernovae, they would have inundated the Galaxy with more heavy elements than astronomers see. And stars born with less than 80 percent of the Sun's mass live so long that they should still be shining, as orange and red dwarfs—which astronomers don't see in the dark halo. Thus, in order to consist solely of white dwarfs, the dark halo must have given birth only to stars having masses between 0.8 and 8 times the Sun's.

"That's an issue," said Alcock, "but I don't know whether to call it a problem. It means that the initial mass function for the dark-halo population had to be very different from the initial mass function that we see today. And that's somewhat peculiar, but you know, we have no theory whatsoever of how the initial mass function today arises, so it seems to me that we don't have any compelling reason to expect it was the same in the past." Long ago, when the dark halo presumably formed its stars, the Galaxy consisted almost entirely of hydrogen and helium, with few heavy elements. That's because stars hadn't yet made any. Such metal-poor material might have manufactured stars with a different range of masses.

"Oh, it might have," said Paczyński, "but nobody can calculate it, so it's put in as a totally ad hoc assumption. I like to call such assumptions 'epicycles': we explain one thing which we do not understand with another thing which we do not understand. So what's the point?"

Aside from requiring a special mass spectrum in the early Galaxy, a white-dwarf dark halo signals another problem. If spent firecrackers litter your street one morning, you can be sure that smoke filled the air overnight. In the same way, a dark halo strewn with white-dwarf cinders implies pollution by their progenitors. White dwarfs evolve from stars like the Sun. Sunlike stars swell into red giants, which forge large amounts of carbon and nitrogen. These elements can seep into the red giants' atmospheres. The red giants become white dwarfs by casting their carbon- and nitrogen-laden atmospheres into space. That's good for us—much of the carbon and nitrogen in our bodies blossomed in red giants—but bad for a white-dwarf dark halo, since the Galaxy doesn't seem to have enough carbon and nitrogen.

Of course, the low metal abundance of that ancient era might have suppressed mass loss from red giant stars. "Again, who knows?" asked Paczyński. "We do not understand mass loss, even for stars around us. Who knows what was the mass loss for zero-metallicity stars? I have a mind block: I simply refuse to speculate on topics which have too many if's along the way. Maybe it's because of my age—I've seen people make horrendous mistakes after going through several if, if, if's. It's much more fun to either get new data or wait for new data."

In 1999 astronomers led by Rodrigo Ibata of the European Southern Observatory in Germany used the Hubble Space Telescope and said that they had spotted actual white dwarfs in the Milky Way's dark halo. There were neither spectra to confirm their white-dwarf nature nor parallaxes to ascertain their distances, but the objects' proper motions—that is, their apparent movements across the sky—were just the right size for stars in the dark halo.

In 2001 white dwarfs received an even stronger boost when Berkeley astronomer Ben R. Oppenheimer and his colleagues reported the dramatic discovery of over three dozen such stars in the dark halo. From this large number, Oppenheimer's team concluded that white dwarfs make up at least 2 percent of the dark halo's mass. The astronomers had spectra of the stars, so no debate erupted over their white-dwarf nature. The white dwarfs reside within just a few hundred light-years of the Sun, but they have high velocities, indicating that they belong to the halo, said Oppenheimer.

However, even before this work reached print, dark-halo white dwarfs suffered a one-two punch. First, Ibata's team retracted its claim of white dwarfs in the dark halo. New proper-motion data showed that the purported white dwarfs had no proper motions at all, so the objects were probably not even stars in the Milky Way but instead galaxies so far beyond that they looked stationary. Second, several astronomers attacked Oppenheimer's work, saying that most of the white dwarfs he had assigned to the dark halo were actually high-velocity white dwarfs belonging to the Milky Way's thick disk.

Whether or not the dark halo has white dwarfs, other MACHO possibilities exist. A priori, the most logical might seem to be black holes, the darkest stars of all. Unfortunately, ordinary black holes pose two problems. First, they arise when large stars collapse and explode, releasing heavy elements such as oxygen and iron in quantities far larger than observed. Second, ordinary black holes are at least three times more massive than the Sun, whereas the MACHO team put the dark-halo MACHOs at only half the Sun's mass.

A split second after the big bang, however, clumps of just that mass might have collapsed and created black holes. "That's a suggestion that I've always found rather appealing," said Alcock. "You could imagine black holes forming in the early universe. Maybe there was some special physics that favored the formation of black holes with this particular mass." While ordinary black holes often orbit other stars—witness the first example, Cygnus X-1—primordial black holes would be single, unbound to other stars. Also, primordial black holes might pack not just baryonic matter but also nonbaryonic matter.

Meanwhile, the MACHO and EROS teams have ruled out two other possible components of the dark halo: free-floating planets and brown dwarfs. "Lower-mass objects make shorter-duration events," said Alcock, "and our experiment is extremely sensitive to shorter-duration events. It was designed to look for brown dwarfs, so the fact that we didn't find any is extremely significant." Also out are neutron stars, the dense remnants of many exploding stars. Ordinary black holes might exist, because they produce such long-lasting microlensing events that the experiments have not had time to see them, but their progenitors would have polluted the Galaxy. Thus, said Alcock, it looks as though about 80 percent of the dark halo consists of exotic subatomic particles. The WIMPs have won.

## DARK DOUBTS

Paczyński, however, said that the dark halo's WIMP fraction could be closer to 100 percent. When observers see a microlensing event toward the Magellanic Clouds, they know only that the responsible party lies somewhere in front of the background star. At first, observers assumed that the unseen objects lay well in front of the Magellanic Clouds, in the Galaxy's dark halo. In 1994, however, Kailash Sahu, an astronomer then at the Canary Islands' Institute of Astrophysics, suggested that the supposed MACHOs were merely ordinary stars in the Magellanic Clouds, microlensing other Magellanic stars. If so, the half-solar-mass objects that the MACHO team claims make up 20 percent of the dark halo don't exist—at least, not in the dark halo.

Sahu submitted his work to *Nature*. "There were at least three negative referee reports," said Paczyński, "and some of the referee reports were signed. The referees thought the paper was so ridiculous that they would put their names in the public domain, and at least one or two of these were EROS people. Somehow, *Nature* took its chance and published the paper. In 1998, faced with new data, the EROS team officially changed its mind and said, 'We don't understand why, but apparently the lenses are in the Magellanic Clouds.' So they reversed their opinion—which I think is a good sign."

The new, crucial data came from microlensing events by binary stars. Such events are rare, but they complicate the light variations of the background star in a way that yields an approximate distance to the lens. The first binary lens toward the Large Magellanic Cloud was discovered by the MACHO team in 1994. Alcock said that the observations were too poor to provide a reliable distance, but Paczyński said that the lens was probably in the Large Magellanic Cloud, not the dark halo. In 1998, a binary microlensed a star in the Small Magellanic Cloud. In this case, the observations were good, and the lens was definitely in the Small Magellanic Cloud, not the dark halo. Thus, of all the microlensing events seen toward the Magellanic Clouds, the only one (according to Alcock) or two (according to Paczyński) whose lenses have known distances were in the Magellanic Clouds. No lens with a known distance was in the Milky Way's dark halo.

To Paczyński, the implication is clear: most if not all of the supposed dark-halo MACHOs don't reside in the dark halo. If so, the dark halo may have no MACHOs at all, and the WIMP fraction could hit 100 percent. Alcock countered that the only distance he's confident about comes from the binary lens toward the Small Magellanic Cloud, which differs from its big brother. The Large Magellanic Cloud is nearly face-on to us, whereas the Small Magellanic Cloud is stretched out along our line of sight. Thus, the Small Magellanic Cloud should exhibit a fair amount of self-lensing.

"From a hard-nosed empirical point of view," said Alcock, "what we really need to do is to come up with a series of observations that will resolve the issue one way or the other. We need to find some way of routinely getting distances to the single lenses toward the Large Magellanic Cloud. When we have that, we'll answer the question."

Despite the dispute, there haven't been enough microlensing events to explain the bulk of the dark halo, the Galaxy's least lustrous but most massive component. Thus, most of the Galaxy we call home consists of mysterious material that no one has ever seen on Earth.

# TWELVE

# THE CONSTANT HUBBLE WAR

*The researches of many commentators have already thrown much darkness on this subject, and it is probable that, if they continue, we shall soon know nothing at all about it.*

—UNKNOWN, BUT OFTEN ATTRIBUTED TO MARK TWAIN

EIGHT decades after Edwin Hubble discovered that the universe expands, astronomers are *still* slugging it out over just how fast it does so. In all of astronomy, no number has provoked more controversy than the one that bears his name. The Hubble constant converts redshifts into distances and yields a rough measure of the universe's age.

For half a century, Hubble's heir, Allan Sandage of the Carnegie Observatories in Pasadena, California, has fathomed the galaxies, near and far, to establish the Hubble constant's numerical value. But for the past quarter century, rival astronomers have attacked his work, claiming that he botched the job—perhaps deliberately, in order to reconcile the big bang theory with the ages of elderly stars.

In 1999, NASA proclaimed the cosmological equivalent of harmonic convergence when it held a press conference to announce that the Hubble constant problem was solved—by its own Hubble Space Telescope—and that all the war-torn Hubble soldiers had finally laid down their arms. Newspapers and magazines dutifully reported the good news.

"Well, I read a lot of nonsense in the newspapers about all sorts of things," said Sandage. "It simply isn't true. NASA had to bring a press conference, because determining the Hubble constant is what they sold the Hubble Space Telescope on. We were denied any say at the press conference. We were disinvited to that conference."

At the press conference, Sandage's rivals reported that they had pinned down the Hubble constant to within 10 percent of its true value—a marked improvement over an earlier era when the Hubble constant had swung wildly between two numbers, one twice as high as the other. Trouble was, the Hubble constant that NASA advertised as good to 10 percent disagreed with Sandage's value by over 20 percent. "We all agree," one of Sandage's foes told *New Scientist*. "We just don't agree that we agree."

"That's a bunch of hooey," said Sandage. "It's war—it's war!"

## A BATTLE-SCARRED UNIVERSE

The Hubble war has blackened much of the universe. Indeed, were it a literal war, the troops would have blasted red supergiants, blue supergiants, Cepheids, RR Lyrae stars, H II regions, exploding white dwarfs, planetary nebulae, the Hyades star cluster, globular star clusters, the Large Magellanic Cloud, the Andromeda Galaxy, the M81 group, the spiral galaxy M101, the Virgo cluster, the Fornax cluster, and the Coma cluster to smithereens. Every Hubble soldier nourishes the hope, the wish, the dream to grasp outward, far outward, hundreds of millions of light-years from Earth, into the so-called Hubble flow, and measure the exact distance of a galaxy. There, unperturbed by the Milky Way, the galaxy's redshift marks the rush of new space bursting open as the universe expands, washing away the tiny Doppler shift induced by the galaxy's random motion. Dividing this far-off galaxy's redshift by its distance then yields the Hubble constant, in units of kilometers per second (the redshift) per megaparsec (the distance). If the Hubble constant's numerical value were known, it would state how fast redshifts rise as one proceeds away from the Milky Way. Do galaxies 100 megaparsecs—326 million light-years—from Earth have redshifts of 5,000 kilometers per second, or instead 10,000 kilometers per second? If astronomers knew the answer, they could reverse the trick and use a galaxy's redshift to infer its distance, a vital astronomical quantity.

With modern technology, measuring galactic redshifts is easy. Measuring distances is not, so it is distances that the Hubble warriors fight over. Life would be easy if all stars and galaxies were the same. Then they'd all have the same *intrinsic* brightness, so their *apparent* brightness would reveal their distance: the fainter a star or galaxy looked, the farther it would be from Earth. Of course, then astronomy would be so boring that there'd be few astronomers around to measure the Hubble constant.

In actuality, stars and galaxies are like snowflakes: no two are the same. The greatest galaxies outshine the least by over a millionfold, and the greatest stars outperform the meekest by over a trillionfold. Fortunately, within a few hundred light-years of Earth, astronomers can measure the distances of stars via parallax, the tiny shift in a star's apparent position that results as the Earth circles the Sun. To reach greater distances, astronomers have constructed an elaborate "distance ladder" that touches farther stars in the Milky Way, then nearby galaxies, and finally more distant ones, in the Hubble flow. A less polite term for the distance ladder would be a house of cards, in which the upper levels rest on all the levels below. If a lower level collapses, so does the whole house of cards.

No galaxy better demonstrates the Hubble war's brutality than a small one named IC 4182. A splotchy swirl of stars and gas beyond the Local Group, it even looks wounded. No one would have ever heard of it, had not the light from one of its dying stars reached the Earth seven decades ago.

In 1937, astronomers little concerned with the Hubble constant watched a star in IC 4182 explode. The supernova happened to be of the most precious variety, a type now labeled "Ia"—which might as well be triple A, as far as Hubble warriors are concerned. Type Ia supernovae arise from stars of nearly identical mass, so they peak at nearly identical luminosities, making them excellent standard candles. Moreover, they shine so brilliantly that astronomers see them exploding in galaxies billions of light-years away, well into the Hubble flow. Thus, if astronomers knew the intrinsic brightness of just one type Ia supernova, they could measure the distances of far-off galaxies that had spawned their own type Ia supernovae. And that would reveal the Hubble constant.

But supernovae are so rare that none of the type Ia variety was seen in our Galaxy or anywhere else in the Local Group during the entire twentieth century. So Allan Sandage and his longtime collaborator Gustav Tammann tried to measure the distance of IC 4182, for that would reveal the intrinsic brightness of its supernova. The galaxy was too far for them to see its Cepheids, the pulsating yellow supergiants that betray distances. Instead, Sandage and Tammann observed IC

4182's red supergiants. These stars outshine Cepheids, and the brightest are roughly the same from galaxy to galaxy. In 1982, Sandage and Tammann deduced a distance for IC 4182 of 14 million light-years, about six times farther than the Andromeda Galaxy. Such a distance implied that the supernova had shone with the strength of several billion Suns. By assuming that type Ia supernovae in remote galaxies had the same luminosity, Sandage and Tammann got those galaxies' distances and derived a low value for the Hubble constant, 50 kilometers per second per megaparsec.

In 1992, however, Hubble warriors led by Michael Pierce, then at Kitt Peak National Observatory, stormed the galaxy. They reobserved its red supergiants at infrared wavelengths, where red stars glow brightest. Pierce's team claimed that the galaxy's actual distance was only 8 million light-years, so its supernova had been less lustrous. Applied to more distant type Ia supernovae, this finding led to a high Hubble constant of 86 kilometers per second per megaparsec—nearly twice the Sandage-Tammann value—suggesting that the universe expanded nearly twice as fast.

A few months later, Sandage's side fired back. Deploying the heavy ammunition of the Hubble Space Telescope, Sandage and his troops took aim at IC 4182 and for the first time glimpsed its Cepheids. These indicated that the galaxy's true distance was 16 million light-years, close to their original estimate but twice that from Pierce's team. The Cepheid distance, coupled with the type Ia supernovae in remote galaxies, yielded a Hubble constant around 45.

The citizens of IC 4182 could hardly relax, however, because the battle wasn't over. After the discovery of Cepheids, no Hubble warrior could question the galaxy's distance; but they did assault the supernova. Because it exploded long ago, astronomers recorded its brightness on photographs, which are cruder than modern detectors. In 1995, after reexamining those photographs, Pierce and George Jacoby at Kitt Peak reported that the supernova had peaked at only 60 percent of the brightness originally claimed. Thus, the intrinsic brightness of that supernova—and by extension *all* type Ia supernovae—was less than what Sandage and Tammann had adopted, so the remote galaxies that had hosted type Ia supernovae must be closer. Pierce and Jacoby ended up with a Hubble constant around 71. Then Yale astronomer Bradley Schaefer attacked their work, saying the supernova had been brighter and the Hubble constant was lower, around 57. Said Jacoby, "That was a very unpleasant experience for Mike and me—and I hope for Brad." Jacoby and Pierce then fired back, saying Schaefer was wrong and reaffirming their Hubble constant of 71.

All this over one small galaxy!

## A Problem of Astronomical Proportions

The war over the Hubble constant achieves epic levels in part because the Hubble constant is the most basic cosmological parameter, quantifying the twentieth century's greatest astronomical discovery—the expansion of the universe. In contrast, the two other cosmological parameters, the matter density (omega) and the cosmological constant (lambda), perturb that expansion, omega slowing it down, lambda speeding it up.

Despite its name, the Hubble constant is not constant for all time. Even if the universe's expansion neither accelerated nor decelerated, the Hubble constant would fall as the universe grew older. That's because the inverse of the Hubble constant provides a rough measure of the universe's age. Thus, when the universe is twice its present age, the Hubble constant will be roughly half its present value.

Actually, astronomical constants rarely hold constant for all time. For example, the amount of sunlight hitting the Earth's atmosphere, the so-called solar constant, is no more constant than the Hubble constant, because the Sun is slowly brightening. In contrast, the constants of physics, such as the speed of light, do appear to be constant. Even some scientists fail to appreciate the distinction between astronomical constants and physical constants, incorrectly calling the Hubble constant the "Hubble parameter."

The Hubble constant affects a host of other astronomical quantities, because it transforms galactic redshifts into galactic distances. During much of the Hubble war, the opposing camps disagreed by a factor of two. Double the Hubble constant and you halve the distance of a galaxy with a particular redshift. As a result, if one astronomer claimed that a galaxy was 300 million light-years from Earth, another astronomer could claim that exactly the same galaxy was 600 million light-years distant.

A lot depends on distance. Start with the galaxy's size. Astronomers can measure the galaxy's apparent diameter, but to convert that into an actual diameter, they need to know its distance—and therefore the Hubble constant. For a galaxy with a given apparent diameter, the farther it is, the greater its true size. The astronomer who said the galaxy was 300 million light-years away might say that it's 100,000 light-years in diameter, while the astronomer claiming a distance of 600 million light-years would say that it's 200,000 light-years across.

The distance affects the galaxy's estimated luminosity even more severely, because that depends on the *square* of the distance. Astronomers can measure the galaxy's apparent brightness, but to convert that into an intrinsic brightness, they must again know the galaxy's distance—and thus the Hubble constant. The astronomer who said the galaxy is 300 million light-years away might say it's as luminous as the Milky Way, while his rival who claimed twice the distance would deduce a luminosity four times greater.

The Hubble constant's numerical value even affects the material content of the universe. That's because primordial nucleosynthesis—the creation of light elements just after the big bang—depended on the density of ordinary, or baryonic, material, as well as on the universe's expansion rate. A high Hubble constant implies a low density of baryonic material (see the discussion at the end of Chapter 7). If the Hubble constant is 100, then baryonic matter contributes a mere 2 percent of the density the universe needs to reverse its expansion. On the other hand, if the Hubble constant is 50, baryonic matter makes up 8 percent of the critical density. Indeed, if the Hubble constant were just 15—which no one thinks it is—then baryonic matter would close the universe.

## A TIMELY TOPIC

The Hubble constant looms large, too, because it could torpedo the big bang theory. The Hubble constant's numerical value largely determines how much time has elapsed since the big bang. The greater the Hubble constant, the faster the universe expands and the younger it must be, because a fast-expanding universe has taken less time to reach its present size. The age of the universe, as derived from the Hubble constant, had better exceed the ages of the oldest stars. From theoretical models of how stars change color and luminosity as they consume their fuel, astronomers can estimate the ages of the ancient stars in the Milky Way's globular clusters. During much of the Hubble war, best estimates for these ages ranged from 15 to 20 billion years. (Current best estimates are lower, 12 to 15 billion years—about three times the age of Earth.) Thus, the universe must be at least this old.

Said the University of Hawaii's R. Brent Tully, "A lot of people who don't work on the Hubble constant say, 'Ah, it's just a number. BORING.' But under the old paradigm, it wasn't just a number—because of this fundamental attack it made on what was then the standard cosmology. If you bought into the evidence that the Hubble constant was anywhere above 60, certainly if it was above 70, certainly if it was in the 80s, you said, 'Um, that world model is *wrong*.'"

The old paradigm Tully referred to assumed that the matter density, omega, had exactly the critical value—1. During the 1980s and much of the 1990s, however, observers and theorists viewed omega differently. Observers tended to believe their observations, that the universe had more matter than met the eye—but not that much more. They therefore favored a low value for omega, around 0.1, which meant the universe would expand forever. In such a lightweight universe, the expansion had decelerated little, and the universe was fairly old. Yet if the Hubble constant was much above 70, the universe seemed younger than the globular clusters. For example, a Hubble constant of 80 and an omega of 0.1 implied that the universe was just 11.0 billion years old, on the wrong side of the globular cluster ages. Still, with uncertainties in all three numbers—the Hubble constant, omega, and the globular cluster ages—observers could squirm out of the problem.

The age conflict was far worse for theorists. Seduced by the beauty of inflationary theory, most theorists believed that the universe was flat. Since they hated the cosmological constant, they believed that the universe was dense and omega precisely 1, placing the cosmos right on the borderline between eternal expansion and collapse. If so, the gravitational pull of all that mass was strongly braking the universe's expansion. Hence, in the past, the universe expanded faster, so it must be younger than if the expansion was not decelerating. If omega was 1, a Hubble constant of 80 yielded a cosmic age of just 8.1 billion years, clearly younger than the globular clusters. After the Hubble war broke out, most theorists therefore sided with those astronomers obtaining *low* values of the Hubble constant. They sometimes claimed this was because the best evidence favored a Hubble constant in the 50s. But the truth was more sinister: a low Hubble constant saved their high-omega universe from the ravages of the ancient globular clusters.

## STEPS TOWARD THE HUBBLE CONSTANT

In 1948, when the mighty 200-inch telescope first peered into the heavens above California's Palomar Mountain, Edwin Hubble planned to use it to photograph other galaxies, hoping to find their Cepheids, discern their distances, and thereby improve the determination of the Hubble constant. Still, the aging Hubble needed an assistant.

"Hubble just asked for the department chairman to send somebody up," said Allan Sandage, then a graduate student at Caltech. "And I was the somebody! Wonderful how fate does its thing! If it hadn't been this, it would have been something else, and I'd have gone to Harvard and been [Bart] Bok's student, and who knows what would've happened? I'd be an accountant by now."

Sandage had first encountered astronomy in third grade, when a neighbor boy showed him the sight through a telescope. Said Sandage, "I've known since that time that there was no other career." In college, at the University of Illinois, he measured magnitudes of stars when he participated in a nationwide star-counting program that Harvard astronomer Bart Bok had organized to probe the Milky Way's structure. Sandage counted a million stars in Perseus—tedious work, but he loved it. As a result of this experience, it was he who was sent up the mountain to meet Hubble.

Fate did its thing again, because Sandage wrote his thesis—on the globular cluster M3—under Walter Baade, the father of the two stellar populations, who in 1952 halved the Hubble constant and doubled the universe's size and age by recognizing the two types of Cepheids. Thus, even as a graduate student, Sandage was working with two of the century's most influential astronomers. In 1953, though, Hubble died, and Sandage carried on his work.

In his six decades as a professional astronomer, Sandage has also contemplated questions that go beyond science. "One seems to be forced to the statement that there is a blueprint that unfolds, and everything seems to have been planned," said Sandage. "I'm very taken with a recent book called *Darwin's Black Box*, by a man by the name of [Michael] Behe. It's an incredible book. He essentially talks about the complications of the human condition. When you cut yourself and the blood clots, so many biochemical things have to happen to prevent you from bleeding to death. But they happen all the time—every time you cut yourself! So since things of that nature feed back upon themselves and are interrelated, you cannot produce that by undirected evolution. You just can't. Something depends on something else, and then a third thing depends upon the first two things being in place, and all up the chain. So that's a mystery that I think will never be solved within science."

A cosmic blueprint might suggest a creator. "I don't know what it suggests," said Sandage. "It just suggests that the world is extraordinarily complicated, and it seems to have to have cooperative phenomena of a very high level. But theology is a mystery; science is a puzzle. The two are different. You can solve a puzzle; you really cannot solve a mystery. And I'm willing to live with the mystery. I wasn't always—scientists want to have solutions, but that's a solution which I think will never come. A scientist would be crazy to work on mysterious problems, because science is all about puzzles, and if you're not trying to solve a puzzle, you're not at the frontier."

During the 1950s, first Baade and then Sandage solved the puzzle of a universe that seemed younger than the Earth—by lowering the Hubble constant and boosting the universe's estimated age. Whereas Hubble had put the Hubble constant around 500 kilometers per second per megaparsec, implying a universe no more than 2 billion years old, by 1958 Sandage had found that the true Hubble constant lay somewhere between 50 and 100 kilometers per second per megaparsec, suggesting a universe 10 to 20 billion years old.

The determination of the Hubble constant centers on distances to other galaxies, and no distance indicator is more sacred than the Cepheids, one of the few things that nearly all Hubble warriors agree on. Dennis Overbye's book *Lonely Hearts of the Cosmos* recounted this memorable scene: "Sandage talked faster and faster. 'Let me identify these for you: IC 1613, member of Local Group; Andromeda nebula; 6822, member of Local Group; M33; John Graham's NGC 300, which you've just seen, at a Cepheid distance.' His finger started thunking datum points. 'Okay, Cepheid distance, Cepheid distance, Cepheid, Cepheid, Cepheid, Cepheid, Cepheid, Cepheid, Cepheid,' as if it were a Fort Knox stamp, this was the good stuff, primo distances."

In the early 1960s, to help him find and measure the periods of Cepheids, Sandage enlisted Swiss astronomer Gustav Tammann. "I am an easygoing person," said Tammann, who called Sandage the twentieth century's best astronomer. "Allan is a *serious* person. Allan wonders what our existence is about, and he believes our existence is justified only if you do something reasonable with your abilities and your life—that is your obligation. If you miss this obligation, you are a sinner. And as a scientist, you look for truth. Allan believes basically that being wrong is failing your obligation. It is sinful to be wrong.

"Life is extremely difficult for him—he suffers through it, he always sees obligations, he is *forced* to find the truth; otherwise, he is a sinner. But he must admit that he is sometimes wrong, that he is sometimes a sinner, and that makes life only harder for him."

Sandage's best-known mistake involved not the Hubble constant but another cosmological parameter, omega, which represents the density of matter that slows the universe's expansion. In the 1960s, Sandage used the brightest galaxy in each of several galaxy clusters to measure how fast the universe's expansion was decelerating. If each such galaxy had the same intrinsic brightness, its apparent brightness would indicate its distance. Sandage then found that the most distant galaxy clusters, billions of light-years away, were closer than they should have been, indicating that the universe's expansion had decelerated greatly since the time they emitted their light. He said that omega was around 2 and that the universe would someday collapse.

However, a young astronomer in Texas, Beatrice Tinsley, challenged Sandage's work. "That was a very personal debate between the two," said Tammann. "It was so bad that at one conference, Beatrice Tinsley got up before a talk by Sandage and said, 'What you will hear next is all wrong.'"

At stake was nothing less than the fate of the cosmos. The debate focused on the stars in the bright galaxies that Sandage had used as standard candles. These galaxies were old—giant ellipticals, like the Virgo cluster's M87—and glowed yellow-orange. Both Sandage and Tinsley knew that these galaxies might fade as their stars aged. But if their yellow-orange light emerged from long-lived main-sequence stars like the Sun, as Sandage thought, then their luminosity should not change much, even over billions of years, so the galaxies should be good standard candles. On the other hand, if the yellow-orange light came from fast-evolving yellow, orange, and red giants, then the galaxies were hardly standard candles, dooming Sandage's program to measure the universe's deceleration.

Other astronomers observed elliptical galaxies at infrared wavelengths and detected the spectral signature of giant stars. "This observation was decisive," said Tammann, "and Beatrice Tinsley was right." Tammann was about to give a talk at an astronomy conference. "I phoned Allan and said, 'Can I say the universe is open?' And he said, 'Of course.' From one second to the other. So Allan is totally capable of switching his opinion, if the evidence requires it."

In the 1960s, Sandage and Tammann embarked on a program they eventually called "Steps Toward the Hubble Constant." Their first target: a small spiral galaxy in the constellation Camelopardalis named NGC 2403. It resembled the Local Group's M33 but belonged to another galaxy group, the M81 group, named for a large spiral in Ursa Major. The crispness of NGC 2403 on photographic plates suggested proximity. No Cepheid had ever been seen outside the Local Group, but the ones that Sandage and Tammann glimpsed in NGC 2403 shocked them: they were faint, so the galaxy was much farther than had been thought, nearly five times farther than the Andromeda Galaxy—11 million light-years from Earth. Tammann and Sandage published this distance in 1968.

They then proceeded to another crisp-looking spiral galaxy, M101. "That was the next stepping stone," said Tammann. "It's a gigantic spiral, very beautiful. For months and months, we looked to find M101's Cepheids. The published distance of M101 should have made it only slightly more difficult than in NGC 2403—but it was impossible to find Cepheids." Even though M101 resided in the same constellation as M81, it had to lie well beyond. Since Sandage and Tammann couldn't find M101's Cepheids, they used cruder distance indicators, such as its brightest stars, to place the striking spiral galaxy 24 million light-years from Earth, over twice as far as the M81 group. This distance, published in 1974, and M101's redshift yielded the Hubble constant: 56 kilometers per second per megaparsec. Such a low Hubble constant meant the universe was old. With omega between 1 and 0, the universe's age was 12 to 17 billion years.

Sandage and Tammann then took aim at the most crucial object of all, the Virgo cluster. By comparing M101 and other spiral galaxies with their counterparts in Virgo, Sandage and Tammann established the cluster's distance, eventually settling on a figure of 71 million light-years, thrice M101's distance. This distance, coupled with Virgo's redshift, again gave a Hubble constant in the 50s, as did the distances and redshifts of more remote galaxies. Thus, the universe was comfortably old—old enough to accommodate the ancient globular clusters. The first and most basic cosmological parameter seemed to have been measured.

Little did Sandage and Tammann then know that their work was about to come under fierce attack, led by maverick Gérard de Vaucouleurs, the self-described "pugnacious" French-born astronomer who two decades earlier had championed the existence of the Local Supercluster. By the 1970s, de Vaucouleurs was at the University of Texas at Austin, the same place where Sandage critic Tinsley had earned her doctorate.

When the Hubble constant controversy erupted, it hit Sandage far harder than Tammann. Said Tammann, "When you talk to Allan, he will say, 'I live in a monastery, and I don't mind what the outside world does. I do my job, and the rest of the world can do what it wants.' But he cannot stick completely to the rules. He still looks over his monastery walls."

In contrast, Tammann said he welcomed the controversy. "It saved my life," he said. "Honestly, in 1974, when I gave a talk at de Vaucouleurs' place in Austin, nobody objected. De Vaucouleurs listened very attentively and was very positive, and I thought, 'Horror, now we have this value of the Hubble constant, and it will never be quoted, and nobody will ever care about it.'" As a result, Tammann thought he would have to find something else to do.

"When the opposition came," he said, "it was stimulating. I never had to think, 'What shall I do next week? What shall I do next year?' I had not the possibility, the restriction, to suffer something new, because de Vaucouleurs saved our life, giving us every week new work to do. So I had a very easy life. The controversy was the red flag which I had simply to follow, and I didn't have to think much."

## DOUBLE HUBBLE TROUBLE

The late Gérard de Vaucouleurs pictured himself the outsider, the crusader fighting an enemy entrenched behind castle walls. When MIT physicist Alan Lightman asked him why astronomers had resisted his idea of the Local Supercluster, de Vaucouleurs replied, "Number one, because it did not come from a member of the establishment. As one of them told me years later, 'If it doesn't come from us, I don't believe it.' There is only one true church."

Later de Vaucouleurs again compared the behavior of the scientific establishment to the worst excesses of religion. "I have a question that I ask my students to illustrate my position here," he said. "'What was God doing during the time before [the big bang]?' I know what the answer of physicists is: 'This is not a question you should ask because this is a mystery. There was no time and there was no space.' The theologians used to make it even simpler. 'You should not ask those questions, brother, because otherwise the Inquisition will start being interested in you.'"

Two decades before he initiated the war over the Hubble constant, de Vaucouleurs first crossed swords with Sandage. At the time, Sandage was working on *The Hubble Atlas of Galaxies*, which would feature photographs of nearly two hundred galaxies—spirals, ellipticals, irregulars. Sandage had lent some of the photographs to de Vaucouleurs, who published them himself, scooping Sandage with his own material. "He never asked my permission," said Sandage. "That was the first indication that there was going to be a strong rivalry." Sandage believes that de Vaucouleurs deliberately set out to get a different number for the Hubble constant. "Now that's a terrible thing to say," said Sandage, "but I felt it at the time, that if we had gotten 75, he would have gotten 50. And indeed, we *did* get 75 at one time, and he *did* get 50." That was in the late 1960s.

In 1976, de Vaucouleurs launched the Hubble war. He assaulted Sandage and Tammann's work, claiming they had bungled the distances to galaxies. Employing all manner of distance indicators—novae, Cepheids, RR Lyrae stars, eclipsing binaries, blue and white supergiants, red supergiants, diameters of ionized hydrogen regions, globular clusters, supernovae, and more—de Vaucouleurs constructed an elaborate distance ladder that he said fathomed the universe far more securely. His philosophy was as simple as his methodology was complex: spread the risks by using as many distance indicators as possible, since any one distance indicator, such as Cepheids, may fail. He also argued that Sandage and Tammann had failed to recognize the gravitational pull of the Virgo cluster, which slowed the universe's expansion among nearby galaxies like M101, leading to a Hubble constant that was too low.

By the time his distance ladder reached the Hubble flow, de Vaucouleurs ended up with a Hubble constant nearly twice the Sandage-Tammann value: 100 kilometers per second per megaparsec, implying an extremely youthful universe. For omega between 1 and 0, the universe was just 6.5 to 9.7 billion years old—not much older than the 4.6-billion-year-old Earth, and far younger than the globular clusters.

De Vaucouleurs' attack on Sandage and Tammann dazzled other astronomers. Up to that time, most had believed Sandage's work on the Hubble constant. He was, after all, Hubble's honorary son. Yet de Vaucouleurs' sophisticated array of distance indicators made Sandage and Tammann's work seem simplistic and naive. Why use one primary distance indicator when you could use five?

Thus did Sandage and de Vaucouleurs polarize their profession by staking out two radically different values for the Hubble constant. "This dichotomy affecting one of the most basic parameters of the universe is intolerable," wrote de Vaucouleurs; "it is as if some physicists believed the speed of light to be 300,000 km per second while others insisted on 600,000!" De Vaucouleurs insinuated that Sandage and Tammann had arrived at their low Hubble constant of 55 only because it made the universe old enough to accommodate the globular clusters.

"He went around the world saying we were deceitful," said Sandage. "But where in our papers, when we had to make a decision, did we cheat, did we 'cook' the data? We simply didn't. I know that's the claim—that's not only his claim, but many other people have said, we will always get 55 regardless of what the data say. That's because it is 55!"

## SPINNING SPIRALS

As the Hubble warriors stormed into action, another astronomer developed a new method for measuring the distances of spiral galaxies. "I was not one of these guys that was really interested in astronomy as a youth," said Brent Tully. "Basically, the decision I made at each point was, 'Where is the biggest area of the unknown?' I said, 'Gee, astronomy—they sure don't know much there!' And when I went into astronomy, I said, 'Oh, they sure don't know much in extragalactic astronomy.'" The thing that hooked Tully on astronomy: "The perception that there were vast areas of ignorance, so there were vast opportunities to find things out without having to be too smart."

With radio astronomer J. Richard Fisher, Tully investigated how a spiral galaxy's luminosity matched its rotation rate. "We had the idea of the relationship from a very simple thought experiment," he said. "Consider two galaxies in the Local Group, the Andromeda Galaxy and M33. Even back then, we knew the distances to those galaxies pretty well, so we certainly knew that Andromeda was a giant galaxy and M33 was a middle-range galaxy. And we also knew that Andromeda had a very rapid rotation speed, something like 250 kilometers per second, and M33 was rotating at something like 120 kilometers per second—much more leisurely. So giant galaxies should be rotating fast and smaller galaxies should be rotating slower.

"The thought experiment was: suppose, out of an erroneous measurement, we moved Andromeda in by a factor of two and M33 out by a factor of two. Then the luminosities we would assign to those galaxies would have said that M33 was a giant galaxy and Andromeda was a more modest galaxy. But that wouldn't have made any sense, given their rotation velocities."

To test this luminosity-rotation relation, Tully and Fisher observed galaxies in the Virgo cluster. Since the Virgo galaxies all had about the same distance from Earth, their apparent brightness served as a surrogate for their intrinsic brightness. To measure the galaxies' rotation, the astronomers exploited the Doppler shift. As a spiral spins, one side whirls toward the Earth and exhibits a smaller redshift than the other side, which whirls away. The more the two redshifts differ, the faster the spiral spins. Fisher measured the redshifts of the radio waves emitted by the spirals' hydrogen gas, and sure enough, the bright spirals spun faster than the faint ones.

Now Tully and Fisher could deduce the Hubble constant. Anchoring their luminosity-rotation relation to galaxies of known distance and hence known luminosity—Andromeda and M33, the M81 group, the M101 group—they used the Virgo galaxies' rotation rates to derive their luminosities and thereby the distance to the Virgo cluster: just 43 million light-years, much less than the 71 million light-years Sandage and Tammann had said. Dividing Virgo's redshift by this short distance led to a high Hubble constant of around 80. After Tully and Fisher published their work in 1977, other astronomers used the Tully-Fisher relation and also found a high Hubble constant. Sandage reportedly referred to the Tully-Fisher relation as "Fishy-Tuller."

Such a high Hubble constant implied that the universe was young and spelled a serious age conflict with the globular clusters—especially if the matter density, omega, was 1, in which case a Hubble constant of 80 led to a universe that was only 8 billion years old. Said Tully, "I myself and de Vaucouleurs can point to papers in 1988 or so where each of us was saying, 'By golly, there's a real problem here, and you theorists should be thinking about models with a cosmological constant.'" A cosmological constant elevates the universe's age. "Okay," said Tully, "we don't know anything about the physics of the cosmological constant, and that's for theorists to figure out, but from an observational standpoint, omega equals 1 is just a nonstarter. Even a low-density universe is just not comfortable, and the only real models that made sense to us had a cosmological constant." At the time, though, because of the furor over the Hubble constant and the perceived ugliness of the cosmological constant, few astronomers entertained the thought. "That shows," said Tully, "the nonsense that theorists do. Theorists are great, they drive the whole field; but you know, that whole thing that lambda is 0 because it's ugly was absurd."

## NEW TECHNIQUES

During the 1980s, astronomers developed new methods for measuring the Hubble constant that also fell on the high side of the Hubble divide. One employed planetary nebulae, gas shells cast off and set aglow by dying stars. This method was spearheaded by George Jacoby, an astronomer at Kitt Peak National Observatory in Arizona.

"The Vietnam War threw me into astronomy," said Jacoby, who graduated with a major in aeronautical engineering in 1969. "My options were to find a job in the aerospace industry, for which I would get a job deferment from the Army; or I could go to Vietnam; or I could leave the country." He chose the first course. "And what I learned," he said, "is that engineering, for me, was deadly dull. Furthermore, I was being asked to design rocket engines which were delivering bombs—so although I wasn't involved in the Vietnam War, I was somehow involved in what I would call inhumane practices. I was so appalled by some of the things I was being asked to do, I decided to look for a new career in science that was guaranteed not to hurt anybody. And I couldn't think of a science less likely to hurt somebody than astronomy."

Jacoby began observing the spectra of planetary nebulae in other galaxies, in order to track how the abundance of chemical elements varied from a galaxy's center to its edge. Different planetary nebulae shine at different brightnesses, but Jacoby and his colleagues noticed that the overall luminosity pattern was the same from galaxy to galaxy. By assuming that the planetary nebulae in the Virgo galaxies followed the same distribution, Jacoby measured the distance to the Virgo cluster and obtained the Hubble constant.

"I actually thought the Hubble constant was going to be in the 50s," said Jacoby. "I was convinced by Sandage's arguments that the age-of-the-universe problem was more easily understood if the Hubble constant was small. I was a little bit surprised that the Hubble constant was higher than I was inclined to expect. But it didn't bother me."

Jacoby and his colleagues reported the news in 1990. As had Tully, they found the Virgo cluster much closer than Sandage and Tammann had said: 48 million light-years versus 71 million. That distance led to a high Hubble constant, around 88, much closer to Gérard de Vaucouleurs' value of 100 than to Allan Sandage's value of 55.

Jacoby soon heard from the two chief combatants. "In my opinion," said Jacoby, "Gérard was very encouraging and I think he was more ready to admit a change in his value than Allan Sandage was. Allan's messages to me were always quite strident. One of his earliest letters was something like, 'Congratulations on a new technique; I hope you find your errors soon.'"

Planetary nebulae shine only so brightly, preventing Jacoby from reaching galaxies much beyond the Virgo cluster. In the same year, however, another technique emerged that probed galaxies at greater distances. Developed by John Tonry, now at the University of Hawaii, this new technique sought fluctuations in the surface brightness over the face of an elliptical galaxy. The farther an elliptical galaxy is, the smoother it looks, so the less are these fluctuations. In 1990, Tonry and his colleagues reported that the surface brightness fluctuation technique yielded a distance to the Virgo cluster of 55 million light-years and a Hubble constant of 78.

"I got two phone calls," said Tonry. "One was from Gérard de Vaucouleurs. He was practically in tears, because I'd betrayed him and I was coming in with a sinfully low answer. And I got a phone call from Allan Sandage, who was charming as usual. He said, 'If there's something wrong with what you're doing, what could it be?'"

## VICTORY IN VIRGO

The year 1990 was momentous for another reason: NASA finally deployed its high-priced Hubble Space Telescope, whose launch the shuttle explosion four years earlier had delayed. The telescope's chief raison d'être was to measure the Hubble constant—in particular, to discern Cepheids in the Virgo cluster and establish its distance. At the time, the farthest detected Cepheids resided in the giant spiral galaxy that had anchored the Sandage-Tammann distance scale: M101, where, in an ironic twist, Sandage's rivals had used them to confirm the Sandage-Tammann distance of 24 million light-years and repudiate de Vaucouleurs' much shorter distance.

One astronomer who hoped to use the Hubble Space Telescope to find Virgo's Cepheids was Wendy Freedman, who is now Sandage's most formidable opponent—even though she works at the same observatory. As did Sandage, she first encountered astronomy at a young age. "The virus was passed on by my father," Freedman said. "The first memory I have of becoming interested in astronomy is my father describing to me that because of the finite travel time of light, the objects we see might not be there any more. We were seeing them as they were when the light left them. That just fascinated me."

During the 1980s, while still in graduate school, Freedman studied bright stars in nearby galaxies such as M33. In order to deduce the stars' exact luminosities, she needed to know those galaxies' distances, so she examined their Cepheids. With her advisor, Barry Madore, she tackled the problem of dust. Dust dims Cepheids and makes them look fainter and farther than they really are. Fortunately, dust grabs more blue light than yellow or red, so by observing Cepheids at these colors, plus infrared wavelengths, Freedman and Madore could quantify the amount of intervening dust and establish accurate distances to nearby galaxies. Cepheids brought Freedman and Madore together in more than just scientific ways; they married in 1985.

Ten years earlier, just before de Vaucouleurs launched his crusade against Sandage, Madore had fired an opening shot. He challenged the distance to NGC 2403, the small spiral galaxy in the M81 group that Sandage and Tammann had said was 11 million light-years away. Madore claimed that dust dimmed the galaxy's Cepheids so much that it was actually just 7 million light-years away.

Sandage is still steamed over the incident. "I turned over all the plates to him and told him why he couldn't do what he was doing," said Sandage. "He used our colors and made the crazy assumption that the colors should be used exactly as measured, when we argued that you could not do that because of the errors." Those colors suggested that the Cepheids were reddened by large amounts of dust, like the setting Sun.

Ironically, in 1988, Freedman and Madore themselves observed NGC 2403's Cepheids at dust-defying infrared wavelengths and proved the Sandage-Tammann distance correct. "Either we were exceedingly lucky," said Sandage, "or exceedingly smart." But he was annoyed that the Freedman-Madore paper never explicitly said that he had been right. Freedman countered that Sandage and Tammann had been right for the wrong reason—errors in their photographic data compensated for the effects of dust.

After the 1990 launch of the Hubble Space Telescope, Freedman's team—including Madore but not Sandage—hoped to solve the Hubble constant problem once and for all. The Hubble Space Telescope should detect Cepheids in dozens of galaxies, thereby better calibrating techniques such as the Tully-Fisher relation, the method that uses spirals' spins to derive distances. The telescope should also find the first Cepheids in the Virgo cluster as well as the Fornax cluster, on the other side of the Milky Way from Virgo.

It didn't work out that way. Although NASA spent over a billion dollars on the Hubble Space Telescope, it launched it with a defective mirror—incapable of seeing Virgo's Cepheids. "It was an enormously depressing time," said Freedman. "I can't overemphasize how *awful* it was. People had waited so long for this telescope, the expectations had been high—and they were shattered. [Princeton astronomer] Jim Gunn was originally in our group, and eventually walked away from anything having to do with Space Telescope. He felt he had spent thirteen years of his life pouring his energy and attention into it, and he was devastated."

What to do? Why, use the "Hobbled" Space Telescope to observe Cepheids in *nearby* galaxies—and wait. Freedman's team studied M81's Cepheids, reconfirming that galaxy's distance. Also, Sandage's team spotted Cepheids in IC 4182, the small galaxy that had spawned the type Ia supernova decades earlier.

Late in 1993, astronauts aboard the space shuttle repaired the Hubble Space Telescope, conferring upon it a dubious honor: it cost more to *fix* than any telescope on Earth had ever cost to *build*. Meanwhile, a team led by Michael Pierce in 1994 scooped the pricey scope by arming a ground-based telescope with advanced optics and detecting three Cepheids in a Virgo spiral named NGC 4571. THE HUBBLE CONSTANT PINNED DOWN, claimed the cover of *Nature*. Virgo's distance was just 49 million light-years, far less than the 71 million light-years that Sandage and Tammann had claimed, implying a high Hubble constant of 87. Shortly afterward, Freedman's team reported that the Hubble Space Telescope had spotted Cepheids in another Virgo spiral, the attractive galaxy M100. Its distance was also small—

56 million light-years—and yielded a high Hubble constant of 80. Newspapers announced the result, and the age crisis worsened: the universe seemed younger than its oldest stars. Even if omega were 0, the age of a universe with a Hubble constant of 80 was a mere 12 billion years. And if omega was 1, the universe was just 8 billion years old.

Sandage stood his ground: "I hope you're not believing all the nonsense in the press." The Cepheid distances to the two Virgo spirals were irrelevant, he said, because they were in front of the cluster's core. "No spiral lies in the Virgo cluster's core," said Sandage. "The spirals are spread in a 15-degree diameter about the core, so the distance to any given spiral does not of and by itself tell you the distance to the core." Instead, he said, the cluster's core houses M87 and the other giant ellipticals, but these old galaxies lack the Cepheids that would resolve the dispute. Einstein once said that the Lord was subtle, not malicious; however, he never measured the Hubble constant.

Freedman acknowledged that M100 might lie in front of the Virgo cluster's core. Indeed, her team's estimate of the Hubble constant carried a stated uncertainty of 21 percent, primarily because M100's exact location relative to the Virgo core was unknown. However, she and her colleagues have since measured Cepheid distances to other Virgo spirals, and all have distances matching M100's. In the meantime, Sandage's team observed one spiral in the constellation Virgo—NGC 4639—because it hosted a type Ia supernova, and this galaxy's Cepheids indicate a much greater distance, 83 million light-years. Sandage therefore argues that the Virgo cluster core sits between M100 and NGC 4639, consistent with his preferred value of 71 million light-years. Freedman countered that even though NGC 4639 lies in the constellation Virgo, it has nothing to do with the Virgo cluster.

The Virgo cluster is north of the Milky Way's disk. South of the Galactic disk is the more obscure Fornax cluster, smaller and tighter, so its galaxies might be more concentrated. Freedman's team used the Hubble Space Telescope to observe the Fornax galaxy NGC 1365, a stunning barred spiral. Its Cepheids yielded a distance of 61 million light-years and a Hubble constant of 72. In 2001, Freedman's team published work that combined several techniques, including the Tully-Fisher relation, surface brightness fluctuations, and type Ia supernovae, to give the same Hubble constant—72—which she said was good to 10 percent.

## AN EXPLOSIVE SITUATION

Meanwhile, Sandage's team used the Hubble Space Telescope's Cepheid-hunting powers to pin down the distances of nearby galaxies that have sprouted type Ia supernovae. "They're the best standard candles we know," he said. Unlike most supernovae, which arise from big stars, type Ia supernovae shatter stars only the size of Earth: white dwarfs, the cinders of former Sunlike stars. If a white dwarf accretes enough material from another star to exceed the Sun's mass by 40 percent, a runaway nuclear explosion annihilates the white dwarf, frying much of it into radioactive nickel, which decays into radioactive cobalt, which decays into nonradioactive iron, which flows through your veins right now. The radioactive decay unleashes so much light—about 5 billion Suns' worth—that a single explosion outshines a modest spiral galaxy like M33. Moreover, because type Ia supernovae arise from the same type of star with the same amount of mass, they should all peak at the same luminosity. Indeed, when two type Ia supernovae erupt in the same galaxy, observers see that they have nearly the same apparent brightness and hence nearly the same intrinsic brightness. During the 1990s, astronomers further tightened the luminosity dispersion. For reasons not yet understood, the slower a type Ia supernova fades, the brighter its peak luminosity. By measuring this decline rate, astronomers can marshal type Ia supernovae into standard candles with luminosities good to 20 percent and thus distances good to 10 percent—excellent, by astronomical standards. When Sandage and his colleagues use type Ia supernovae to measure the distances of remote galaxies, as they reported in 2000, they find a Hubble constant of 58. However, Freedman and her colleagues use type Ia supernovae, too—but as they reported in 2001, they find a Hubble constant of 71.

How can this be? How can two teams use the same method yet disagree by 20 percent? For one thing, to derive the intrinsic brightness of type Ia supernovae, the two teams choose different nearby type Ia supernovae. Some of these exploded long ago, so they were not observed as well as modern supernovae. Freedman throws them out, saying they are not reliable. Sandage, however, retains them, arguing that what counts is not just the uncertainty in the supernova's observed brightness—which he acknowledges is larger—but also the uncertainty in the Cepheid distance. "The unfortunate thing," said Tammann, "is the wickedness of nature: the less-well-observed supernovae tend to have better Cepheid distances." That's because those old supernovae exploded in nearer galaxies, such as IC 4182.

Also, until recently, Freedman added type Ia supernovae that had exploded in elliptical galaxies of the Fornax cluster. Since elliptical galaxies have no Cepheids, Freedman assumed they had the same distance as Fornax's barred spiral, NGC 1365, in which Cepheids had been seen. Sandage blasted the practice, saying NGC 1365 lies in front of the Fornax ellipticals and the supernovae they produced. Freedman no longer uses those supernovae.

"I must say, I stopped reading the opponents' papers some few years ago," said Sandage, "so I don't know what Freedman actually does. But up until 1999, in a conference where Freedman and I were on the same platform, she used that argument." Freedman said that even then, the Fornax supernovae carried little weight.

Furthermore, Freedman prefers to employ a large array of distance indicators, not just type Ia supernovae. "Type Ia supernovae are one of the most promising means of measuring the Hubble constant," she said. "But would I put all of my eggs in that basket? Absolutely not." She would rather hedge her bets by adopting a suite of the best distance indicators and taking an average. To which Sandage said: "Well, okay, you have a few rotten eggs, and you ruin your soufflé."

## MALMQUIST BIAS

Actually, although Sandage and Tammann emphasize type Ia supernovae, they contend that other techniques, if properly used, agree with their low Hubble constant of 58. Take, for example, the Tully-Fisher relation, the correlation between a spiral's spin and luminosity. When Tully uses the Tully-Fisher relation, he gets a Hubble constant of 77. When Freedman and her colleagues do so, they get a similar value, 71. However, when Sandage and Tammann use the Tully-Fisher relation, they get only 50. Fishy indeed.

Sandage and Tammann argue that their opponents neglect an insidious effect known as Malmquist bias. Named for a Swedish astronomer who analyzed star counts early in the twentieth century, Malmquist bias can falsely raise measured values of the Hubble constant. It works like this. Suppose an observer examines dozens of distant spiral galaxies that all spin exactly as fast as the Milky Way. According to the Tully-Fisher relation, each galaxy should shine just as brightly as the Milky Way, allowing the observer to deduce each galaxy's distance from how faint it looks. In reality, however, the Tully-Fisher relation has scatter: a galaxy that spins as fast as the Milky Way may be somewhat brighter or somewhat fainter. No problem, you think—on average, they're the same luminosity as the Milky Way. But as observers look farther and farther away, they tend to see just galaxies more luminous than the Milky Way. An observer who incorrectly assumed these galaxies shone only as brightly as the Milky Way would think the galaxies were closer than they are, boosting the estimated Hubble constant. Furthermore, say Sandage and Tammann, other subtle biases afflict their opponents' work, again pushing the Hubble constant too high.

"It's impossible to talk with these people," said Sandage. "I have talked with them a few times, and their eyes just glaze over." Tammann mentioned a 2000 paper by Freedman's team that used the Tully-Fisher method to obtain the Hubble constant. "It's senseless!" he said. "It is *the* illustration of Malmquist bias." Tammann delivered what may be the worst insult one astronomer can give another: "For me, this is more astrology than astronomy."

Sandage's critics claim that they know all about Malmquist bias, which can be corrected if the size of the scatter is known. The greater the scatter, the worse is the Malmquist bias. Type Ia supernovae show little Malmquist bias, because their luminosities are so uniform. The Tully-Fisher relation suffers more, because it has greater scatter. Sandage's critics contend that he adopts too large a scatter for the Tully-Fisher relation, causing him to *over*correct for Malmquist bias and leading him to too *low* a Hubble constant.

## A New Focus

As the Hubble warriors duke it out over Malmquist bias, other astronomers have recently used methods to measure the Hubble constant that leapfrog over the Cepheid-based distance ladder altogether. One of the most promising exploits gravitational lensing, whereby gravity bends light. In the 1990s, dark matter hunters used this phenomenon to detect MACHOs.

Long before that, in 1979, astronomers had discovered what seemed to be two quasars near the Big Dipper. Both quasars had identical redshifts, 1.41, so their light had taken some 10 billion years to reach the Earth. But the two quasars were really just one quasar, whose light was split into two by an intervening giant elliptical galaxy at a redshift of 0.36. The two light beams follow different routes from the quasar to

Earth. One beam reaches Earth before the other, so one path is shorter than the other. By measuring the time delay between the two light beams, astronomers could hope to measure the difference in path length, in light-years. The longer the time delay, the greater the difference in path length—and the farther the lensing galaxy from Earth. Comparing its distance with its redshift would then yield the Hubble constant.

First, though, astronomers had to measure the time delay between the two images. "It was a long process," said Rudolph Schild of the Harvard-Smithsonian Center for Astrophysics, "and I faced a lot of opposition. It has been very hard for me, in a kind of psychological way, to know that for six or seven years the world was on a wrong tack—that the literature was being piled up with heaps and heaps of garbage."

Still, Schild was lucky to be alive for the dispute. Three years before the double quasar's discovery, while atop Mount Hopkins in Arizona, he was struck by lightning. Minutes later, his colleagues noticed him missing, searched the mountain, and soon found him, unconscious. Fortunately, because thunderstorms spark forest fires, the Forest Service had readied a helicopter to search for trouble, so the pilot headed for the mountain, landing amid thick fog, heavy rain, and increasing wind, with a mere six feet of clearance for the rotor blades. The helicopter whisked Schild, still unconscious, to the hospital.

Quasars can flicker, so to establish the double quasar's time delay, Schild monitored both images, hoping to see one image wax or wane and then see the other image repeat the feat. Caltech astronomer Peter Young and his colleagues had discovered the intervening galaxy and modeled how its gravity bent the quasar's light. Young estimated that the galaxy's gravity delayed one light beam by approximately 3 years, the other by approximately 2 years.

Said Schild, "Somehow, in the mathematics of doing all this for the first time, he ended up adding the 3 and 2 together, instead of subtracting them, as we know to do today. Things were not so obvious then. And so people thought that the time delay would be 3 + 2, or about 5 years, instead of 3 – 2, or about 1 year." So Schild observed the quasar only once a month. Around the time the mistake was discovered, Young committed suicide.

"I have always felt," said Schild, "that this amazing mistake of his largely contributed to his suicide. He was a young scientist at Caltech, and he might have feared that that's what he would be remembered for.

I had the greatest admiration for Peter Young, because it was very beautiful work he did. It's fairly easy for us to go back and say, 'Oh, you subtract the one time from the other to get the time delay.' At that time, he saw a maze of mathematical calculations, and just one minus sign nailed him."

Schild began observing the quasar more intensely, and in 1986 he and a student, Bryan Cholfin, published what they believed to be the correct time delay: around 1.1 years. Soon, though, Schild's work came under fire, from radio astronomers and other optical astronomers observing the quasar. Their data suggested a longer time delay. William Press, a smart but little-liked theorist, then devised an elegant method to analyze the quasar data. During part of the 1980s, Press had chaired the Harvard astronomy department, until faculty complaints booted him out of the position. Press and his colleagues analyzed the other astronomers' data, but not Schild's, and concluded that Schild was wrong; the true time delay was 1.5 years. The difference in time delay affected the derived Hubble constant. Schild's shorter time delay implied a Hubble constant around 65, Press's time delay around 45.

"Bill Press then went around giving a lot of colloquia, saying that my time delay value was wrong," said Schild, "and I got a very bad reputation. He would say, 'Schild is off his rocker; he doesn't know what he's doing; he's just a dumb observer.' And he would take any occasion to trash me." Schild contended that his data—which Press refused to analyze—were better than the other data. "Bill Press was part of this Princeton astrophysical mob," said Schild, "and they always speak with one voice. They sit around and have teas and meetings together, and then effectively vote on the truth. So what's the truth on the time delay matter? Well, they quickly established that Bill Press's time delay value was right and Rudy's was wrong, and then they started to write scientific papers. So there are these absurd papers by [Princeton's] Ed Turner describing a universe with a variable Hubble constant—variable at different places in the universe." The Hubble constant derived from the double quasar applied at enormous distances, because the gravitational lens was billions of light-years from Earth. Since the Press time delay gave such a low value for the Hubble constant, it differed from determinations of the Hubble constant closer to Earth, suggesting that the Hubble constant varied from place to place.

"I also know," said Schild, "because it was reported to me, that there was a symposium in Aspen—I think it was in 1992—and for some reason people were laughing at me: 'Oh, yeah, Rudy—you know, he measures the two quasar images, and then he measures five standard stars on the same image frames, and then he compares the quasar brightness to them. Isn't that silly? Ha, ha, ha.' Well, you know, it's a damned good way to do it. So I was made to be the laughingstock of that community. Psychologically, I found it extremely difficult to have people thinking that I was probably a crank. It's very hard to shake that, so even today, there are a lot of people who find it hard to take me as a serious scientist." Press offered bets, at ten-to-one odds, that he was right. Schild took him up on the offer, giving Press one dollar, confident he'd someday receive ten back.

Then it happened. In December 1994 the lead image of the double quasar dimmed 10 percent. The other image should repeat the trick—in February 1996, if the time delay was 1.1 years, as Schild had said; or in June 1996, if the time delay was 1.5 years, as Press had said. Astronomers observed the quasar, and sure enough, the second image faded 10 percent—in February, just as Schild said it would. Schild thought, "Thank God, that chapter is behind me." Actually it isn't entirely: Press never paid him the ten bucks.

Soon after Schild's vindication, astronomers succeeded in measuring the time delays of other gravitationally lensed quasars. "Any one lens may always do something to stab you in the back," said Christopher Kochanek, also at the Harvard-Smithsonian Center for Astrophysics. "What you want to do is use a basket of lenses." Taken together, he said, those lenses give a Hubble constant in the low 60s, although uncertainties allow values from the high 40s to the high 70s. "My feeling on the Hubble constant," said Kochanek, "is that you should never think about the value; you should think only about the error bar. Because when you think about the value, you succumb to Hubble constant disease." Sounds terrible—what is it? "Hubble constant disease," he explained, "is what drives everyone to ignore or lowball systematic errors as soon as they start talking about the Hubble constant."

In the case of a Hubble constant derived from gravitational lenses, the chief troublemaker is the galaxy whose gravity lenses the quasar's light. A Hubble constant in the low 60s follows if the intervening galaxies have standard dark halos with standard flat rotation curves. If the galaxies' mass is less concentrated—that is, if they have *rising* rotation curves—then the Hubble constant is lower. If the galaxies' mass is more concentrated—if they have *falling* rotation curves—then the Hubble constant is higher. Furthermore, the derived Hubble constant also depends somewhat on the cosmological parameters omega and lambda, introducing additional uncertainty.

Despite these problems, Schild said that gravitational lensing is now the best route to the Hubble constant, far surpassing the traditional Cepheid-based distance ladder. The Hubble constant, he said, is around 65. Schild then uttered a statement that managed to achieve the impossible: uniting, albeit briefly, both sides of the raging Hubble war. He said, "Today we know effectively nothing about the Hubble constant from the Cepheid business. I consider that to be a monumental failure."

Allan Sandage said, "Schild's dead wrong." Wendy Freedman said, "We know effectively nothing about the Hubble constant from Cepheids? Well, he's entitled to his opinion." Brent Tully said, "That's ridiculous! And if he's suggesting he gets the Hubble constant out of gravitational lensing, that's just dumb. What're we talking about here? We're talking about light being bent as it passes by a whole bunch of gravitational potential wells—not just an individual galaxy but other galaxies around that galaxy, a complicated situation—and he's telling me he understands that sufficiently well to give me a Hubble constant?"

## HOT GAS AGAINST A COLD SKY

In recent years, astronomers have tried to exploit another technique for measuring the Hubble constant that sails over the Cepheid-based distance ladder. Named for two Russian scientists, the Sunyaev-Zel'dovich effect occurs when the cold photons from the cosmic microwave background zip through galaxy clusters harboring hot gas. The hot gas boosts the energy of the cold photons, distorting the cosmic microwave background's spectrum in a way that radio astronomers can observe. It would be as if some of a rainbow's red light acquired energy and became orange: the rainbow's red light would be duller, its orange light more vibrant.

The bigger a galaxy cluster, the more hot gas the cold photons encounter and the bigger is the boost they receive from the Sunyaev-Zel'dovich effect. Thus, from the size of the Sunyaev-Zel'dovich effect, astronomers can deduce the cluster's true diameter. By comparing this with its apparent diameter, and assuming the cluster is spherical, astronomers can derive the cluster's distance. Dividing the cluster's redshift by its distance then yields the Hubble constant.

In practice, astronomers must also know the gas's temperature and density, since the hotter and denser the gas, the greater is the Sunyaev-Zel'dovich effect. Hot gas emits x-rays, which can be observed by satellites such as Chandra, launched in 1999. The x-ray spectrum reveals the gas temperature, and the x-ray brightness gives information about the gas density.

In the early 1990s, when astronomers began deriving the Hubble constant in this way, the number came out low—in the 30s, 40s, and 50s—but with large uncertainties. More recently, astronomers using the Sunyaev-Zel'dovich effect have found values for the Hubble constant in the 60s and low 70s. The largest uncertainty is the exact nature of the hot gas pervading the cluster. If the gas clumps, it emits more x-rays, fooling astronomers into thinking the cluster has more gas than it actually does.

## GROUND ZERO: THE LARGE MAGELLANIC CLOUD

During the late 1990s and early 2000s, more traditional Hubble warriors faced a battle erupting in a most disturbing place: close to home, in the Large Magellanic Cloud, the brightest of the Milky Way's satellite galaxies. "It is annoying," said Freedman. "But that's where a significant amount of the total uncertainty lies."

The Large Magellanic Cloud looms large because it anchors the Cepheid distance scale, displaying the closest collection of Cepheids all having the same distance from Earth. Thus, the distance to the Large Magellanic Cloud reveals the intrinsic brightness of Cepheids. So if the Large Magellanic Cloud's estimated distance were to decrease by 10 percent, then all Cepheid-based distances to other galaxies would likewise decrease, and the Hubble constant would *in*crease by 10 percent. Fighting over such a nearby galaxy, Hubble warriors must feel like world conquerors returning home—and crashing on the rocks of their native land.

During the 1990s, the two warring sides used similar distances for the Large Magellanic Cloud. Freedman's team used a standard distance of 163,000 light-years. Sandage's team used at most a slightly greater distance, 169,000 light-years. That's not even a 4 percent difference.

In 1997, with much fanfare, other astronomers announced that the European satellite Hipparcos had measured the parallaxes of Cepheids near the Sun, finding them more luminous than had been thought and implying a *greater* distance to the Large Magellanic

Cloud, by 10 percent. But hardly anyone else agreed. "I don't take the Cepheid result that seriously," said Michael Bolte of Lick Observatory in California. "They had lots of measurements that were right on the hairy edge of what Hipparcos could do, and it's hard to average lots of lousy measurements to get the right answer." Because Cepheids are distant—even the nearest, Polaris, is 430 light-years from Earth—the errors on the Hipparcos parallaxes sometimes exceeded the parallaxes themselves.

Since then, astronomers have found evidence that the Large Magellanic Cloud is *closer* than the standard distance of 163,000 light-years. One team used so-called red clump stars, which are yellow or orange giants whose centers burn helium rather than hydrogen. "They are quite bright and quite common," said Krzysztof Stanek of the Harvard-Smithsonian Center for Astrophysics, "so we see a lot of them." The cooler of the two giant stars in the nearby Capella system—a brilliant yellow star in the constellation Auriga—is a red clump star.

Because red clump stars are common, many are nearby, so the Hipparcos parallaxes were good—and revealed that red clump stars all had nearly the same intrinsic brightness, making them good standard candles. However, when this absolute brightness was applied to red clump stars in the Large Magellanic Cloud, the distance came out to be only 145,000 light-years, or 12 percent less than the standard distance of 163,000 light-years.

Another method has also given a short distance. "Cosmology isn't my field," said Villanova University's Edward Guinan, who studies stars. "I was more interested in the properties of stars beyond our Galaxy. We did measure the first mass of a star outside our Galaxy, but no one cares about that. Only when I realized that this was such a hot issue did we change the title of the proposals to things like 'Distance to the Magellanic Clouds' and 'Cosmic Yardstick.' It was dumb of me not to think of it first!" That got his team time on big telescopes.

Guinan and his colleagues observed double stars in the Large Magellanic Cloud. As the stars swing around each other, one eclipses the other, dimming its light. The longer the eclipse, the bigger the eclipsing star must be. The bigger and hotter the star, the greater is its intrinsic brightness, so knowing the star's size and temperature lets astronomers deduce its luminosity—and hence its distance. Astronomers have long used this technique to measure stellar distances inside the Milky Way.

Before Guinan began his work, he preferred a long distance to the Large Magellanic Cloud—around 169,000 light-years, which followed from a presumably circular ring of gas lit up by a 1987 supernova in the Cloud. "When the first binary came out to be at 149,000 light-years, I thought it was a mistake," he said. "We weren't happy. And when the second one came out to be near that same number, then I started to believe it. Now with the third one, I'm getting more convinced." He and his colleagues published the result for the second system in 2001. "The people who like the short distance are all very happy," said Guinan. "And the people who like the long distance haven't said much!"

In response to these developments, Freedman's team first lowered its adopted distance to the Large Magellanic Cloud from 163,000 light-years to 160,000 light-years, then raised it back to 163,000 light-years, leading to the Hubble constant of 72, published in 2001. "These things tend to take some time to shake out," said Freedman. "My strong preference is to say, there are many methods, they all have strengths and weaknesses, and we need to average over them, because there isn't a strong reason for preferring one over another." Freedman's team used seven different methods.

"Oh yeah, sure. Well, why not fifteen different methods?" Sandage asked sarcastically. He and Tammann are sticking with their distance of 169,000 light-years, based on what he claims are the best methods, including RR Lyrae stars, Cepheids in Milky Way clusters of known distance, and the supernova ring. As for the red clump stars, he said, "Oh, the red clump! You know the scatter in the red clump?" It's large, he said, and the stars' luminosities depend on both age and metallicity, making them poor standard candles, especially because the stars in the Large Magellanic Cloud tend to be younger and more metal-poor than those near the Sun. Nevertheless, if the Large Magellanic Cloud's true distance is only 150,000 light-years, then his Hubble constant will rise from 58 to 66, and Freedman's from 72 to 78.

## OLD SOLDIERS NEVER DIE

A factor of two no longer separates the two sides battling over the Hubble constant, but the Hubble war rages as fiercely as ever. Sandage and Tammann insist the Hubble constant is in the upper 50s; Freedman and her team insist it is in the lower 70s.

"Sorry, this sounds psychological," said Gustav Tammann, "but

Wendy Freedman hates to go below 70. I think she has the feeling she is yielding to us if she gets under 70; so she still gives surprisingly much weight to routes to the Hubble constant—to the Tully-Fisher relation, surface brightness fluctuations, and distances to field galaxies—which give a higher value."

Freedman laughed at the accusation. "I have no problem with going below 70," she said. "I have no problem if it's 50. I just want to see data that give evidence for that, and if that's where it ends up, that's where it ends up."

Meanwhile, other astronomers have accused Sandage and Tammann of a fixation with a Hubble constant in the 50s. "I discount Allan Sandage's value immediately," said Brent Tully. "He gets his value whatever. I mean, it isn't a measurement."

In recent years, many astronomers, not knowing whom to believe, have adopted a Hubble constant midway between Sandage and Freedman: 65. Unless otherwise stated, all distances in this book, as well as other quantities that depend on the Hubble constant, use a Hubble constant of 65. This number gives a Hubble time of 15.0 billion years. In a low-omega universe, this is just able to accommodate the ages of the oldest globular clusters, implying that the universe didn't waste any time forming them. If the Hubble constant is indeed 65, then it is closer to Sandage's value than to that of de Vaucouleurs, his arch rival.

So can Sandage live with a Hubble constant of 65? "No!" he shouted. "Why do you want me to do that? Because of the precepts that give that? The precepts that give that are absolutely wrong. You can't average a good value with a bad value and get a correct value. Our opponents want to correct our absolute magnitude calibration—Calm down, Allan, calm down, Allan—of type Ia supernovae by second-parameter corrections which don't exist. So you're asking if I can live with 65. Of course not; it's just dead wrong!" The Hubble constant, he said, is 58.

Nor does Freedman like a Hubble constant of 65. "The best results are coming out in the low 70s," she said. "The results are actually all quite consistent. There may be some systematic error that's identified at some point, and it may be necessary to move toward 65, but at the present time the best available evidence suggests the low 70s."

Meanwhile, Sandage has reached agreement with Freedman on at least one astronomical topic. "I think she believes that the Sun comes up in the east and sets in the west," he said.

# THIRTEEN

# OMEGA: THE WEIGHT OF THE UNIVERSE

OF THE three parameters that rule the cosmos, the one that carries the greatest weight, literally, is omega. Just as the Hubble constant serves as the universe's speedometer, quantifying the universe's expansion, so omega serves as the universe's accountant, adding up all the material, seen and unseen, in all the galaxies, galaxy clusters, superclusters, and even voids throughout the universe. Since matter tugs other matter via its gravity, omega tries to restrain the universe's expansion. If the matter density is great enough—if omega exceeds 1—then it will first slow, then halt, and finally reverse the expansion, until the cosmos crashes together in a fiery grand finale. On the other hand, if the matter density is lower—if omega is less than or equal to 1—then the universe will expand forever, growing still colder and darker. Which end you prefer, fire or ice, is a matter of taste—although as one astronomer noted, "If you think about it, neither one of them is very good."

With the destiny of the universe hinging on its numerical value, omega ought to be the most controversial cosmological parameter, provoking far more passion than either the Hubble constant or lambda, the cosmological constant. Furthermore, whereas the Hubble constant has generated furor by oscillating between numbers that differ by a factor of two, omega has vacillated by over a factor of *ten*.

In actuality, however, omega is the *least* controversial cosmological parameter. "There's never been a huge emotional investment in omega," said Kitt Peak's George Jacoby, veteran of the Hubble wars. "Some theorists have advocated a value, but they haven't been as vocal as some of the observers working on the Hubble constant, which has had a tremendous history of being controversial. And the lambda thing has always been a wild-eyed concept, and anybody crazy enough to say it's not 0 was treated as kind of nuts."

Also, the omega debate differs from the Hubble constant dispute sociologically. "You see, there's two kinds of people out in the community," said the University of Hawaii's Brent Tully. "There're people who see the beauty of theoretical concepts; and then there are observers who are more nuts-and-bolts—they say, 'Let's make a measurement here.'" The Hubble constant war is bitter in part because it is internecine, between observers who accuse each other of bungling the measurements.

In contrast, the omega debate pitted theorists steeped in physics, who preferred an omega of 1, against astronomical observers scanning the heavens, who thought omega was closer to 0. Theorists wanted omega to be 1 because this value followed from what they regarded as a beautiful and elegant theory: inflation, in which the hyperexpansion a split second after the big bang flattened the universe and drove omega to 1. Moreover, the flatness problem also convinced theorists that omega must be 1. If it wasn't, omega would drift farther and farther from this value, so that today, billions of years after the big bang, omega should be much farther from 1 than it actually is. Such grandiose theoretical concepts, however, underwhelmed most astronomical observers. Instead, they trusted their observations, which suggested that omega was low, closer to 0.

If omega was exactly 1, as most theorists said, the universe was so dense that the expansion was slowing greatly. If instead omega was low, as most observers said, the universe's matter barely braked the expansion at all. Either way, however, the universe would expand forever. Thus, the debate over omega didn't alter estimates of how the universe would end, another reason the subject engendered little passion.

## THE HEART OF THE MATTER

To estimate the numerical value of omega, astronomers naturally looked to the great galactic gatherings that spangle constellations such as Virgo and Coma Berenices. "Galaxy clusters are like the Mount Everests in the world—the highest mass concentrations that we see in the universe," said Neta Bahcall of Princeton University. "They are the largest, most massive systems that are in equilibrium—not falling together or spreading apart—so they are one of the best tools in constraining the mass density of the universe." As a result, the motions of member galaxies reflect the cluster's gravity and thereby indicate its mass. Said Bahcall, "It's just like our Earth goes around the Sun: it's in equilibrium, so from the motion of the Earth around the Sun, we can measure the mass of the Sun."

If the universe consisted solely of Sunlike stars, the feeble glow of the galaxies would translate into a paltry omega—roughly 0.001, a mere thousandth of the amount needed to make the universe flat. Of course, most material in the universe emits no light, and back in the 1930s Fritz Zwicky found the first strong evidence for this dark matter: if the Coma galaxy cluster didn't have far more mass than met the eye, its high-flying galaxies would have shot out of the cluster. Today astronomers can use the same technique for measuring cluster masses and then extrapolate to the universe at large. If the cosmic mass-to-light ratio matches those of galaxy clusters, omega is hundreds of times greater than 0.001—but far shy of 1. Instead, it's only 0.2 or 0.3, which means that the universe has only 20 to 30 percent of the matter density it needs to be flat.

As the University of Arizona's Simon White and his colleagues found in 1993, galaxy clusters yield another measure of omega that confirms this low figure. In addition to boasting hundreds or thousands of galaxies, clusters teem with hot, x-ray-emitting gas—the same gas that causes the Sunyaev-Zel'dovich effect, by boosting the energy of photons from the cosmic microwave background. The hot gas outweighs all the stars in all the cluster's galaxies. Still, as do the stars, the gas consists of ordinary, baryonic matter. By adding up the stars and gas, astronomers can ascertain the total amount of baryonic matter in the cluster. The total amount of baryonic matter in the universe they can infer from the abundances of the light elements—hydrogen, helium, lithium—manufactured moments after the big bang. If the Hubble constant is 65, these indicate that baryonic matter contributes an omega of about 0.05. Thus, if omega is truly 1, then baryons constitute only 5 percent of the universe's matter; and if a galaxy cluster faithfully represents the universe's composition, then the baryons in its stars and gas should likewise constitute only 5 percent of the cluster's total mass. In fact, they constitute about 3 times that percentage—suggesting, again, that omega is 1/3 the critical amount.

Other astronomers have probed galaxy clusters through gravitational lensing, because a cluster's mass—visible or invisible, baryonic or nonbaryonic—distorts the light from galaxies and quasars behind it. "When we started getting lensing results of clusters of galaxies," said J. Anthony Tyson of Bell Labs, "sure enough, there was all this huge mass." The maverick Zwicky had been right: clusters indeed harbor large quantities of dark matter. Tyson and his colleagues studied a cluster in the constellation Pisces bearing the unimaginative name CL 0024+1654—"CL" for *cluster*, the numbers the equivalent of celestial longitude and latitude. As they found in 1997, this cluster's contents, if typical of the universe, imply an omega around 0.2.

Galaxy clusters, though, house only a few percent of all galaxies. Most galaxies, such as our own, reside in lesser gatherings, like the Local Group. Thus, theorists argued, galaxy clusters need not represent the entire cosmos, and omega could be higher than the clusters suggest. During the 1980s, theorists concocted a concept they called bias—a fitting term, since its inventors were themselves biased toward a high omega. According to this idea, dense regions of the universe, such as galaxy clusters, preferentially emitted light, while sparser regions, such as the voids, emitted none, even though they had matter. In individual galaxies, bias undeniably exists. Most of the Milky Way's light streams out of the disk and bulge, the densest parts, but most of its mass lumbers in the surrounding dark halo. For omega to be 1, the seemingly empty voids between the superclusters must harbor huge quantities of hidden material. Galaxies formed and stars shone, so the argument went, only where matter was densest, in galaxy clusters and galaxy groups. Thus, light was like snow on mountaintops, crowning only the tallest. To press their case, theorists argued that the larger the scale on which astronomers looked, the more the estimated omega rose. For example, the mass of the Galactic disk implied an omega around only 0.01; but the dark halo surrounding the disk implied an omega around 0.1; and the galaxy clusters implied a still higher omega, around 0.2 or 0.3. Thus, on the scale of the entire universe, omega was surely 1. Observers tended to dismiss this argument, deeming the belief in an omega of 1 "religion."

Furthermore, the presence of distant galaxy clusters argued against such a high omega. "If the universe's mass density is high," said Bahcall, "you expect to find almost no massive clusters at very high redshift; whereas if the mass density is low, you expect to find quite a few."

Clusters form through gravity, as galaxies attract one another. Over time, more and more galaxies join clusters, so today the universe has more clusters than ever before, and they're greater than ever before, too. Exactly how many clusters the universe had in the past depends on the numerical value of omega. In a low-density, low-omega universe, clusters must form early in the universe's life. Otherwise, the universe's expansion would separate the galaxies from their peers before they could nestle into clusters like Coma or Virgo. Thus, if omega is low, the Coma cluster formed shortly after the big bang. In contrast, a high-density universe has lots of matter. As a result, few clusters formed early on, because if they had, they would have overrun the present universe. If omega is 1, the Coma cluster formed only a few billion years ago. The difference is enormous: at half its present age, a low-omega universe had thousands of times more Coma-sized clusters than a high-omega universe.

Unfortunately, observers can't date the epoch when the Coma cluster formed. However, they can achieve a similar goal by seeking far-off clusters. If they see a galaxy cluster 5 billion light-years away, it must have existed 5 billion years ago. The presence of just one Coma-sized cluster at such a large distance would nearly rule out an omega of 1. And such clusters exist. As Bahcall and Xiaohui Fan, also at Princeton, noted in 1998, one named MS 1054–03 resides in the constellation Leo with a redshift of 0.83, corresponding to a distance of some 8 billion light-years. It's at least twice as massive as the Coma cluster.

Galaxy clusters join forces to create superclusters. In fact, clusters cluster much more strongly than galaxies themselves do. "That was a big, big surprise," said Bahcall. "That's very important for understanding how the large-scale structure formed and also for cosmology." Using computers, astronomers can simulate how those superclusters arose, as galaxies and clusters attracted one another. These simulations match the observed properties of the superclusters, but only if omega is low.

Since mass warps space, astronomers have also sought to measure the numerical value of omega by studying the universe's geometry. One method, which involves radio galaxies, was developed in the 1990s by Pennsylvania State University's Ruth Daly, who came to astronomy through a circuitous route. "I've always been fascinated by stars and the sky, ever since childhood," she said, "but I didn't realize that you could do astronomy as a profession until I had finished my undergraduate degree. So I actually obtained my undergraduate degree in English and psychology." The latter, she said, helps her understand cosmologists.

"Cosmology asks the ultimate questions," said Daly. "What are the global properties of the universe? And if we ever figure out what those properties are, we can ask, why does the universe have those properties? It's not a lot of nitty-gritty detailed modeling, where you get down to the fourth decimal place. Cosmology is still very much broad brush stroke."

The denser the universe, the more its expansion should be slowing. Thus, in a high-omega universe, a galaxy with a particular redshift should be closer to Earth and look larger. Nature neglected to stamp out galaxies with uniform diameters, so Daly and her colleagues used another measure of their size. Radio galaxies shoot out jets of material whose intrinsic lengths and widths can be determined from models. By seeing how large these jets look at different redshifts, Daly's team deduced that omega is low, agreeing with other determinations of the parameter.

## THE SHADOW UNIVERSE

Although an omega of 0.3 falls far short of theorists' dreams, it still implies a substantial quantity of exotic, nonbaryonic material. If the omega in baryonic material is only 0.05, yet omega in all matter is 0.30, then the vast majority of material in the universe must be nonbaryonic. Indeed, nonbaryonic matter must outweigh baryonic matter roughly five to one.

Despite its prevalence, nonbaryonic matter has eluded every attempt to identify it. "I think that's the most important and mysterious question in cosmology today," said Bahcall. "And the answer is: we don't know. The popular models all assume it is some kind of cold dark matter, and it may well be. But we don't know for sure. There may be something else that we have not thought about."

Whatever they are, the sluggish particles composing cold dark matter should be not just up there but down here, too. If such particles make up the Milky Way's dark halo, then as the Sun orbits the Galaxy, it should drag the Earth—and our bodies—through a sea of these ghostly particles. However, because nonbaryonic particles interact at best weakly with ordinary matter, physicists who have searched for them have yet to find any.

Daly painted an intriguing picture: imagine creatures who are the reverse of ourselves, consisting of cold dark matter and unable to see the baryonic matter that makes up stars, planets, and terrestrial people. "If we couldn't see baryons," Daly said, "we would never guess that all of these things are happening in the baryon universe. So there could be lots of things happening in that other world. There could be things as interesting as nucleosynthesis in stars and supernova explosions. It's so easy, if you don't have any data, to not understand what's happening."

In the late 1990s, physicists discovered that one nonbaryonic particle—the neutrino—had mass and thus contributed to omega. Earlier claims of a neutrino mass had proved spurious, but the new evidence was almost too much for one scientist. "It's been an exciting week," Fermilab's Rocky Kolb told *Astronomy*. "First the Spice Girls break up, and now we have the first experimental indication that neutrinos have mass." Nevertheless, the mass in a single neutrino turned out to be so piddly that these fleet particles make up only a sliver of the universe's total mass.

Actually, astronomers had already figured that out. Neutrinos are *hot* dark matter, shooting swiftly through space. "If most of the dark matter were made of neutrinos," said Tyson, "they would stream away—and galaxies wouldn't have formed. It's sort of like building sand castles at high tide. Since galaxies exist, neutrinos aren't the answer for the dark matter."

## LOOKING OUT FOR NUMBER 1

As the 1990s progressed, theorists longing for an omega of 1 faced a dilemma. Previously, they could argue that the observations suggesting a low omega were faulty. After all, cosmological parameters are notoriously difficult to measure: Hubble himself had overestimated the most basic one—the Hubble constant—by a factor of ten. Thus, said theorists, once observers did their job right, measured values of omega would rise toward the magic number of 1. Instead, as observations improved, the case for a low omega only strengthened.

As a consequence, some theorists began toying with alternatives. In 1990, for example, British cosmologist George Efstathiou and his colleagues advocated the virtues of little-liked lambda, the cosmological constant. Even if omega was low, noted Efstathiou's team, lambda's energy density could warp space enough to make it flat, as inflation predicted. Furthermore, lambda could increase the universe's age, so that even if the Hubble constant was high, the universe would be old enough to accommodate the ancient globular clusters. Moreover, Efstathiou and his colleagues said that a lambda-dominated universe would give rise to galaxy superclusters resembling those which actually cross the cosmos. However, if lambda existed, then empty space bristled with repulsive energy—a notion that most cosmologists themselves found repulsive.

Other theorists explored "open inflation" models, in which the inflationary episode just after the big bang somehow failed to flatten the universe completely. If anything, though, such models were even less popular than those invoking lambda. After all, one of inflationary theory's firm predictions had been that the universe be flat.

Thus, even as observational evidence mounted that omega was low, many theorists held out for high values of omega. "One cannot exclude omega of 1," University of Chicago cosmologist David Schramm insisted during a Berkeley talk in 1997. Furthermore, he said, there was no need for lambda. "There is no serious age problem yet!" Predictably, he emphasized determinations of the Hubble constant that fell on the low side, so that a high-omega universe could be older than the globular clusters.

Late that year, Schramm died when a plane he was piloting crashed in Colorado. At nearly the same time, two rival teams seeking the universe's deceleration were shocked to find there wasn't any. Instead, the universe's expansion was *ac*celerating, caught in the grip of the cosmological parameter Einstein himself had damned as his biggest blunder—the evil cosmological constant.

# FOURTEEN

# LAMBDA: EINSTEIN'S CURSE

NO ONE much liked the cosmological constant, the funny force that might pervade empty space, counteract gravity, and accelerate the universe's expansion. Indeed, the scientists who stumbled across it weren't even looking for it. Instead, they were stalking the material content of the cosmos, seeking the universe's deceleration, hoping to quantify omega, the cosmic matter density. True, most matter failed to glow through telescopes; most matter consisted of unearthly material; and most matter might hide in the dark voids between superclusters. But all matter, visible or invisible, baryonic or nonbaryonic, galactic or ungalactic, exerted gravitational force that should slow the universe's expansion. To designate the universe's deceleration rate, cosmologists had long reserved a letter of the alphabet, $q$; but through the decades, $q$ remained only a letter, for no one had pinned it to an actual number.

Measuring the universe's deceleration challenged astronomers to fathom galaxies *billions* of light-years distant, to witness an era billions of years earlier, when the universe expanded noticeably faster than today. The denser the universe and the more it decelerated, the closer high-redshift objects should be to Earth and the brighter they should look to terrestrial observers. In the same way, if two people on two different planets simultaneously hurled glowing lightbulbs of equal luminosity into the air at equal speeds, the lightbulb launched from the planet of greater gravity would slow down more, stay closer to the ground, and therefore look brighter to a ground-based observer. By measuring each lightbulb's apparent brightness, each observer could deduce the planet's gravity. Decades ago, Allan Sandage had grasped billions of light-years outward by using the brightest galaxies in clusters, but questions over their evolution impeded the measurement.

Still, an actual measurement of the deceleration parameter would constrain omega far better than motions of galaxies in clusters, which sampled the universe's density over much smaller distances. Theorists might concoct schemes in which dark material littered the voids, but even the most ingenious couldn't prevent that matter from braking the universe's expansion.

What those who embarked on this herculean quest never considered was that the deceleration parameter might turn out to be an *ac*celeration parameter.

## SEARCHING FOR SUPERNOVAE

As a graduate student, Saul Perlmutter of the Lawrence Berkeley Laboratory in California had been chasing supernovae, albeit in nearby galaxies that didn't probe the universe's deceleration. Even then, though, he had greater things on his mind. "I was always fascinated by the fundamental questions of how this universe works and how it all happened," said Perlmutter. "As a child, I was amazed that here we were, put in this life, and nobody had the owner's manual to the place. Why don't we all fall through the floor? What is doable and what is not doable? I always thought that we all needed to know these things, and as I grew up, I was a little bit surprised to realize that not everybody felt the same way."

The supernovae that Perlmutter and his colleagues were discovering in nearby galaxies came in sundry types, classified Ia, Ib, Ic, and II. All supernovae mark the violent deaths of stars, but the cumbersome nomenclature obscures the key point. Type Ia supernovae arise from white dwarf stars, whereas type Ib, Ic, and II supernovae conjure up what most people envision a supernova to be: a bloated, massive star like Betelgeuse reaches the end of its life, runs out of fuel, and explodes, its core collapsing into a neutron star or black hole. The Roman numeral—I or II—denotes the absence or presence, respectively, of hydrogen, the most common cosmic element. Most white dwarf stars have

SUPERNOVAE

| Type | Exploding Star | Hydrogen? | Helium? | Remnant | In Which Galaxies? | Nearly Uniform Peak Luminosity? | Examples | |
|---|---|---|---|---|---|---|---|---|
| | | | | | | | Supernova | Galaxy |
| Ia | White dwarf | No | No | Nothing | Ellipticals; spirals; irregulars | Yes | SN 1937C | IC 4182 |
| Ib | Supergiant without hydrogen | No | Yes | Neutron star or black hole | Spirals; irregulars | No | SN 1983N | M83 |
| Ic | Supergiant without hydrogen or helium | No | No | Neutron star or black hole | Spirals; irregulars | No | SN 1994I | Whirlpool Galaxy |
| II | Supergiant | Yes | Yes | Neutron star or black hole | Spirals; irregulars | No | SN 1987A | Large Magellanic Cloud |

no hydrogen, because they consist chiefly of carbon and oxygen; therefore, they erupt as the hydrogen-poor type I supernovae. In contrast, massive stars usually harbor huge hydrogen atmospheres, so they produce the hydrogen-rich type II supernovae. So far, so good. However, complications infest the nomenclature because some massive stars, before exploding, lose their hydrogen atmospheres and thus produce the hydrogen-poor type I supernovae. Such supernovae are designated type Ib, or, if the stars have also lost their helium, type Ic.

To cosmologists pursuing the universe's deceleration rate, the key supernovae are those labeled type Ia, the exploding white dwarfs that also star in the war over the Hubble constant. Although type Ia supernovae come from small stars, they outshine their bigger brethren by bursting forth with the brilliance of 5 billion Suns. That's good, because they can be seen in galaxies billions of light-years away, far enough back in time that cosmologists can discern how fast the universe's expansion is decelerating. Of course, the greatest type Ia virtue is their homogeneity: they all come from the same type of star with nearly the same amount of mass, a white dwarf which slips over the fatal 1.4-solar-mass threshold that triggers a runaway nuclear explosion. Thus, the explosions peak at nearly identical luminosities. In contrast, the other types of supernovae come from stars with various masses and therefore explode at various luminosities, making them poor standard candles.

During the 1980s, Perlmutter and his colleagues began exploring how to find supernovae not just nearby but faraway, too. "With this project," he said, "we realized that we had the possibility of going much, much, much farther. We were thinking about redshifts around 0.3. We were after measuring how much gravity was slowing the universe, and in that sense we would be weighing the universe, because we were finding out how much matter density there was in the universe that could slow it down. It would also tell us whether the universe was flat or curved. So it sounded like the perfect experiment to do, because you could get these very fundamental questions—which is the sort of science that has always interested me. And you could do it with just a simple measurement."

Finding remote supernovae wasn't so simple, however. When Perlmutter's team had sought nearby supernovae, the scientists snapped pictures of individual galaxies, night after night, and searched each for an exploding star. Now they would cast a broader net, imaging hundreds of distant galaxies at a time, hoping at least one would give birth to a dying star. If a new spot of light appeared, the scientists could obtain spectra to verify its supernova nature and determine its exact type.

However, obtaining telescope time for this project wasn't easy. "Everybody was completely convinced it wouldn't work," said Perlmutter. "They all felt that the supernovae were too far away, that we wouldn't be able to find them; and even if we found them, we wouldn't be able to get useful data on them, because the calibration for type Ia supernovae was at their *peak* brightness, so if we caught them a few weeks later, we wouldn't know how bright they were at their top."

Nevertheless, Perlmutter and his colleagues did receive twelve nights of time on the large Anglo-Australian Telescope in Australia. Unfortunately, on all but two and a half of those nights, clouds obscured the sky, and the scientists discovered not a single supernova. Meanwhile, they sought other large telescopes around the world.

"Our very first discovery was at the Canary Islands in 1992," he said. "That was a *key* discovery to allow us to go ahead. It was a rather dramatic run, because it was three years into the project, and we'd already had such terrible luck in Australia that the review panels and the various funding groups that were helping to support the project were all saying, 'Well, you see—we thought this wasn't going to work.'"

THE SUPERNOVA AT THE END OF THE UNIVERSE, said *New Scientist*, the first to report the discovery: "The most distant supernova ever found has raised hopes of determining the ultimate fate of the Universe. Because of its enormous distance—about 5 billion light years—it may help to reveal how quickly the expansion of the Universe is slowing, and so whether the Universe will expand forever or eventually collapse." The far-off supernova had erupted inside a dim galaxy with no name in the constellation Hercules. Its redshift, 0.46, bested the previous supernova champion by nearly 50 percent. Before reaching Earth, its light had crossed one third of the observable universe, perfect for determining the numerical value of omega.

When he began the search, Perlmutter didn't prefer any particular value of omega. "We already understood that the theorists felt it should come out near 1," he said. "But coming from an experimentalists' background, we were much more interested that so far there had been no observational evidence that was indicating numbers that high. So we didn't know what it would turn out to be, and we were rather excited about getting to try and find out."

The brighter the supernova looked, the closer it was to Earth, and the more the universe's expansion was decelerating; hence, the greater omega must be. If omega was 1, the 1992 supernova should have looked 25 percent brighter than if omega was 0. It actually looked rather faint, implying that the universe's expansion had suffered little deceleration and that omega was low, around 0.1—in accord with what most other observers had been finding. However, not even the scientists themselves believed this result, because any one type Ia supernova could be slightly more or less luminous than the average. When the scientists analyzed the uncertainties, they concluded that omega could be anywhere from 0 to 2, covering the full suite of open, flat, and closed universes. And the supernova's dimness even agreed with a *negative* deceleration—that is, an acceleration.

However, Perlmutter barely thought about that. "It hadn't occurred to us at all," he said. "It was only after we had already gotten through all the very rough stages where nothing was working, in 1993 or 1994, that we started paying attention to a change in the culture of cosmology. Theorists were just beginning to say, 'We are so sure that the universe should be flat, that even if the observationalists keep coming in with omega being low, then we might actually want to start throwing in a cosmological constant.'"

## COSMOLOGY'S UGLY DUCKLING

The cosmological constant, designated lambda, is a repulsive force pervading empty space, just what Albert Einstein had needed to keep his static universe from collapsing back in 1917. Lambda therefore resembles an "antigravity" force—but only in quotes. Gravity is attractive, so an antigravity force should be repulsive, as the cosmological constant indeed is; but gravity arises from mass, so a true antigravity force should also arise from mass, whereas the cosmological constant arises from space.

Because omega attracts and lambda repels, the two oppose each other and fight to control the universe. Omega tries to slow down the universe's expansion, lambda tries to speed it up. Nevertheless, both omega and lambda warp space in the same way. That's because lambda's energy density can be translated into a matter density via Einstein's equation $E = mc^2$. Ordinarily, if omega is less than 1, the universe is open. But if lambda exists, omega *plus* lambda can still be 1—and the universe be flat, as the flatness problem suggested and as inflationary theory predicted.

Still, lambda lovers were scarce, for the cosmological constant meant that empty space not only exerted force but might even dictate the universe's destiny. Furthermore, lambda seemed a fudge factor, invoked to patch up defective work. Einstein had invented it because he mistakenly thought the universe static. After Edwin Hubble discovered the universe's expansion, Einstein disowned the cosmological constant, calling it the biggest blunder of his life.

"That expresses the feeling people have about it," said Princeton University's Edwin Turner. "It's certainly an inelegant thing to have a situation in which the laws of physics don't have consistent zeros. In a sense, that's what a nonzero cosmological constant is saying: when you go down to zero content, you don't go down to zero gravitational effect. It's as though the scales on these two things were not properly connected or normalized. It just doesn't have theoretical elegance or beauty; it doesn't make much sense."

Indeed, intuitively, empty space should exert no force nor possess any energy whatsoever. If so, lambda is 0, and the cosmological constant does not exist. However, physicists can calculate an expected value for lambda, because quantum mechanics says that the vacuum is anything but vacuous. Instead, it abounds with pairs of particles and antiparticles popping into existence, only to annihilate each other and disappear. These so-called virtual particles contribute an energy density equaling an enormous lambda of roughly $10^{120}$—a 1 followed by 120 zeros. Such a superlambda would have driven the cosmos into a hyperexpansion like that which puffed up the universe during the inflationary epoch. This hyperexpansion would have prevented the galaxies from forming. If Earth still managed to arise, this lambda would interfere with daily life: from a short distance, traffic lights would be so redshifted that their green lights would look red and their red lights invisible, shifted to infrared wavelengths. Thus, physicists argued, some unknown factor must cancel this huge number so that the true numerical value of the ugly parameter was a perfect 0.

Despite its unpopularity, lambda does do one good thing for the universe: it boosts its estimated age. If lambda accelerates the universe's expansion, then the expansion in the past was slower; thus, the universe took longer to reach its present size. As a result, if the Hubble constant is high, lambda can free age-conscious astronomers from their bind, making the universe old enough to accommodate the ancient globular clusters. A universe with an omega of 0.3 and a lambda of 0.7 is 19 percent older than a lambda-less universe with an omega of 0.3, and 45 percent older than a lambda-less universe with an omega of 1. This amounts to an age increase of billions of years. For example, if the Hubble constant is 65, then a universe with an omega of 1 and no lambda is just 10.0 billion years old; a universe with an omega of 0.3 and no lambda is 12.2 billion years old; but a universe with an omega of 0.3 and a lambda of 0.7 is 14.5 billion years old.

In the early 1990s, lambda loathers were pleased to hear that observations of gravitationally lensed quasars suggested the cosmological constant did not exist. A Japanese team led by Masataka Fukugita of Kyoto University, and independently Turner, had realized that the number of lensed quasars could reveal the numerical value of lambda. Because lambda accelerates the universe's expansion, more space lies between Earth and a quasar having a particular redshift; thus, the greater lambda is, the greater is the chance that a galaxy sits in front of the quasar and lenses its light. So lambda ups the number of gravitationally lensed quasars.

"Want to impress your friends?" asked *Sky and Telescope* in 1992. "Drop this on them next time there's a lull in the conversation: 'Did you hear it's not a lambda-dominated universe?'" The Hubble Space Telescope had searched for gravitationally lensed quasars but come up mostly empty-handed, implying a low lambda, consistent with 0. What the article failed to mention, however, was that the constraint was sufficiently weak that any value of lambda up to 0.6 or 0.7 or possibly 0.8 was permitted.

In 1995, gravitational lens expert Christopher Kochanek, of the Harvard-Smithsonian Center for Astrophysics, was asked about the cosmological constant. "I'd bet *your* life there isn't one," he said. "Notice I said your life, not my life. You do the best job you can on the calculation. But you can't include every conceivable systematic error. Look through the history of astronomy—how many other great ideas for measuring cosmological parameters have sunk without a trace because you made the wrong assumptions? So realistically, you'd be foolish to rule out the possibility that one of the assumptions in the lensing calculation is wrong."

## THE DECELERATING UNIVERSE

Initially, the distant supernovae that Saul Perlmutter and his colleagues were finding backed up the lensing result, arguing against lambda. "Every single time we got another supernova, it was very exciting," said Perlmutter. "We were starting to get these beautiful spectra at these great distances which nobody thought we'd be able to do, and eventually we had enough supernovae that we were ready to start making the measurements of omega."

The first analysis, performed in 1996, comforted not only the theorists but also lambdaphobes everywhere: the supernovae were bright, so the universe seemed to be strongly decelerating, with omega around 1 and lambda at 0. Said Perlmutter, "A number of theorists took it as a nice indicator that things were moving their way." Perlmutter and his colleagues studied seven supernovae at redshifts greater than 0.35, with the one at redshift 0.46 remaining the champ. Although Perlmutter's team put omega close to 1, uncertainties permitted any value from about 0.3 to 1.6. The scientists also ruled out the cosmological constant: "The results . . . are inconsistent with [lambda]-dominated, low-density, flat cosmologies that have been proposed to reconcile the ages of globular cluster stars with higher Hubble constant values," they wrote.

"We were rather excited that we could do the measurements at all," said Perlmutter. "Everybody saw it coming in on the side of omega of 1. Now with the error bars we had at that point, you really couldn't tell apart an omega of 1 from an omega of 0. People tend to look at a number, and if that's what they're expecting, then they don't pay quite as much attention to the error bar." The theorists, who all along had been insisting that omega was 1, pointed to the supernova result as vindication. Unlike lesser measures of omega, such as those from galaxy clusters, the supernova study probed the entire cosmos, sensing even the gravitational pull of the matter hiding in voids, the very regions theorists had claimed swarmed with material that the observers pushing for a low omega couldn't see. So, it seemed, the theorists were doubly right: omega was 1, and the universe was biased, having lit up only where densest.

## THE ACCELERATING UNIVERSE

By now, however, Perlmutter's team had competition. Inspired by his success, a second group of observers was hunting for distant type Ia supernovae, too. They also wanted to use them to measure the universe's deceleration rate and determine the numerical value of omega. Yet their very first discovery raised troubling questions, for it showed no sign that the universe's expansion was slowing.

The second team's leader was Brian Schmidt, an American-born astronomer in Australia. He had been interested in astronomy as a child, but had planned to go into meteorology and climatology instead. "One summer in high school," said Schmidt, "I went and worked a bit for the Weather Service, and I don't know—I just didn't like it as well as I thought I might, and so I had to make a decision. I went to some speech some guy gave, and he said, 'Go and do something with your life which you would do for free.' So I said, 'Well, I'd do *astronomy* for free.' Astronomy has big questions—that's why I like it."

While a graduate student at Harvard University, Schmidt had studied supernovae, of the type II variety. He doubted that the type Ia supernovae could betray the universe's deceleration. "Quite frankly," he said, "I had been extremely skeptical of what the guys at Berkeley were doing—not because I didn't think they could find supernovae, but because without knowledge of how to actually use them, it was sort of wasting time. I just felt that there was a lot of dirty laundry." Type Ia supernovae, he thought, weren't the great standard candles they were cracked up to be.

In 1993, however, Mark Phillips in Chile harnessed type Ia supernovae into outstanding distance indicators. By comparing the apparent brightnesses of nearby type Ia supernovae at different redshifts, Phillips recognized that the longer a type Ia supernova took to fade, the brighter it shone. Thus, by measuring any type Ia supernova's decline rate, astronomers could deduce the supernova's luminosity, allowing accurate measurements of distances to galaxies, near or far, that had spawned the explosions. When Schmidt heard that, and realized that new technology allowed the discovery of distant supernovae, he decided to start looking for himself.

"It was a program to measure $q$, the deceleration parameter," said Schmidt. "That's related to omega. I certainly was not thinking of the cosmological constant. Indeed, I remember joking once with a friend who's a biologist, saying, 'Ah, yeah, I'm working on something that could potentially lead to the Nobel prize. If the universe were really crazy and did something I wasn't expecting—like there was a cosmological constant—geez, that would be great.'" To him, though, lambda wasn't really a possibility. "It was something that we were going to show didn't exist," he said. At the time, he favored a low omega, around 0.2, because of the other observations that indicated this value.

Schmidt and his colleagues scored a success with their first supernova. When they discovered it in 1995, it broke the supernova distance record: its redshift was 0.48, edging out the farthest supernova from Perlmutter's group. Said Schmidt, "I felt suddenly confident that we would be able to get telescope time again."

That same supernova, however, posed trouble. It looked faint, and therefore it was farther than it should have been if omega was 1, or even 0. In fact, the supernova's faintness suggested that the universe's expansion was speeding up, not slowing down. If lambda was 0, omega was, impossibly, *below* 0—a *negative* 0.6.

"Then Saul Perlmutter's team published their distances to their first seven supernovae," said Schmidt. Those indicated omega was 1. "We had one observation; they had seven; but ours was completely different from theirs. At this point, I was concerned. Maybe we had screwed up somewhere." Still, he never thought the universe's expansion was really accelerating. "I honestly did not worry about that," said Schmidt. "I figured, you know, it was still consistent with a very-close-to-0-omega universe."

His team applied for more telescope time. Said Schmidt, "One person who was on a time allocation committee said, 'You know, you've got to do something about that supernova—it can't be right.' They knew Saul was putting in proposals that said omega was 1, and they just assumed we had screwed up. Interestingly enough, we didn't get any time from this time allocation committee, and Saul did. And I have to admit: if I was going to decide between one team or the other, I probably would have picked them, too." After all, Perlmutter's answer made sense; Schmidt's didn't.

Soon, however, Perlmutter's team began to face the same disturbing prospect. Said Perlmutter, "Probably in the spring and summer of 1997, we were starting to make a plot as the different supernovae were being analyzed, and we were getting odd results, making it look as if omega was coming way, way down. As the error bars got better, the mass density got much, much lower." Good-bye to a universe with an omega of 1.

"At that point," said Perlmutter, "I wasn't too concerned. It was a little bit surprising, but there're so many steps in the analysis and so many places where we had to check the calibrations, that I figured, as we completed the calibrations, things would sharpen up, and then we'd know what the final answer was. But as we kept doing more and more of these calibrations, and more and more of these analyses, it looked like we were still getting these low numbers. And we started to joke that if this went on, the mass density would be 0, and all of us would disappear!"

As 1997 proceeded, Perlmutter and his colleagues found omega dropping not just to 0 but below—in other words, the universe's expansion was not decelerating but *ac*celerating. During the autumn, they realized the startling truth: space was pervaded by the repulsive force of the cosmological constant, the very parameter Einstein himself had condemned.

"It was probably when I gave a talk at the University of California at Santa Cruz that it first hit me," said Perlmutter. "When you're working with it that long, you get used to seeing the data, and you get used to the surprise—and it's only when you start saying it out loud to people that you start realizing: this is actually a really bizarre result. At the Santa Cruz talk, one of the more respected theorists in cosmology, Joel Primack, stood up and said that the audience may not realize what an important result this is, but this could mean that we've got a new fundamental part in our physics—that we now have a possible cosmological constant."

Meanwhile, in Australia, Schmidt and his colleagues had begun to acquire precise data on the brightness of distant supernovae by using the Hubble Space Telescope. But there was trouble. "We couldn't figure out why we still had a negative deceleration," said Schmidt. "At this point, I wasn't really sure what was going on. I was figuring maybe we had some—I don't know—bad statistical luck or something." He thought Perlmutter's team still had omega pegged at 1.

In October 1997, however, Schmidt's team submitted a paper to *The Astrophysical Journal Letters* reporting on four distant type Ia supernovae. One of them had a redshift of 0.97, corresponding to a lookback distance of 8 billion light-years, over halfway to the edge of the observable universe. When that star exploded, the universe was only half its present size. From the faintness of the four supernovae, Schmidt's team concluded that if lambda was 0, then omega was –0.1. Still, uncertainties permitted an omega as high as a positive 0.4, consistent with an open universe having a low omega and no lambda. Such a low omega, of course, fit what most observers had been saying and refuted theorists' claims that omega must be 1. Indeed, Schmidt's team was 95 percent confident that omega was *not* 1—"matter alone is insufficient to produce a flat universe," they wrote—and noted that this conclusion conflicted with the omega-equals-1 result from Perlmutter's group. The universe, said Schmidt's team, was either open or flat; but if it was flat, then "a cosmological constant makes a considerable contribution."

Schmidt and the other astronomers discovered still more far-off supernovae. "At the end of 1997," he said, "it was very clear that, like it or not, every one of these objects was coming in at the negative part of the deceleration diagram." He reanalyzed the data; a colleague did, too. The result stood. "So at this point," he said, "I had to believe it. I was still shaking my head, but we had checked everything, and we had to go forth and tell the world about it—no matter what they're going to think about us. And I was very reluctant about telling people, because I truly thought that we were going to get massacred."

In January 1998, at a meeting of the American Astronomical Society, Perlmutter's group announced its find: the universe's expansion was accelerating and lambda not only existed but also dominated omega. Said Perlmutter, "Lambda had been in and out of the cosmology for years and years, since Einstein first threw it in. Ever since, people have realized that it could be there, but nobody took it seriously, because they figured it was a complication and the story was complicated enough without it." Furthermore, no one liked it. He said, "It certainly made us very aware of why we had to do this measurement so carefully—because if you're going to suggest something new in the story that's very ugly, you'd better be confident about it. Most of what being a scientist is, is trying to figure out how you're wrong, not trying to convince people you're right." Better to prove yourself wrong than for someone else to do it for you.

When Perlmutter's team announced the result, Schmidt heard about it via e-mail. Said Schmidt, "Well, needless to say, it suddenly makes you much more confident when you've independently got the same answer, but you've been dragging your feet because you thought the other team had a different answer." Soon after, Schmidt and his colleagues released their result. "I certainly was fearing the worst," said Schmidt. "Even after six months, I had expected everyone basically to say no—but in fact, the community as a whole rapidly accepted the observations, much to my surprise." Indeed, the double whammy from the two competing groups convinced most astronomers that the universe was indeed accelerating.

Because both Perlmutter and Schmidt measure the universe's acceleration, which omega opposes and lambda fosters, the astronomers cannot determine either parameter separate from the other. Instead, what they measure is akin to lambda *minus* omega. The greater lambda is, and the less omega is, the faster the universe accelerates. If the universe is flat—so that lambda *plus* omega equals 1—then the supernova data suggest that omega is 0.3 and lambda is 0.7. The number for omega agrees with the matter density that observers had long lobbied for: the universe has only 30 percent of the matter needed to reverse its expansion. However, matter doesn't call the shots. Because lambda is over twice as large as omega, ours is actually a lambda-dominated universe, one that will expand faster and faster.

The cosmological constant, though, hasn't always ruled the cosmos. Billions of years ago, the universe was denser and omega greater. Indeed, early in the universe's life, when galaxies were closer together, the universe's expansion was actually slowing down. But as the universe expanded, the galaxies raced apart, the matter density got diluted, omega fell—and lambda grew greater and greater, eventually pushing the universe from deceleration to acceleration. If lambda is now 0.7, then it already governs the universe and will only strengthen as the universe expands further, since lambda derives its power from empty space. Each year, the expanding universe gives birth to more empty space, which fuels lambda's power, which gives birth to yet more empty space, which further fuels lambda's power, and so on. Ironically, although Einstein introduced the cosmological constant to stabilize the universe, lambda will actually do the opposite, destabilizing the cosmos and driving it toward ever faster expansion, until it becomes a runaway.

## NOT SO FAST

Although two independent teams arrived at the same shocking conclusion, both did so by using the same objects—type Ia supernovae—raising the specter that these objects might be flawed. "The most worrisome thing, the thing that has really wiped out other attempts to do this in the past, is evolution," said Schmidt. "That is, are the supernovae back then the same as today?"

Allan Sandage had failed in his attempt to measure the universe's deceleration because his objects—the brightest galaxies in clusters—had changed their luminosity over billions of years, so he could not easily compare nearby galaxies with distant ones. Likewise, when Perlmutter and Schmidt see type Ia supernovae exploding in far-off galaxies, those explosions actually occurred billions of years ago. If such ancient type Ia supernovae underperform modern ones, they could fool supernova hunters into thinking they are farther than they really are, mimicking lambda. In particular, if ancient type Ia supernovae were 25 percent less powerful than their modern counterparts, then omega could be low and lambda 0. And if those ancient type Ia supernovae were a full 50 percent less luminous than their modern peers, then omega could be 1 and lambda 0.

"Fundamentally," said Schmidt, "we have a real problem because we don't understand these explosions that well. We don't know what the progenitors are." Although type Ia supernovae come from white dwarf stars that receive material from a companion star, the companion's exact nature is unknown. It could be a giant star, a main-sequence star, or even another white dwarf star. Furthermore, white dwarfs billions of years ago may have differed from those today. In particular, the proportions of their two chief elements—carbon and oxygen—may have differed, altering the supernova luminosities. Schmidt doesn't think such problems will torpedo his work, but he added, "We do need to keep on investigating this, since so much of cosmology relies on these observations."

In 1999, supernova hunters got a scare when reports surfaced that distant type Ia supernovae did differ from those closer to home. Berkeley's Adam Riess, a member of Schmidt's team, claimed that the distant explosions reached their peak brightness faster than did nearby explosions. This feature alone didn't necessarily mean that the ancient explosions were less powerful, but it raised the alarming possibility that the two epochs spawned different breeds of supernovae. However, Riess based his report on preliminary data. Later data, published in 2000, showed no significant difference in rise times, removing the problem.

One supernova expert thinks all will be well. "Type Ia's don't lie," said the University of Oklahoma's David Branch, who belongs to neither Perlmutter's team nor Schmidt's but has long advocated using type Ia supernovae as distance indicators. "I'll admit that the supernova results really surprised me. I was one of those who sort of made fun of the lambda—until the type Ia's said that it's there. We have questions about metallicity and evolution and gray dust, but really, my bet is that all that stuff will turn out to be okay."

As Branch mentioned, dust could also masquerade as lambda. Sprinkled somewhere between Earth and the distant galaxies, it could dim the supernovae and make them seem farther than they really are, creating the illusion of an accelerating universe. Wouldn't that be just like the universe: astronomers think they've made a great discovery, only to get tripped up by dust.

It's happened before. For more than a century, astronomers were convinced that the Sun and the Earth inhabited the Milky Way's center. That's because in every direction they looked, the number of stars fell off, suggesting that we were in the densest thicket of the Galaxy. Only in the twentieth century did astronomers learn that the Galaxy's disk contains dust particles that absorb starlight, so we only seem to be at the center, just as fog-shrouded suburbanites seem to be at the center of the little they can see.

Fortunately, dust normally betrays its presence, by absorbing more red light than blue. The light of the distant supernovae wasn't reddened, suggesting no dust. In 1999, however, Anthony Aguirre of the Harvard-Smithsonian Center for Astrophysics published work suggesting that "gray" dust—which absorbs light of all colors equally—might exist, dimming the supernovae and fooling their observers.

To rule out this possibility, both supernova teams are finding extremely distant supernovae, those exceeding a redshift of 1. "If you can go far enough away," said Perlmutter, "you get back to a time when the universe's expansion was slowing down, because the mass density was much greater then; therefore, you can find a time when the supernovae would *not* have been fainter than you would have expected in a zero-cosmological-constant universe." In contrast, if dust dims the supernova light, the supernovae should keep getting fainter and fainter the farther they are.

In 2001, Schmidt's team reported the farthest supernova yet seen—whose brightness confirmed that the universe's expansion is accelerating, because its light emerged during the ancient epoch *before* the acceleration began. The supernova had a redshift of 1.7, corresponding to a lookback distance of over 10 billion light-years. The star therefore exploded just 4 billion years after the big bang. Unlike the other supernovae, this ancient explosion appeared brighter than it would have looked if the universe had always expanded at the same rate. That's because the supernova's light set out when the universe was so dense that gravity overwhelmed the cosmological constant, causing the universe at that time to *de*celerate.

The year before this announcement, in 2000, two other scientific teams announced results that strengthened the case for lambda. Lofting balloons into the atmosphere, the BOOMERANG and MAXIMA experiments observed the ripples in the cosmic microwave background that the COBE satellite had first detected a decade earlier. (BOOMERANG stands for Balloon Observations of Millimetric Extragalactic Radiation and Geophysics; MAXIMA stands for Millimeter Anisotropy Experiment Imaging Array.) The exact pattern of these cosmic ripples depends on the geometry of space and thereby signals whether the universe is open, flat, or closed. Both BOOMERANG and MAXIMA indicated that the universe was flat—omega plus lambda equals 1. Since the supernova work had ruled out a high omega, this result implies that lambda must exist. In fact, the two observations complement each other: the supernova data measure the acceleration of the universe—lambda *minus* omega—whereas the cosmic microwave background results measure the universe's geometry, or lambda *plus* omega. Neither experiment gave omega or lambda independently, but together they yield an omega around 0.3 and a lambda around 0.7.

Indeed, the ripples in the cosmic microwave background promise such a treasure that on June 30, 2001, NASA launched a satellite to mine it. Named MAP, for Microwave Anisotropy Probe, the satellite is COBE's successor and was designed to determine omega, lambda, the Hubble constant, and the baryonic omega. And the Europeans plan to launch PLANCK, another satellite devoted to the cosmic microwave background.

## QUINTESSENCE

Confronted by a cosmological parameter that few liked, scientists are struggling to understand the problems it presents. As if the notion that completely empty space teems with energy that dictates the universe's fate isn't troublesome enough, lambda poses a further difficulty. Right now, omega seems to be around 0.3 and lambda around 0.7, so the two opposing parameters are roughly equal. True, lambda has about twice omega's strength, but over the course of the universe's life, omega plummets and lambda rockets upward; yet right now, they're of comparable strength.

"That's nasty," said Schmidt, "because why is the universe so finely tuned so that we can just right now see lambda? If we had been born at a time when the Earth had just formed, the effect of the cosmological constant would be much smaller than it is right now. It would have almost been unmeasurable. So that coincidence is telling people we need to come up with a way to make the accelerating universe always sort of look like lambda is 0.7."

In the young universe, omega dominated lambda. The universe then was dense and decelerating, and any astronomers would have had a hard time detecting lambda. However, fueled by nothing but the energy of empty space, lambda was lying in wait, ready to seize control. As the universe got larger, and the matter density smaller, lambda rose and rose. At some point—when omega fell to 0.67 and lambda rose to 0.33—the universe switched from deceleration to acceleration. If omega is presently 0.3 and lambda 0.7, then this switch-over occurred just a few billion years ago. Cosmologists consider that recent, implying that we inhabit an unusual era. That makes them uncomfortable, since they don't like our time or place to be special.

Actually, a similar transition occurred once before, in the cosmological stone age. Shortly after the big bang, radiation dominated the universe. The expansion of space, however, stretched and weakened the radiation, until matter took control. This transition doesn't bother cosmologists, because it happened so long ago—a few tens of thousands of years after the big bang. In contrast, the omega-to-lambda switch-over seems to have happened recently.

One possible solution: quintessence, which resembles a cosmological constant, only it isn't constant. The cosmological constant is a constant that pervades empty space; thus, as the universe's expansion creates more empty space, omega gets diluted and lambda takes over. Quintes-

LAMBDA'S REIGN

| | | Acceleration Began | | | Lambda Equaled Omega | | |
|---|---|---|---|---|---|---|---|
| Present Omega | Present Lambda | At Redshift | Billions of Years Ago* | Percentage of Universe's Age | At Redshift | Billions of Years Ago* | Percentage of Universe's Age |
| 0.1 | 0.9 | 1.62 | 12.3 | 36 | 1.08 | 9.9 | 48 |
| 0.2 | 0.8 | 1.00 | 8.8 | 46 | 0.59 | 6.3 | 61 |
| 0.3 | 0.7 | 0.67 | 6.6 | 54 | 0.33 | 3.9 | 73 |
| 0.4 | 0.6 | 0.44 | 4.8 | 64 | 0.14 | 1.9 | 85 |
| 0.5 | 0.5 | 0.26 | 3.2 | 75 | 0.00 | 0.0 | 100 |

*For a Hubble constant of 65 kilometers per second per megaparsec.*

sence, in contrast, varies as the universe expands and stays linked to omega. If quintessence exists, then future astronomers will still see the universe caught in the grip of two more-or-less equal and opposing forces, omega and a quasi lambda. In contrast, without quintessence, omega will drop to nearly 0 and lambda will rise to nearly 1.

Schmidt doubts, though, whether he can detect such a thing. He said, "Quintessence is sort of this mystery matter which has this weird form that sort of tracks the universe as it expands, and in essence it gives you a different expansion history of the universe. Unfortunately, it's *not* much different from a cosmological constant model, so it's going to be extraordinarily difficult to show that the universe is significantly different, and interestingly different, from a cosmological constant model." Unless quintessence exerts a large influence, Schmidt doesn't think he'll be able to distinguish it from a vanilla lambda.

In addition, the coincidence between the numerical values of omega and lambda may be just that, a coincidence. Astronomers have long been aware of a startling noncosmological coincidence: the two brightest objects in the sky, the Sun and the Moon, look equally large, so that during solar eclipses the Moon can cover the Sun in dramatic fashion. The Sun's actual diameter is four hundred times greater than the Moon's, but it's also four hundred times farther. Furthermore, this solar-lunar equivalence is unique to our time, because the Moon slowly recedes from the Earth. In former times, the Moon looked larger than the Sun, because it was closer to the Earth; and in future times, the Moon will look smaller. Just as we happen to live at the time when the Moon's size matches the Sun's, perhaps we also happen to live at the time when omega equals lambda.

In any case, physicists want to know what causes empty space to exert such force. Quantum physics, after all, suggests that the nothingness of space should lead to a lambda roughly $10^{120}$ times greater than observed. As British cosmologist John Peacock wrote in 1999, "It is perhaps just as well that the average taxpayer, who funds research in physics, is unaware of the trouble we have in understanding even nothing at all."

## FROM FUDGE FACTOR TO FUNDAMENTAL PARAMETER

Although no one much liked it—indeed, the two supernova teams weren't even looking for it—the universe seems to have something like a cosmological constant. If the recent evidence is correct, then the observers who favored a low omega, on the basis of data from galaxy clusters, were right. The theorists who insisted that omega equaled 1 were on a wild-goose hunt for far more dark matter than actually existed. Furthermore, the dark voids between the superclusters are just as empty as they look: the universe is *not* biased; only the theorists were. Still, the theorists do get the flat universe they fought so hard for, but at the cost of a new parameter that most everyone deemed ugly.

Ugly though it may be, lambda nevertheless increases the universe's estimated age by a couple billion years. If the Hubble constant is 65 kilometers per second per megaparsec, if omega is 0.3, and if lambda is 0.7, then the big bang occurred 14.5 billion years ago. Thus, the universe is indeed older than the oldest globular clusters, which probably formed 12 to 15 billion years ago.

Said Swiss astronomer Gustav Tammann, "I am impressed with what fantastic shape cosmology is in at the moment, if you accept the solution for lambda. I know about the possible objections, but I am very, very impressed by Saul Perlmutter's work and also by Brian Schmidt's group."

Modern cosmology may indeed be in fantastic shape, but lambda alters the universe's future in ways that few ever foresaw. No longer need astronomers debate whether the universe ends in fire or ice. Instead, it appears that the universe's expansion will inexorably speed up, pushing most other galaxies out of terrestrial view, until future astronomers are left with very few sights on which to train their telescopes.

# FIFTEEN

# THE FUTURE OF THE UNIVERSE

BY EXPLOITING and extrapolating the latest cosmological discoveries, astronomers can predict the universe's future anew. Previous generations debated whether the universe might collapse under its own weight, with the gravitational pull of the galaxies halting and then reversing the cosmic expansion, turning their redshifts blue and reheating the microwave background until it fried the stars away, right before the entire cosmos headed toward a fiery big crunch.

No more. The universe will never collapse. Its galaxies can't muster the gravitational strength to stop the expansion, and space itself teems with a repulsive force that accelerates the expansion. Thus, not only will the universe expand forever, but it will also expand forever faster, rushing headlong into the cold and dark. As a result, astronomers face a bleak prospect: during the next 150 billion years, over 99.99999 percent of the galaxies they now see will slip out of sight.

## THE FUTURE OF THE EARTH

First, though, astronomers and the taxpayers who support them will confront problems closer to home. Since the Sun's birth 4.6 billion years ago, it has faithfully showered the Earth with light. The Sun's center generates this energy by converting hydrogen, the most potent stellar fuel, into helium. The newborn Sun consisted mostly of hydrogen, but as its center depleted the element, the Sun gradually brightened. Today it shines 40 percent brighter than when it first stoked its nuclear flame.

Even as it remains a main-sequence star—burning hydrogen into helium at its center—the Sun will continue to brighten. In a billion years, the Sun will outshine its present self by 10 percent. Conventional wisdom, which wallows in the pessimistic, holds that such intense sunlight will begin to boil the Earth's oceans, transforming the solar system's lushest planet into a desert world like Venus, whose surface sears at 860 degrees Fahrenheit. Thus, the very light that now sustains terrestrial life will someday kill it off.

Conventional wisdom, however, overlooks the Earth's best defense: the ingenuity it has spawned. As long as the Earth bears intelligent life, the planet should survive far beyond a billion years. In fact, if its inhabitants arm themselves with the right technology, the Earth can thrive for at least a *quintillion* years—a billion billion years. To retain the oceans, our descendants could pursue two different strategies. First, they could construct a large shield between the Sun and the Earth. When the Sun shines 10 percent brighter, a shield that deflected 10 percent of the sunlight from Earth would keep the world at its present pleasant temperature. Building such a shield exceeds present technology but should be trivial to a race a billion years more advanced.

Second, to lessen the amount of sunlight striking Earth, our descendants could slowly move the world away from the Sun. If they nudge the Earth outward by just twenty-four feet a year, the planet would keep its cool even as the Sun brightens. They could do so by hurling asteroids past the Earth. The gravity of each passing asteroid would alter the Earth's orbit slightly. Interplanetary spacecraft already employ a similar trick. When the Pioneer and Voyager spacecraft flew past Jupiter, Jupiter's gravity sped up the spacecraft, and Jupiter fell ever so slightly sunward. Although catapulting asteroids into new orbits far outstrips present technology, our descendants have a billion years—a million times the interval that separated the Middle Ages from the Apollo Moon landing. As long as people get smarter faster than the Sun gets brighter, the Earth should thrive.

Alas, 6 billion years from now, the Sun will suffer more drastic changes, threatening not just the oceans but the Earth's very survival. The Sun's center will finally run out of hydrogen, and the Sun will start burning the element in a layer surrounding the core. As a result, the

Sun will expand, brighten, and cool, swelling into a red giant star like Aldebaran that outshines the present Sun a hundred times. The expanding red giant Sun will first swallow Mercury, then possibly Venus. And Earth? If our descendants fail to move it, the Earth may or may not survive, but either way the blast furnace of the red giant Sun will scorch the planet with heat worse than presently bakes Mercury. To maintain its current climate, the Earth would have to be placed in orbit around Saturn.

Then the red giant Sun will cast off its atmosphere, encircling itself in a rainbow-colored smoke ring that resembles the beautiful Ring Nebula in the constellation Lyra. At this time, the Sun's hot, hydrogen-depleted core will emerge into view. The core is really a white dwarf star that burns no fuel, so over millions and billions of years it will cool and fade. Our descendants could haul the Earth back inward again, so that the Earth intercepts a greater fraction of the Sun's dwindling light, but this will merely postpone the inevitable, because the Sun will eventually fade so much that it can no longer warm the Earth. In order to live, the Earth must find another star.

Fortunately, the Milky Way abounds with stars, which continually whisk past the solar system. During the several billion years before the Sun bloats into a red giant, our descendants could snare a passing star into orbit around the Sun, then jettison the Earth from its solar orbit into an orbit around the new star. Deflecting a star into orbit around the Sun vastly exceeds the capabilities of modern technology and would require roughly a million times more energy than the Sun radiates in a year. (The obvious alternative—transporting everyone from Earth to some other star's planet—would also require large amounts of energy but far greater delicacy, since people are more fragile than stars. It would also require that another Earthlike world exist nearby, and would probably sacrifice much of the Earth's rich heritage.) Once in orbit around the new star, the Earth's residents would watch from afar as the old Sun ballooned into a red giant; but they would receive nearly all their light from the new star, which should support them for billions of years. In fact, if the new star had less mass than the Sun, it would live longer, since the smaller a main-sequence star, the more it hoards its fuel. Before the new star dies, our descendants could repeat the process by lassoing another star from the Milky Way. They could keep doing this as long as the Milky Way has main-sequence stars.

Sooner or later, though, the Milky Way will run out of gas—literally. Gas clouds give birth to new stars, so when the gas runs out, so will the new stars, and trillions of years later all the Galaxy's main-sequence stars will have shriveled into stellar cinders—white dwarfs, neutron stars, black holes—that will do the Earth little good. Elliptical galaxies have already ceased star formation, because they exhausted their gas. Even the Milky Way harbors just 5 to 10 billion Suns' worth of this star-spawning material. As stars die, however, they shed gas and dust into the Galaxy, prolonging its star-forming career.

Unfortunately, a collision with the Andromeda Galaxy will likely change the Milky Way into a gas-depleted elliptical galaxy. But no matter. Even when all the Galaxy's main-sequence stars die, the Earth can still live. Roaming among the stellar cinders will be billions of brown dwarfs, stars too small to sustain the hydrogen-to-helium fusion reaction that powers stars like the Sun. Astronomers discovered the first definite brown dwarf in 1995, a mere nineteen light-years from Earth, and brown dwarfs may outnumber all other stars put together.

Brown dwarfs will be especially useful to our distant descendants. Unlike ordinary stars, they do not burn their hydrogen; thus, trillions of years, quadrillions of years, even quintillions of years from now, they will still possess untapped reservoirs of hydrogen. If our descendants smash several brown dwarfs together, they will spark a nuclear flame in the resulting star that could warm the Earth. Because the Milky Way has so many brown dwarfs, the Earth could thrive for quintillions of years.

Ironically, although the human race has driven numerous species into extinction, it may ultimately be terrestrial life's savior. Indeed, without the human race, or some other form of intelligent life, the Earth will inevitably die, a victim of the harsh rules of stellar evolution. True, the technology to save the Earth is advanced, but the physics requires no exotica such as wormholes or parallel universes. Fortunately, the human race has plenty of time to develop the necessary technology. The immediate task is much simpler: to survive. If through nuclear war, overpopulation, or some other catastrophe the human race wipes itself out, the Earth may never again give birth to another intelligent species capable of saving the planet. Furthermore, if Earth is the only life-bearing world in the cosmos, then our self-destruction would be the most supreme tragedy imaginable, for it would ultimately snuff out all life in the universe.

## THE FUTURE OF THE GALAXY

Long before the Earth's residents have to harvest brown dwarfs, their Galaxy will entertain several visitors. The Milky Way's two brightest satellite galaxies, the Large and Small Magellanic Clouds, hurtle through its dark halo. As these lesser galaxies rub against the Milky Way's dark matter, they lose orbital energy, slipping closer and closer to the Milky Way. Billions of years from now, they will spiral into the Milky Way, splattering some 10 billion stars onto their master and augmenting its already mighty gravitational power.

As they rain down on the Milky Way, the Magellanic stars won't collide with any Galactic stars, because stars are small compared with the gulfs between. However, the Magellanic Clouds bear more than just stars; they also harbor gas clouds that span dozens of light-years of space. These will smack the Milky Way's, compress the gas, and trigger an avalanche of new star formation. For a time, as the Galaxy gobbles its two most lustrous colonies, the Milky Way may outshine its rival in Andromeda.

Meanwhile, the Andromeda Galaxy may crash the party. Since the work of astronomer Vesto Slipher early in the twentieth century, astronomers have known that Andromeda exhibits a blueshift of about 300 kilometers per second. Most of this blueshift, though, arises from the Sun's orbital motion around the Galaxy of 230 kilometers per second, in a direction roughly toward Andromeda. Subtracting this solar motion gives the Andromeda Galaxy's approach velocity toward the Milky Way: 120 kilometers per second, or 6 million miles per day.

This velocity alone, however, does not determine whether Andromeda will hit the Milky Way—just as a ball thrown toward you may sail over your head rather than into your hand. To know whether Andromeda will hit the Milky Way, astronomers must also measure Andromeda's *sideways* velocity. If the sideways velocity is large, Andromeda will miss us; if it's small, Andromeda is heading straight for us, and the collision should occur even before the Sun becomes a red giant. Within the next twenty years, astronomers should be able to measure Andromeda's proper motion—its apparent movement across the sky—and thus its sideways velocity, allowing them to determine just how soon our Galaxy will meet Andromeda.

Even if Andromeda misses the Milky Way on this pass, sooner or later the two galaxies will collide. In fact, their dark halos may already brush each other. As the two giants approach, tides will strip some of the stars off the spirals and hurl them into the black depths of intergalactic space. Such tidal tails decorate the Antennae, a pair of colliding galaxies where two arcing strings of cast-off stars resemble the antennae of some celestial insect.

Most stars, however, will remain bound to the growing galactic conglomeration. Enormous gravitational forces will shuffle the stars in all directions, tearing the beautiful spirals asunder, and merging the two galaxies into a single giant. Gas clouds, but not stars, will smash together, unleashing a horde of new stars. Already, both Andromeda and the Milky Way outshine most other galaxies. But the combined Andromeda-Milky Way superpower, glistening with a rash of new stars, may outdo every other galaxy in the Local Supercluster.

This starry spectacle carries a price. It will consume most of the Galaxy's star-forming material. Indeed, the supergalaxy that emerges from the collision will be a gas-depleted giant elliptical like the Virgo cluster's M87, unable to sculpt nebulae into stars.

## THE FUTURE OF THE UNIVERSE

Meanwhile, as the Galaxy metamorphoses from a beautiful spiral into an ugly elliptical, cosmologists may grow alarmed—not at the diminishing splendor of their home Galaxy, but at the vanishing sights beyond.

Blame lambda. Without the cosmological constant, the gravitational tugs from the galaxies would retard the universe's expansion. Even though these tugs wouldn't reverse the expansion—the universe isn't dense enough, since omega seems to be around 0.3—the deceleration would confer on future astronomers a nice treasure: they'd see more galaxies each year. Of course, astronomers see more *space* each year, because as the universe ages, they see light from greater distances. As a result, in any direction, astronomers see one light-year farther each year. If the universe's expansion were decelerating, previously unseen distant galaxies would pop into view as their light finally reached the Earth.

An *ac*celerating universe, however, forces future astronomers to face the opposite scenario: galaxies they once saw will vanish across the horizon. So much space will open up between those galaxies and Earth that their light will no longer make it to Earth and they will disappear from terrestrial view. Worse, this predicament won't be long in coming. After the discovery of the cosmological constant, Lawrence Krauss and Glenn Starkman at Case Western Reserve University in Cleveland peered into the future, and in 2000 they reported just how bad it will be. According to their study, in 150 billion years—just ten times longer than the universe's present age—all galaxies beyond the Local Supercluster will vanish from the sky. Whereas the presently observable universe harbors some 100 billion galaxies, astronomers then will see only the few thousand galaxies in the Local Supercluster. Beyond those galaxies will yawn 150 billion light-years of empty space.

Actually, things might be even worse. Krauss and Starkman assumed that the gravity of the Local Supercluster's galaxies will hold it together, even as the universe's expansion accelerates. In particular, they assumed that the Milky Way, and the rest of the Local Group, will remain bound to the Local Supercluster. If instead the Local Group goes its own way, then all galaxies beyond it will disappear, and future terrestrial astronomers will see only thirty-six galaxies in the entire universe. If deprived of modern knowledge, observers would reach radically different conclusions about the cosmos. They might not sense that the universe is expanding, because the Local Group doesn't; they'd never know the plethora of galaxies scattered throughout the universe; and they might not even realize that the universe started with the big bang. On the positive side, present-day astronomers have a whole new argument to convince politicians to fund their research *now*, before the objects of their study slip out of sight.

As the universe continues to expand, it will continue to cool. Already the universe's expansion has so attenuated the heat of creation that the cosmic thermometer reads just 2.73 Kelvin. If the expansion neither accelerated nor decelerated, then in 150 billion years the temperature would drop to 0.273 Kelvin. But in an accelerating universe, the universe faces its frosty fate faster: the thermometer at that time will plummet to a mere 0.0005 degrees above absolute zero. What is now the cosmic microwave background will become the cosmic *radio* background, but its intensity may wither to undetectability, robbing astronomers of a vital big bang relic.

Even as it turns the universe into a runaway, lambda's energy could conceivably produce new forces and new phenomena, unforeseen by present science, that counteract the universe's future poverty. Indeed, new universes could blossom out of ours. If observers had witnessed the immediate aftermath of the universe's creation, when matter raced madly about and light scattered to and fro, they might have reached equally pessimistic conclusions about the future; yet that initial chaos sprouted into a universe resplendent with stars, galaxies, planets, and people. Perhaps, unbeknownst to us, the accelerating universe holds future surprises for its inhabitants.

Still, the straightforward extrapolation of the best cosmological knowledge suggests an eternal night, against which the human spirit will stand as a brilliant flame, possibly unique in the cosmos. With sufficient ingenuity, the Earth can survive and even thrive, but it will do so in a universe that grows ever colder, ever darker, ever emptier.

*The standard model of the universe at the end of the nineteenth century was unlike the standard model at the end of the twentieth century in almost every respect. This prompts the question: Is it possible that the standard model of the universe at the end of the twenty-first century will be totally unlike that at the end of the twentieth century? The Victorians were confident that they were close to the truth. What are we to make of the fact that today there is a similar attitude?*

—EDWARD HARRISON (*COSMOLOGY*, 2000 EDITION, PAGE 83)

TABLE 1
THE OBSERVABLE UNIVERSE

| | |
|---|---|
| Temperature | 2.73 Kelvin (-454.76 Fahrenheit) |
| Expansion rate (Hubble constant) | 65 kilometers per second per megaparsec? |
| Matter density (Omega—Ω) | 0.3? |
| Cosmological Constant (Lambda—λ) | 0.7? |
| Omega plus Lambda | 1.0? |
| Geometry | Flat? |
| Age | 14.5 billion years? |
| Radius | 14.5 billion light-years? |
| Fate | Forever expanding |
| Matter Density | $2.4 \times 10^{-30}$ grams per cubic centimeter? |
| Mass-to-Light Ratio | 300? |
| Omega in Baryonic Matter | 0.05? |
| Omega in Nonbaryonic Matter | 0.25? |
| Number of Galaxies | 100 billion? |
| Number of Stars | 1 billion trillion? |
| Questionable Entries Above | 87 percent |

## TABLE 2
## THE CRITICAL DENSITY

The critical density is the minimum density that the universe must have in order to collapse, if lambda is 0. It depends on the square of the Hubble constant: the higher the Hubble constant, the faster the universe expands, so the greater the density must be to stop it.

| HUBBLE CONSTANT (km/sec/Mpc) | CRITICAL DENSITY (grams per cubic centimeter) |
|---|---|
| 40 | $0.30 \times 10^{-29}$ |
| 45 | $0.38 \times 10^{-29}$ |
| 50 | $0.47 \times 10^{-29}$ |
| 55 | $0.57 \times 10^{-29}$ |
| 60 | $0.68 \times 10^{-29}$ |
| 65 | $0.79 \times 10^{-29}$ |
| 70 | $0.92 \times 10^{-29}$ |
| 75 | $1.06 \times 10^{-29}$ |
| 80 | $1.20 \times 10^{-29}$ |
| 85 | $1.36 \times 10^{-29}$ |
| 90 | $1.52 \times 10^{-29}$ |
| 95 | $1.70 \times 10^{-29}$ |
| 100 | $1.88 \times 10^{-29}$ |

## TABLE 3
## LAMBDA AS THE UNIVERSE'S SAVIOR

If the universe is denser than the critical density, then omega exceeds 1 and the universe will someday collapse. However, if lambda, the cosmological constant, exists, it can prevent such a collapse.

| OMEGA | MINIMUM VALUE OF LAMBDA TO PREVENT COLLAPSE |
|---|---|
| 1.0 | 0.000 |
| 1.5 | 0.009 |
| 2.0 | 0.042 |
| 2.5 | 0.096 |
| 3.0 | 0.168 |
| 4.0 | 0.347 |
| 5.0 | 0.563 |
| 6.0 | 0.805 |
| 7.0 | 1.067 |
| 8.0 | 1.345 |
| 9.0 | 1.637 |
| 10.0 | 1.939 |

## TABLE 4
## LOOKBACK DISTANCES

As Edwin Hubble discovered, the greater a galaxy's redshift, the greater is its distance. The following tables convert a galaxy's redshift into its lookback distance—the distance its light has traveled before reaching Earth. Thus, a galaxy whose lookback distance is 8 billion light-years appears to terrestrial observers as it was 8 billion years ago.

The redshifts in these tables indicate how much the expansion of space has stretched a galaxy's light waves. For example, a galaxy whose spectral lines are shifted to the red by 10 percent has a redshift of 0.1; a galaxy whose spectral lines are shifted to the red by 100 percent has a redshift of 1; and so on. If the galaxy's redshift is in velocity units, divide by the speed of light (300,000 kilometers per second, or 186,000 miles per second) to get the redshift. For example, a galaxy with a redshift of 30,000 kilometers per second (18,600 miles per second) has a redshift of 0.1.

Lookback distances depend on three cosmological parameters: the Hubble constant, the universe's present expansion rate; omega (Ω), the universe's matter density; and lambda (λ), the strength of the cosmological constant. The following tables give lookback distances for a Hubble constant of 65 kilometers per second per megaparsec and for over 100 different combinations of omega and lambda. The numbers were calculated using Ned Wright's cosmological calculator (http://www.astro.ucla.edu/~wright/CosmoCalc.html), which includes the energy of the cosmic microwave background, even when omega is 0.

To use these tables:

1. Choose a value for omega, and turn to the page headed by that number.
2. Look in the far left-hand column for the redshift you want.
3. Choose a value for lambda. Follow the row rightward until you reach the column headed by that value. There you will find a lookback distance, in billions of light-years.
4. Choose a value for the Hubble constant, in kilometers per second per megaparsec. Divide 65 by your value for the Hubble constant. Multiply the resulting number by the distance in the table, and you will have the lookback distance you want.

For example, to compute the lookback distance to an object with a redshift of 0.6, for a Hubble constant of 50, omega of 0.3, and lambda of 0.7, turn to the table that says "Omega = 0.3." Then look down the column that says "Redshift" until you find the row that starts with "0.6." Then move along that row to the right until you reach the column headed by a lambda of "0.7." There the table says "6.1." This would be the lookback distance if the Hubble constant were 65. However, in this example, the Hubble constant is 50, so divide 65 by 50 to get 65/50 = 1.3. Then multiply 6.1 by 1.3 to get the answer, 7.9 billion light-years.

Flat universes are in **bold**. Open universes lie to the left of the bold line, closed universes to the right. Open universes expand forever. Closed universes with lambda = 0 collapse, but most closed universes with nonzero lambda expand forever—see Table 3.

*(CONTINUED)*

## LOOKBACK DISTANCES*

HUBBLE CONSTANT = 65 OMEGA = 0.0

| REDSHIFT | LAMBDA 0.0 | 0.1 | 0.2 | 0.3 | 0.4 | 0.5 | 0.6 | 0.7 | 0.8 | 0.9 | **1.0** |
|---|---|---|---|---|---|---|---|---|---|---|---|
| 0.0 | 0.0 | 0.0 | 0.0 | 0.0 | 0.0 | 0.0 | 0.0 | 0.0 | 0.0 | 0.0 | **0.0** |
| 0.1 | 1.4 | 1.4 | 1.4 | 1.4 | 1.4 | 1.4 | 1.4 | 1.4 | 1.4 | 1.4 | **1.4** |
| 0.2 | 2.5 | 2.5 | 2.5 | 2.6 | 2.6 | 2.6 | 2.6 | 2.7 | 2.7 | 2.7 | **2.7** |
| 0.3 | 3.5 | 3.5 | 3.5 | 3.6 | 3.6 | 3.7 | 3.7 | 3.8 | 3.8 | 3.9 | **3.9** |
| 0.4 | 4.3 | 4.4 | 4.4 | 4.5 | 4.5 | 4.6 | 4.7 | 4.8 | 4.9 | 5.0 | **5.1** |
| 0.5 | 5.0 | 5.1 | 5.2 | 5.3 | 5.4 | 5.5 | 5.6 | 5.7 | 5.8 | 5.9 | **6.1** |
| 0.6 | 5.6 | 5.7 | 5.8 | 5.9 | 6.1 | 6.2 | 6.3 | 6.5 | 6.7 | 6.9 | **7.1** |
| 0.7 | 6.2 | 6.3 | 6.4 | 6.6 | 6.7 | 6.9 | 7.0 | 7.2 | 7.4 | 7.7 | **8.0** |
| 0.8 | 6.7 | 6.8 | 7.0 | 7.1 | 7.3 | 7.5 | 7.7 | 7.9 | 8.2 | 8.5 | **8.8** |
| 0.9 | 7.1 | 7.3 | 7.4 | 7.6 | 7.8 | 8.0 | 8.3 | 8.5 | 8.8 | 9.2 | **9.7** |
| 1.0 | 7.5 | 7.7 | 7.9 | 8.1 | 8.3 | 8.5 | 8.8 | 9.1 | 9.5 | 9.9 | **10.4** |
| 1.1 | 7.9 | 8.1 | 8.3 | 8.5 | 8.7 | 9.0 | 9.3 | 9.6 | 10.0 | 10.5 | **11.2** |
| 1.2 | 8.2 | 8.4 | 8.6 | 8.8 | 9.1 | 9.4 | 9.7 | 10.1 | 10.6 | 11.1 | **11.9** |
| 1.3 | 8.5 | 8.7 | 8.9 | 9.2 | 9.5 | 9.8 | 10.1 | 10.6 | 11.1 | 11.7 | **12.5** |
| 1.4 | 8.8 | 9.0 | 9.2 | 9.5 | 9.8 | 10.1 | 10.5 | 11.0 | 11.5 | 12.2 | **13.2** |
| 1.5 | 9.0 | 9.3 | 9.5 | 9.8 | 10.1 | 10.5 | 10.9 | 11.4 | 12.0 | 12.7 | **13.8** |
| 1.6 | 9.3 | 9.5 | 9.8 | 10.0 | 10.4 | 10.8 | 11.2 | 11.7 | 12.4 | 13.2 | **14.4** |
| 2 | 10.0 | 10.3 | 10.6 | 10.9 | 11.3 | 11.8 | 12.3 | 13.0 | 13.8 | 14.9 | **16.5** |
| 3 | 11.3 | 11.6 | 12.0 | 12.4 | 12.9 | 13.5 | 14.2 | 15.0 | 16.2 | 17.8 | **20.8** |
| 4 | 12.0 | 12.4 | 12.8 | 13.3 | 13.9 | 14.5 | 15.3 | 16.3 | 17.7 | 19.8 | **24.1** |
| 5 | 12.5 | 12.9 | 13.4 | 13.9 | 14.5 | 15.2 | 16.1 | 17.2 | 18.8 | 21.2 | **26.7** |
| 6 | 12.9 | 13.3 | 13.8 | 14.3 | 15.0 | 15.7 | 16.6 | 17.8 | 19.5 | 22.2 | **28.9** |
| 7 | 13.2 | 13.6 | 14.1 | 14.6 | 15.3 | 16.1 | 17.1 | 18.3 | 20.1 | 22.9 | **30.6** |
| 8 | 13.4 | 13.8 | 14.3 | 14.9 | 15.6 | 16.4 | 17.4 | 18.7 | 20.5 | 23.5 | **32.1** |
| 9 | 13.5 | 14.0 | 14.5 | 15.1 | 15.8 | 16.6 | 17.6 | 19.0 | 20.9 | 24.0 | **33.3** |
| 10 | 13.7 | 14.1 | 14.6 | 15.2 | 16.0 | 16.8 | 17.9 | 19.2 | 21.2 | 24.4 | **34.2** |
| 15 | 14.1 | 14.6 | 15.1 | 15.8 | 16.5 | 17.4 | 18.5 | 20.0 | 22.1 | 25.6 | **37.1** |
| 20 | 14.3 | 14.8 | 15.4 | 16.0 | 16.8 | 17.7 | 18.9 | 20.4 | 22.5 | 26.3 | **38.2** |
| 40 | 14.6 | 15.2 | 15.7 | 16.4 | 17.2 | 18.2 | 19.4 | 20.9 | 23.2 | 27.1 | **39.5** |
| 100 | 14.8 | 15.4 | 15.9 | 16.6 | 17.4 | 18.4 | 19.6 | 21.2 | 23.5 | 27.4 | **39.9** |
| 1000 | 14.9 | 15.4 | 16.0 | 16.7 | 17.5 | 18.5 | 19.7 | 21.3 | 23.6 | 27.5 | **39.9** |

***In billions of light-years; flat universes in *bold*.**

## LOOKBACK DISTANCES*

| | HUBBLE CONSTANT = 65 | | | | | | OMEGA = 0.1 | | | | |
|---|---|---|---|---|---|---|---|---|---|---|---|
| | LAMBDA | | | | | | | | | | |
| REDSHIFT | 0.0 | 0.1 | 0.2 | 0.3 | 0.4 | 0.5 | 0.6 | 0.7 | 0.8 | **0.9** | 1.0 |
| 0.0 | 0.0 | 0.0 | 0.0 | 0.0 | 0.0 | 0.0 | 0.0 | 0.0 | 0.0 | **0.0** | 0.0 |
| 0.1 | 1.4 | 1.4 | 1.4 | 1.4 | 1.4 | 1.4 | 1.4 | 1.4 | 1.4 | **1.4** | 1.4 |
| 0.2 | 2.5 | 2.5 | 2.5 | 2.6 | 2.6 | 2.6 | 2.6 | 2.6 | 2.7 | **2.7** | 2.7 |
| 0.3 | 3.4 | 3.5 | 3.5 | 3.6 | 3.6 | 3.6 | 3.7 | 3.7 | 3.8 | **3.8** | 3.9 |
| 0.4 | 4.3 | 4.3 | 4.4 | 4.4 | 4.5 | 4.6 | 4.6 | 4.7 | 4.8 | **4.9** | 5.0 |
| 0.5 | 5.0 | 5.0 | 5.1 | 5.2 | 5.3 | 5.4 | 5.5 | 5.6 | 5.7 | **5.8** | 6.0 |
| 0.6 | 5.6 | 5.7 | 5.8 | 5.9 | 6.0 | 6.1 | 6.2 | 6.4 | 6.5 | **6.7** | 6.9 |
| 0.7 | 6.1 | 6.2 | 6.3 | 6.5 | 6.6 | 6.7 | 6.9 | 7.1 | 7.3 | **7.5** | 7.7 |
| 0.8 | 6.6 | 6.7 | 6.8 | 7.0 | 7.1 | 7.3 | 7.5 | 7.7 | 7.9 | **8.2** | 8.5 |
| 0.9 | 7.0 | 7.1 | 7.3 | 7.4 | 7.6 | 7.8 | 8.0 | 8.3 | 8.5 | **8.9** | 9.2 |
| 1.0 | 7.4 | 7.5 | 7.7 | 7.9 | 8.1 | 8.3 | 8.5 | 8.8 | 9.1 | **9.5** | 9.9 |
| 1.1 | 7.7 | 7.9 | 8.1 | 8.3 | 8.5 | 8.7 | 9.0 | 9.3 | 9.6 | **10.0** | 10.5 |
| 1.2 | 8.0 | 8.2 | 8.4 | 8.6 | 8.8 | 9.1 | 9.4 | 9.7 | 10.1 | **10.5** | 11.1 |
| 1.3 | 8.3 | 8.5 | 8.7 | 8.9 | 9.2 | 9.4 | 9.7 | 10.1 | 10.5 | **11.0** | 11.6 |
| 1.4 | 8.6 | 8.8 | 9.0 | 9.2 | 9.5 | 9.8 | 10.1 | 10.5 | 10.9 | **11.4** | 12.1 |
| 1.5 | 8.8 | 9.0 | 9.2 | 9.5 | 9.7 | 10.0 | 10.4 | 10.8 | 11.3 | **11.8** | 12.5 |
| 1.6 | 9.0 | 9.2 | 9.5 | 9.7 | 10.0 | 10.3 | 10.7 | 11.1 | 11.6 | **12.2** | 12.9 |
| 2 | 9.7 | 10.0 | 10.2 | 10.5 | 10.8 | 11.2 | 11.6 | 12.1 | 12.7 | **13.4** | 14.3 |
| 3 | 10.8 | 11.1 | 11.4 | 11.8 | 12.2 | 12.6 | 13.1 | 13.7 | 14.4 | **15.3** | 16.5 |
| 4 | 11.5 | 11.8 | 12.1 | 12.5 | 12.9 | 13.4 | 14.0 | 14.6 | 15.4 | **16.4** | 17.7 |
| 5 | 11.9 | 12.2 | 12.6 | 13.0 | 13.4 | 13.9 | 14.5 | 15.2 | 16.0 | **17.1** | 18.4 |
| 6 | 12.2 | 12.5 | 12.9 | 13.3 | 13.8 | 14.3 | 14.9 | 15.6 | 16.4 | **17.5** | 18.9 |
| 7 | 12.4 | 12.7 | 13.1 | 13.5 | 14.0 | 14.5 | 15.2 | 15.9 | 16.7 | **17.8** | 19.2 |
| 8 | 12.6 | 12.9 | 13.3 | 13.7 | 14.2 | 14.7 | 15.3 | 16.1 | 17.0 | **18.0** | 19.5 |
| 9 | 12.7 | 13.0 | 13.4 | 13.8 | 14.3 | 14.9 | 15.5 | 16.2 | 17.1 | **18.2** | 19.6 |
| 10 | 12.8 | 13.1 | 13.5 | 14.0 | 14.4 | 15.0 | 15.6 | 16.4 | 17.2 | **18.3** | 19.8 |
| 15 | 13.1 | 13.4 | 13.8 | 14.3 | 14.7 | 15.3 | 15.9 | 16.7 | 17.6 | **18.7** | 20.2 |
| 20 | 13.2 | 13.6 | 14.0 | 14.4 | 14.9 | 15.5 | 16.1 | 16.9 | 17.8 | **18.9** | 20.3 |
| 40 | 13.4 | 13.8 | 14.1 | 14.6 | 15.1 | 15.6 | 16.3 | 17.1 | 18.0 | **19.1** | 20.5 |
| 100 | 13.5 | 13.8 | 14.2 | 14.7 | 15.2 | 15.7 | 16.4 | 17.1 | 18.0 | **19.2** | 20.6 |
| 1000 | 13.5 | 13.9 | 14.3 | 14.7 | 15.2 | 15.8 | 16.4 | 17.2 | 18.1 | **19.2** | 20.6 |

**In billions of light-years; flat universes in* ***bold.***

## LOOKBACK DISTANCES*

| | HUBBLE CONSTANT = 65 | | | | | | | OMEGA = 0.2 | | | |
|---|---|---|---|---|---|---|---|---|---|---|---|
| | LAMBDA | | | | | | | | | | |
| REDSHIFT | 0.0 | 0.1 | 0.2 | 0.3 | 0.4 | 0.5 | 0.6 | 0.7 | **0.8** | 0.9 | 1.0 |
| 0.0 | 0.0 | 0.0 | 0.0 | 0.0 | 0.0 | 0.0 | 0.0 | 0.0 | **0.0** | 0.0 | 0.0 |
| 0.1 | 1.4 | 1.4 | 1.4 | 1.4 | 1.4 | 1.4 | 1.4 | 1.4 | **1.4** | 1.4 | 1.4 |
| 0.2 | 2.5 | 2.5 | 2.5 | 2.5 | 2.6 | 2.6 | 2.6 | 2.6 | **2.7** | 2.7 | 2.7 |
| 0.3 | 3.4 | 3.5 | 3.5 | 3.5 | 3.6 | 3.6 | 3.7 | 3.7 | **3.8** | 3.8 | 3.9 |
| 0.4 | 4.2 | 4.3 | 4.3 | 4.4 | 4.5 | 4.5 | 4.6 | 4.7 | **4.7** | 4.8 | 4.9 |
| 0.5 | 4.9 | 5.0 | 5.1 | 5.1 | 5.2 | 5.3 | 5.4 | 5.5 | **5.6** | 5.7 | 5.9 |
| 0.6 | 5.5 | 5.6 | 5.7 | 5.8 | 5.9 | 6.0 | 6.1 | 6.3 | **6.4** | 6.6 | 6.7 |
| 0.7 | 6.0 | 6.1 | 6.2 | 6.4 | 6.5 | 6.6 | 6.8 | 6.9 | **7.1** | 7.3 | 7.5 |
| 0.8 | 6.5 | 6.6 | 6.7 | 6.9 | 7.0 | 7.2 | 7.3 | 7.5 | **7.7** | 8.0 | 8.2 |
| 0.9 | 6.9 | 7.0 | 7.2 | 7.3 | 7.5 | 7.6 | 7.8 | 8.1 | **8.3** | 8.6 | 8.9 |
| 1.0 | 7.3 | 7.4 | 7.5 | 7.7 | 7.9 | 8.1 | 8.3 | 8.5 | **8.8** | 9.1 | 9.5 |
| 1.1 | 7.6 | 7.7 | 7.9 | 8.1 | 8.3 | 8.5 | 8.7 | 9.0 | **9.3** | 9.6 | 10.0 |
| 1.2 | 7.9 | 8.0 | 8.2 | 8.4 | 8.6 | 8.8 | 9.1 | 9.4 | **9.7** | 10.1 | 10.5 |
| 1.3 | 8.1 | 8.3 | 8.5 | 8.7 | 8.9 | 9.2 | 9.4 | 9.7 | **10.1** | 10.5 | 10.9 |
| 1.4 | 8.4 | 8.6 | 8.8 | 9.0 | 9.2 | 9.4 | 9.7 | 10.0 | **10.4** | 10.8 | 11.3 |
| 1.5 | 8.6 | 8.8 | 9.0 | 9.2 | 9.4 | 9.7 | 10.0 | 10.3 | **10.7** | 11.2 | 11.7 |
| 1.6 | 8.8 | 9.0 | 9.2 | 9.4 | 9.7 | 10.0 | 10.3 | 10.6 | **11.0** | 11.5 | 12.0 |
| 2 | 9.5 | 9.7 | 9.9 | 10.2 | 10.4 | 10.8 | 11.1 | 11.5 | **12.0** | 12.5 | 13.2 |
| 3 | 10.5 | 10.7 | 11.0 | 11.3 | 11.6 | 12.0 | 12.4 | 12.9 | **13.4** | 14.0 | 14.8 |
| 4 | 11.1 | 11.3 | 11.6 | 11.9 | 12.3 | 12.7 | 13.1 | 13.6 | **14.2** | 14.9 | 15.7 |
| 5 | 11.4 | 11.7 | 12.0 | 12.3 | 12.7 | 13.1 | 13.6 | 14.1 | **14.7** | 15.4 | 16.2 |
| 6 | 11.7 | 12.0 | 12.3 | 12.6 | 13.0 | 13.4 | 13.8 | 14.4 | **15.0** | 15.7 | 16.5 |
| 7 | 11.9 | 12.1 | 12.5 | 12.8 | 13.2 | 13.6 | 14.0 | 14.6 | **15.2** | 15.9 | 16.8 |
| 8 | 12.0 | 12.3 | 12.6 | 12.9 | 13.3 | 13.7 | 14.2 | 14.7 | **15.4** | 16.1 | 16.9 |
| 9 | 12.1 | 12.4 | 12.7 | 13.0 | 13.4 | 13.8 | 14.3 | 14.9 | **15.5** | 16.2 | 17.1 |
| 10 | 12.2 | 12.5 | 12.8 | 13.1 | 13.5 | 13.9 | 14.4 | 14.9 | **15.6** | 16.3 | 17.2 |
| 15 | 12.4 | 12.7 | 13.0 | 13.4 | 13.8 | 14.2 | 14.7 | 15.2 | **15.8** | 16.6 | 17.4 |
| 20 | 12.5 | 12.8 | 13.1 | 13.5 | 13.9 | 14.3 | 14.8 | 15.3 | **15.9** | 16.7 | 17.6 |
| 40 | 12.6 | 12.9 | 13.3 | 13.6 | 14.0 | 14.4 | 14.9 | 15.5 | **16.1** | 16.8 | 17.7 |
| 100 | 12.7 | 13.0 | 13.3 | 13.7 | 14.1 | 14.5 | 15.0 | 15.5 | **16.2** | 16.9 | 17.8 |
| 1000 | 12.7 | 13.0 | 13.3 | 13.7 | 14.1 | 14.5 | 15.0 | 15.5 | **16.2** | 16.9 | 17.8 |

**In billions of light-years; flat universes in bold.*

## LOOKBACK DISTANCES*

HUBBLE CONSTANT = 65 OMEGA = 0.3

| REDSHIFT | LAMBDA 0.0 | 0.1 | 0.2 | 0.3 | 0.4 | 0.5 | 0.6 | **0.7** | 0.8 | 0.9 | 1.0 |
|---|---|---|---|---|---|---|---|---|---|---|---|
| 0.0 | 0.0 | 0.0 | 0.0 | 0.0 | 0.0 | 0.0 | 0.0 | **0.0** | 0.0 | 0.0 | 0.0 |
| 0.1 | 1.4 | 1.4 | 1.4 | 1.4 | 1.4 | 1.4 | 1.4 | **1.4** | 1.4 | 1.4 | 1.4 |
| 0.2 | 2.5 | 2.5 | 2.5 | 2.5 | 2.6 | 2.6 | 2.6 | **2.6** | 2.6 | 2.7 | 2.7 |
| 0.3 | 3.4 | 3.4 | 3.5 | 3.5 | 3.6 | 3.6 | 3.6 | **3.7** | 3.7 | 3.8 | 3.8 |
| 0.4 | 4.2 | 4.2 | 4.3 | 4.4 | 4.4 | 4.5 | 4.5 | **4.6** | 4.7 | 4.8 | 4.9 |
| 0.5 | 4.9 | 4.9 | 5.0 | 5.1 | 5.2 | 5.2 | 5.3 | **5.4** | 5.5 | 5.6 | 5.8 |
| 0.6 | 5.4 | 5.5 | 5.6 | 5.7 | 5.8 | 5.9 | 6.0 | **6.1** | 6.3 | 6.4 | 6.6 |
| 0.7 | 5.9 | 6.0 | 6.1 | 6.3 | 6.4 | 6.5 | 6.6 | **6.8** | 6.9 | 7.1 | 7.3 |
| 0.8 | 6.4 | 6.5 | 6.6 | 6.7 | 6.9 | 7.0 | 7.2 | **7.4** | 7.5 | 7.8 | 8.0 |
| 0.9 | 6.8 | 6.9 | 7.0 | 7.2 | 7.3 | 7.5 | 7.7 | **7.9** | 8.1 | 8.3 | 8.6 |
| 1.0 | 7.1 | 7.3 | 7.4 | 7.6 | 7.7 | 7.9 | 8.1 | **8.3** | 8.5 | 8.8 | 9.1 |
| 1.1 | 7.4 | 7.6 | 7.7 | 7.9 | 8.1 | 8.3 | 8.5 | **8.7** | 9.0 | 9.3 | 9.6 |
| 1.2 | 7.7 | 7.9 | 8.0 | 8.2 | 8.4 | 8.6 | 8.8 | **9.1** | 9.4 | 9.7 | 10.0 |
| 1.3 | 8.0 | 8.1 | 8.3 | 8.5 | 8.7 | 8.9 | 9.1 | **9.4** | 9.7 | 10.0 | 10.4 |
| 1.4 | 8.2 | 8.4 | 8.6 | 8.7 | 9.0 | 9.2 | 9.4 | **9.7** | 10.0 | 10.4 | 10.8 |
| 1.5 | 8.4 | 8.6 | 8.8 | 9.0 | 9.2 | 9.4 | 9.7 | **10.0** | 10.3 | 10.7 | 11.1 |
| 1.6 | 8.6 | 8.8 | 9.0 | 9.2 | 9.4 | 9.7 | 9.9 | **10.2** | 10.6 | 11.0 | 11.4 |
| 2 | 9.2 | 9.4 | 9.6 | 9.9 | 10.1 | 10.4 | 10.7 | **11.0** | 11.4 | 11.8 | 12.3 |
| 3 | 10.2 | 10.4 | 10.6 | 10.9 | 11.2 | 11.5 | 11.8 | **12.2** | 12.7 | 13.2 | 13.7 |
| 4 | 10.7 | 10.9 | 11.2 | 11.5 | 11.8 | 12.1 | 12.5 | **12.9** | 13.3 | 13.8 | 14.5 |
| 5 | 11.0 | 11.3 | 11.5 | 11.8 | 12.1 | 12.5 | 12.8 | **13.3** | 13.7 | 14.3 | 14.9 |
| 6 | 11.3 | 11.5 | 11.8 | 12.1 | 12.4 | 12.7 | 13.1 | **13.5** | 14.0 | 14.5 | 15.2 |
| 7 | 11.4 | 11.7 | 11.9 | 12.2 | 12.5 | 12.9 | 13.3 | **13.7** | 14.2 | 14.7 | 15.4 |
| 8 | 11.5 | 11.8 | 12.1 | 12.3 | 12.7 | 13.0 | 13.4 | **13.8** | 14.3 | 14.9 | 15.5 |
| 9 | 11.6 | 11.9 | 12.1 | 12.4 | 12.8 | 13.1 | 13.5 | **13.9** | 14.4 | 15.0 | 15.6 |
| 10 | 11.7 | 11.9 | 12.2 | 12.5 | 12.8 | 13.2 | 13.6 | **14.0** | 14.5 | 15.0 | 15.7 |
| 15 | 11.9 | 12.1 | 12.4 | 12.7 | 13.0 | 13.4 | 13.8 | **14.2** | 14.7 | 15.3 | 15.9 |
| 20 | 12.0 | 12.2 | 12.5 | 12.8 | 13.1 | 13.5 | 13.9 | **14.3** | 14.8 | 15.4 | 16.0 |
| 40 | 12.1 | 12.4 | 12.6 | 12.9 | 13.2 | 13.6 | 14.0 | **14.4** | 14.9 | 15.5 | 16.1 |
| 100 | 12.1 | 12.4 | 12.7 | 13.0 | 13.3 | 13.7 | 14.0 | **14.5** | 15.0 | 15.5 | 16.2 |
| 1000 | 12.2 | 12.4 | 12.7 | 13.0 | 13.3 | 13.7 | 14.1 | **14.5** | 15.0 | 15.5 | 16.2 |

**In billions of light-years; flat universes in* ***bold.***

## LOOKBACK DISTANCES*

HUBBLE CONSTANT= 65 OMEGA = 0.4

| | LAMBDA | | | | | | | | | | |
|---|---|---|---|---|---|---|---|---|---|---|---|
| REDSHIFT | 0.0 | 0.1 | 0.2 | 0.3 | 0.4 | 0.5 | **0.6** | 0.7 | 0.8 | 0.9 | 1.0 |
| 0.0 | 0.0 | 0.0 | 0.0 | 0.0 | 0.0 | 0.0 | **0.0** | 0.0 | 0.0 | 0.0 | 0.0 |
| 0.1 | 1.4 | 1.4 | 1.4 | 1.4 | 1.4 | 1.4 | **1.4** | 1.4 | 1.4 | 1.4 | 1.4 |
| 0.2 | 2.5 | 2.5 | 2.5 | 2.5 | 2.5 | 2.6 | **2.6** | 2.6 | 2.6 | 2.7 | 2.7 |
| 0.3 | 3.4 | 3.4 | 3.5 | 3.5 | 3.5 | 3.6 | **3.6** | 3.7 | 3.7 | 3.7 | 3.8 |
| 0.4 | 4.2 | 4.2 | 4.3 | 4.3 | 4.4 | 4.4 | **4.5** | 4.6 | 4.6 | 4.7 | 4.8 |
| 0.5 | 4.8 | 4.9 | 4.9 | 5.0 | 5.1 | 5.2 | **5.3** | 5.4 | 5.5 | 5.6 | 5.7 |
| 0.6 | 5.4 | 5.5 | 5.5 | 5.6 | 5.7 | 5.8 | **5.9** | 6.1 | 6.2 | 6.3 | 6.5 |
| 0.7 | 5.9 | 6.0 | 6.1 | 6.2 | 6.3 | 6.4 | **6.5** | 6.7 | 6.8 | 7.0 | 7.2 |
| 0.8 | 6.3 | 6.4 | 6.5 | 6.6 | 6.8 | 6.9 | **7.0** | 7.2 | 7.4 | 7.6 | 7.8 |
| 0.9 | 6.7 | 6.8 | 6.9 | 7.1 | 7.2 | 7.3 | **7.5** | 7.7 | 7.9 | 8.1 | 8.3 |
| 1.0 | 7.0 | 7.1 | 7.3 | 7.4 | 7.6 | 7.7 | **7.9** | 8.1 | 8.3 | 8.6 | 8.8 |
| 1.1 | 7.3 | 7.5 | 7.6 | 7.7 | 7.9 | 8.1 | **8.3** | 8.5 | 8.7 | 9.0 | 9.3 |
| 1.2 | 7.6 | 7.7 | 7.9 | 8.0 | 8.2 | 8.4 | **8.6** | 8.8 | 9.1 | 9.4 | 9.7 |
| 1.3 | 7.8 | 8.0 | 8.1 | 8.3 | 8.5 | 8.7 | **8.9** | 9.1 | 9.4 | 9.7 | 10.0 |
| 1.4 | 8.1 | 8.2 | 8.4 | 8.5 | 8.7 | 8.9 | **9.2** | 9.4 | 9.7 | 10.0 | 10.3 |
| 1.5 | 8.3 | 8.4 | 8.6 | 8.8 | 9.0 | 9.2 | **9.4** | 9.7 | 10.0 | 10.3 | 10.6 |
| 1.6 | 8.4 | 8.6 | 8.8 | 9.0 | 9.2 | 9.4 | **9.6** | 9.9 | 10.2 | 10.5 | 10.9 |
| 2 | 9.0 | 9.2 | 9.4 | 9.6 | 9.8 | 10.1 | **10.3** | 10.6 | 11.0 | 11.3 | 11.7 |
| 3 | 9.9 | 10.1 | 10.3 | 10.6 | 10.8 | 11.1 | **11.4** | 11.7 | 12.1 | 12.5 | 13.0 |
| 4 | 10.4 | 10.6 | 10.8 | 11.1 | 11.3 | 11.6 | **11.9** | 12.3 | 12.7 | 13.1 | 13.6 |
| 5 | 10.7 | 10.9 | 11.2 | 11.4 | 11.7 | 12.0 | **12.3** | 12.6 | 13.0 | 13.5 | 14.0 |
| 6 | 10.9 | 11.1 | 11.4 | 11.6 | 11.9 | 12.2 | **12.5** | 12.9 | 13.3 | 13.7 | 14.2 |
| 7 | 11.0 | 11.3 | 11.5 | 11.8 | 12.0 | 12.3 | **12.7** | 13.0 | 13.4 | 13.9 | 14.4 |
| 8 | 11.2 | 11.4 | 11.6 | 11.9 | 12.1 | 12.4 | **12.8** | 13.1 | 13.5 | 14.0 | 14.5 |
| 9 | 11.2 | 11.5 | 11.7 | 11.9 | 12.2 | 12.5 | **12.9** | 13.2 | 13.6 | 14.1 | 14.6 |
| 10 | 11.3 | 11.5 | 11.8 | 12.0 | 12.3 | 12.6 | **12.9** | 13.3 | 13.7 | 14.1 | 14.6 |
| 15 | 11.5 | 11.7 | 11.9 | 12.2 | 12.5 | 12.8 | **13.1** | 13.5 | 13.9 | 14.3 | 14.8 |
| 20 | 11.6 | 11.8 | 12.0 | 12.3 | 12.6 | 12.9 | **13.2** | 13.6 | 14.0 | 14.4 | 14.9 |
| 40 | 11.7 | 11.9 | 12.1 | 12.4 | 12.7 | 13.0 | **13.3** | 13.7 | 14.1 | 14.5 | 15.0 |
| 100 | 11.7 | 11.9 | 12.2 | 12.4 | 12.7 | 13.0 | **13.3** | 13.7 | 14.1 | 14.6 | 15.1 |
| 1000 | 11.7 | 11.9 | 12.2 | 12.4 | 12.7 | 13.0 | **13.4** | 13.7 | 14.1 | 14.6 | 15.1 |

**In billions of light-years; flat universes in* ***bold***.

## LOOKBACK DISTANCES*

HUBBLE CONSTANT= 65 OMEGA = 0.5

| REDSHIFT | LAMBDA 0.0 | 0.1 | 0.2 | 0.3 | 0.4 | **0.5** | 0.6 | 0.7 | 0.8 | 0.9 | 1.0 |
|---|---|---|---|---|---|---|---|---|---|---|---|
| 0.0 | 0.0 | 0.0 | 0.0 | 0.0 | 0.0 | **0.0** | 0.0 | 0.0 | 0.0 | 0.0 | 0.0 |
| 0.1 | 1.4 | 1.4 | 1.4 | 1.4 | 1.4 | **1.4** | 1.4 | 1.4 | 1.4 | 1.4 | 1.4 |
| 0.2 | 2.5 | 2.5 | 2.5 | 2.5 | 2.5 | **2.5** | 2.6 | 2.6 | 2.6 | 2.6 | 2.7 |
| 0.3 | 3.4 | 3.4 | 3.4 | 3.5 | 3.5 | **3.5** | 3.6 | 3.6 | 3.7 | 3.7 | 3.8 |
| 0.4 | 4.1 | 4.2 | 4.2 | 4.3 | 4.3 | **4.4** | 4.4 | 4.5 | 4.6 | 4.7 | 4.7 |
| 0.5 | 4.8 | 4.8 | 4.9 | 5.0 | 5.0 | **5.1** | 5.2 | 5.3 | 5.4 | 5.5 | 5.6 |
| 0.6 | 5.3 | 5.4 | 5.5 | 5.6 | 5.7 | **5.8** | 5.9 | 6.0 | 6.1 | 6.2 | 6.3 |
| 0.7 | 5.8 | 5.9 | 6.0 | 6.1 | 6.2 | **6.3** | 6.4 | 6.6 | 6.7 | 6.8 | 7.0 |
| 0.8 | 6.2 | 6.3 | 6.4 | 6.5 | 6.7 | **6.8** | 6.9 | 7.1 | 7.2 | 7.4 | 7.6 |
| 0.9 | 6.6 | 6.7 | 6.8 | 6.9 | 7.1 | **7.2** | 7.4 | 7.5 | 7.7 | 7.9 | 8.1 |
| 1.0 | 6.9 | 7.0 | 7.2 | 7.3 | 7.4 | **7.6** | 7.7 | 7.9 | 8.1 | 8.3 | 8.6 |
| 1.1 | 7.2 | 7.3 | 7.5 | 7.6 | 7.8 | **7.9** | 8.1 | 8.3 | 8.5 | 8.7 | 9.0 |
| 1.2 | 7.5 | 7.6 | 7.7 | 7.9 | 8.0 | **8.2** | 8.4 | 8.6 | 8.8 | 9.1 | 9.4 |
| 1.3 | 7.7 | 7.8 | 8.0 | 8.1 | 8.3 | **8.5** | 8.7 | 8.9 | 9.1 | 9.4 | 9.7 |
| 1.4 | 7.9 | 8.1 | 8.2 | 8.4 | 8.5 | **8.7** | 8.9 | 9.2 | 9.4 | 9.7 | 10.0 |
| 1.5 | 8.1 | 8.3 | 8.4 | 8.6 | 8.8 | **8.9** | 9.2 | 9.4 | 9.6 | 9.9 | 10.2 |
| 1.6 | 8.3 | 8.4 | 8.6 | 8.8 | 8.9 | **9.1** | 9.4 | 9.6 | 9.9 | 10.2 | 10.5 |
| 2 | 8.8 | 9.0 | 9.2 | 9.4 | 9.6 | **9.8** | 10.0 | 10.3 | 10.6 | 10.9 | 11.2 |
| 3 | 9.7 | 9.9 | 10.1 | 10.3 | 10.5 | **10.7** | 11.0 | 11.3 | 11.6 | 11.9 | 12.3 |
| 4 | 10.1 | 10.3 | 10.5 | 10.7 | 11.0 | **11.2** | 11.5 | 11.8 | 12.1 | 12.5 | 12.9 |
| 5 | 10.4 | 10.6 | 10.8 | 11.0 | 11.3 | **11.5** | 11.8 | 12.1 | 12.5 | 12.8 | 13.2 |
| 6 | 10.6 | 10.8 | 11.0 | 11.2 | 11.5 | **11.7** | 12.0 | 12.3 | 12.7 | 13.0 | 13.5 |
| 7 | 10.7 | 10.9 | 11.1 | 11.4 | 11.6 | **11.9** | 12.2 | 12.5 | 12.8 | 13.2 | 13.6 |
| 8 | 10.8 | 11.0 | 11.2 | 11.5 | 11.7 | **12.0** | 12.3 | 12.6 | 12.9 | 13.3 | 13.7 |
| 9 | 10.9 | 11.1 | 11.3 | 11.5 | 11.8 | **12.0** | 12.3 | 12.6 | 13.0 | 13.4 | 13.8 |
| 10 | 11.0 | 11.2 | 11.4 | 11.6 | 11.8 | **12.1** | 12.4 | 12.7 | 13.1 | 13.4 | 13.9 |
| 15 | 11.1 | 11.3 | 11.5 | 11.8 | 12.0 | **12.3** | 12.6 | 12.9 | 13.2 | 13.6 | 14.0 |
| 20 | 11.2 | 11.4 | 11.6 | 11.8 | 12.1 | **12.4** | 12.6 | 13.0 | 13.3 | 13.7 | 14.1 |
| 40 | 11.3 | 11.5 | 11.7 | 11.9 | 12.2 | **12.4** | 12.7 | 13.0 | 13.4 | 13.8 | 14.2 |
| 100 | 11.3 | 11.5 | 11.7 | 12.0 | 12.2 | **12.5** | 12.8 | 13.1 | 13.4 | 13.8 | 14.2 |
| 1000 | 11.3 | 11.5 | 11.8 | 12.0 | 12.2 | **12.5** | 12.8 | 13.1 | 13.4 | 13.8 | 14.3 |

**In billions of light-years; flat universes in **bold**.*

## LOOKBACK DISTANCES*

HUBBLE CONSTANT = 65 OMEGA = 0.6

| REDSHIFT | LAMBDA 0.0 | 0.1 | 0.2 | 0.3 | **0.4** | 0.5 | 0.6 | 0.7 | 0.8 | 0.9 | 1.0 |
|---|---|---|---|---|---|---|---|---|---|---|---|
| 0.0 | 0.0 | 0.0 | 0.0 | 0.0 | **0.0** | 0.0 | 0.0 | 0.0 | 0.0 | 0.0 | 0.0 |
| 0.1 | 1.3 | 1.4 | 1.4 | 1.4 | **1.4** | 1.4 | 1.4 | 1.4 | 1.4 | 1.4 | 1.4 |
| 0.2 | 2.4 | 2.5 | 2.5 | 2.5 | **2.5** | 2.5 | 2.6 | 2.6 | 2.6 | 2.6 | 2.6 |
| 0.3 | 3.3 | 3.4 | 3.4 | 3.4 | **3.5** | 3.5 | 3.6 | 3.6 | 3.6 | 3.7 | 3.7 |
| 0.4 | 4.1 | 4.1 | 4.2 | 4.2 | **4.3** | 4.3 | 4.4 | 4.5 | 4.5 | 4.6 | 4.7 |
| 0.5 | 4.7 | 4.8 | 4.9 | 4.9 | **5.0** | 5.1 | 5.1 | 5.2 | 5.3 | 5.4 | 5.5 |
| 0.6 | 5.3 | 5.3 | 5.4 | 5.5 | **5.6** | 5.7 | 5.8 | 5.9 | 6.0 | 6.1 | 6.2 |
| 0.7 | 5.7 | 5.8 | 5.9 | 6.0 | **6.1** | 6.2 | 6.3 | 6.4 | 6.6 | 6.7 | 6.9 |
| 0.8 | 6.1 | 6.2 | 6.3 | 6.4 | **6.6** | 6.7 | 6.8 | 6.9 | 7.1 | 7.3 | 7.4 |
| 0.9 | 6.5 | 6.6 | 6.7 | 6.8 | **7.0** | 7.1 | 7.2 | 7.4 | 7.5 | 7.7 | 7.9 |
| 1.0 | 6.8 | 6.9 | 7.0 | 7.2 | **7.3** | 7.4 | 7.6 | 7.8 | 7.9 | 8.1 | 8.4 |
| 1.1 | 7.1 | 7.2 | 7.3 | 7.5 | **7.6** | 7.8 | 7.9 | 8.1 | 8.3 | 8.5 | 8.7 |
| 1.2 | 7.4 | 7.5 | 7.6 | 7.7 | **7.9** | 8.1 | 8.2 | 8.4 | 8.6 | 8.8 | 9.1 |
| 1.3 | 7.6 | 7.7 | 7.8 | 8.0 | **8.1** | 8.3 | 8.5 | 8.7 | 8.9 | 9.1 | 9.4 |
| 1.4 | 7.8 | 7.9 | 8.1 | 8.2 | **8.4** | 8.5 | 8.7 | 8.9 | 9.1 | 9.4 | 9.7 |
| 1.5 | 8.0 | 8.1 | 8.2 | 8.4 | **8.6** | 8.7 | 8.9 | 9.1 | 9.4 | 9.6 | 9.9 |
| 1.6 | 8.1 | 8.3 | 8.4 | 8.6 | **8.8** | 8.9 | 9.1 | 9.3 | 9.6 | 9.8 | 10.1 |
| 2 | 8.7 | 8.8 | 9.0 | 9.2 | **9.3** | 9.5 | 9.8 | 10.0 | 10.2 | 10.5 | 10.8 |
| 3 | 9.5 | 9.6 | 9.8 | 10.0 | **10.2** | 10.4 | 10.7 | 10.9 | 11.2 | 11.5 | 11.8 |
| 4 | 9.9 | 10.1 | 10.3 | 10.5 | **10.7** | 10.9 | 11.1 | 11.4 | 11.7 | 12.0 | 12.4 |
| 5 | 10.2 | 10.3 | 10.5 | 10.7 | **10.9** | 11.2 | 11.4 | 11.7 | 12.0 | 12.3 | 12.7 |
| 6 | 10.3 | 10.5 | 10.7 | 10.9 | **11.1** | 11.4 | 11.6 | 11.9 | 12.2 | 12.5 | 12.9 |
| 7 | 10.4 | 10.6 | 10.8 | 11.0 | **11.2** | 11.5 | 11.7 | 12.0 | 12.3 | 12.6 | 13.0 |
| 8 | 10.5 | 10.7 | 10.9 | 11.1 | **11.3** | 11.6 | 11.8 | 12.1 | 12.4 | 12.7 | 13.1 |
| 9 | 10.6 | 10.8 | 11.0 | 11.2 | **11.4** | 11.6 | 11.9 | 12.2 | 12.5 | 12.8 | 13.2 |
| 10 | 10.7 | 10.8 | 11.0 | 11.2 | **11.5** | 11.7 | 12.0 | 12.2 | 12.5 | 12.9 | 13.2 |
| 15 | 10.8 | 11.0 | 11.2 | 11.4 | **11.6** | 11.9 | 12.1 | 12.4 | 12.7 | 13.0 | 13.4 |
| 20 | 10.9 | 11.1 | 11.3 | 11.5 | **11.7** | 11.9 | 12.2 | 12.5 | 12.8 | 13.1 | 13.4 |
| 40 | 11.0 | 11.1 | 11.3 | 11.5 | **11.8** | 12.0 | 12.3 | 12.5 | 12.8 | 13.2 | 13.5 |
| 100 | 11.0 | 11.2 | 11.4 | 11.6 | **11.8** | 12.0 | 12.3 | 12.6 | 12.9 | 13.2 | 13.6 |
| 1000 | 11.0 | 11.2 | 11.4 | 11.6 | **11.8** | 12.1 | 12.3 | 12.6 | 12.9 | 13.2 | 13.6 |

**In billions of light-years; flat universes in bold.*

## LOOKBACK DISTANCES*

HUBBLE CONSTANT = 65 OMEGA = 0.7

| REDSHIFT | LAMBDA 0.0 | 0.1 | 0.2 | **0.3** | 0.4 | 0.5 | 0.6 | 0.7 | 0.8 | 0.9 | 1.0 |
|---|---|---|---|---|---|---|---|---|---|---|---|
| 0.0 | 0.0 | 0.0 | 0.0 | **0.0** | 0.0 | 0.0 | 0.0 | 0.0 | 0.0 | 0.0 | 0.0 |
| 0.1 | 1.3 | 1.4 | 1.4 | **1.4** | 1.4 | 1.4 | 1.4 | 1.4 | 1.4 | 1.4 | 1.4 |
| 0.2 | 2.4 | 2.4 | 2.5 | **2.5** | 2.5 | 2.5 | 2.5 | 2.6 | 2.6 | 2.6 | 2.6 |
| 0.3 | 3.3 | 3.4 | 3.4 | **3.4** | 3.5 | 3.5 | 3.5 | 3.6 | 3.6 | 3.7 | 3.7 |
| 0.4 | 4.1 | 4.1 | 4.2 | **4.2** | 4.3 | 4.3 | 4.4 | 4.4 | 4.5 | 4.6 | 4.6 |
| 0.5 | 4.7 | 4.7 | 4.8 | **4.9** | 4.9 | 5.0 | 5.1 | 5.2 | 5.2 | 5.3 | 5.4 |
| 0.6 | 5.2 | 5.3 | 5.4 | **5.4** | 5.5 | 5.6 | 5.7 | 5.8 | 5.9 | 6.0 | 6.1 |
| 0.7 | 5.7 | 5.8 | 5.8 | **5.9** | 6.0 | 6.1 | 6.2 | 6.3 | 6.5 | 6.6 | 6.7 |
| 0.8 | 6.1 | 6.2 | 6.3 | **6.4** | 6.5 | 6.6 | 6.7 | 6.8 | 7.0 | 7.1 | 7.3 |
| 0.9 | 6.4 | 6.5 | 6.6 | **6.7** | 6.8 | 7.0 | 7.1 | 7.2 | 7.4 | 7.6 | 7.7 |
| 1.0 | 6.7 | 6.8 | 6.9 | **7.1** | 7.2 | 7.3 | 7.5 | 7.6 | 7.8 | 8.0 | 8.2 |
| 1.1 | 7.0 | 7.1 | 7.2 | **7.4** | 7.5 | 7.6 | 7.8 | 7.9 | 8.1 | 8.3 | 8.5 |
| 1.2 | 7.2 | 7.4 | 7.5 | **7.6** | 7.8 | 7.9 | 8.1 | 8.2 | 8.4 | 8.6 | 8.8 |
| 1.3 | 7.5 | 7.6 | 7.7 | **7.8** | 8.0 | 8.1 | 8.3 | 8.5 | 8.7 | 8.9 | 9.1 |
| 1.4 | 7.7 | 7.8 | 7.9 | **8.1** | 8.2 | 8.4 | 8.5 | 8.7 | 8.9 | 9.1 | 9.4 |
| 1.5 | 7.8 | 8.0 | 8.1 | **8.2** | 8.4 | 8.6 | 8.7 | 8.9 | 9.1 | 9.4 | 9.6 |
| 1.6 | 8.0 | 8.1 | 8.3 | **8.4** | 8.6 | 8.7 | 8.9 | 9.1 | 9.3 | 9.6 | 9.8 |
| 2 | 8.5 | 8.7 | 8.8 | **9.0** | 9.1 | 9.3 | 9.5 | 9.7 | 10.0 | 10.2 | 10.5 |
| 3 | 9.3 | 9.4 | 9.6 | **9.8** | 10.0 | 10.2 | 10.4 | 10.6 | 10.8 | 11.1 | 11.4 |
| 4 | 9.7 | 9.8 | 10.0 | **10.2** | 10.4 | 10.6 | 10.8 | 11.0 | 11.3 | 11.6 | 11.9 |
| 5 | 9.9 | 10.1 | 10.3 | **10.4** | 10.6 | 10.9 | 11.1 | 11.3 | 11.6 | 11.9 | 12.2 |
| 6 | 10.1 | 10.3 | 10.4 | **10.6** | 10.8 | 11.0 | 11.3 | 11.5 | 11.8 | 12.0 | 12.4 |
| 7 | 10.2 | 10.4 | 10.5 | **10.7** | 10.9 | 11.1 | 11.4 | 11.6 | 11.9 | 12.2 | 12.5 |
| 8 | 10.3 | 10.5 | 10.6 | **10.8** | 11.0 | 11.2 | 11.5 | 11.7 | 12.0 | 12.3 | 12.6 |
| 9 | 10.3 | 10.5 | 10.7 | **10.9** | 11.1 | 11.3 | 11.5 | 11.8 | 12.0 | 12.3 | 12.6 |
| 10 | 10.4 | 10.6 | 10.7 | **10.9** | 11.1 | 11.3 | 11.6 | 11.8 | 12.1 | 12.4 | 12.7 |
| 15 | 10.5 | 10.7 | 10.9 | **11.1** | 11.3 | 11.5 | 11.7 | 12.0 | 12.2 | 12.5 | 12.8 |
| 20 | 10.6 | 10.8 | 10.9 | **11.1** | 11.3 | 11.6 | 11.8 | 12.0 | 12.3 | 12.6 | 12.9 |
| 40 | 10.7 | 10.8 | 11.0 | **11.2** | 11.4 | 11.6 | 11.9 | 12.1 | 12.4 | 12.7 | 13.0 |
| 100 | 10.7 | 10.9 | 11.1 | **11.2** | 11.5 | 11.7 | 11.9 | 12.1 | 12.4 | 12.7 | 13.0 |
| 1000 | 10.7 | 10.9 | 11.1 | **11.3** | 11.5 | 11.7 | 11.9 | 12.2 | 12.4 | 12.7 | 13.0 |

**In billions of light-years; flat universes in* ***bold***.

## LOOKBACK DISTANCES*

HUBBLE CONSTANT = 65 OMEGA = 0.8

| | LAMBDA | | | | | | | | | | |
|---|---|---|---|---|---|---|---|---|---|---|---|
| REDSHIFT | 0.0 | 0.1 | **0.2** | 0.3 | 0.4 | 0.5 | 0.6 | 0.7 | 0.8 | 0.9 | 1.0 |
| 0.0 | 0.0 | 0.0 | **0.0** | 0.0 | 0.0 | 0.0 | 0.0 | 0.0 | 0.0 | 0.0 | 0.0 |
| 0.1 | 1.3 | 1.3 | **1.4** | 1.4 | 1.4 | 1.4 | 1.4 | 1.4 | 1.4 | 1.4 | 1.4 |
| 0.2 | 2.4 | 2.4 | **2.5** | 2.5 | 2.5 | 2.5 | 2.5 | 2.6 | 2.6 | 2.6 | 2.6 |
| 0.3 | 3.3 | 3.3 | **3.4** | 3.4 | 3.4 | 3.5 | 3.5 | 3.5 | 3.6 | 3.6 | 3.7 |
| 0.4 | 4.0 | 4.1 | **4.1** | 4.2 | 4.2 | 4.3 | 4.3 | 4.4 | 4.4 | 4.5 | 4.6 |
| 0.5 | 4.6 | 4.7 | **4.8** | 4.8 | 4.9 | 5.0 | 5.0 | 5.1 | 5.2 | 5.3 | 5.4 |
| 0.6 | 5.2 | 5.2 | **5.3** | 5.4 | 5.5 | 5.5 | 5.6 | 5.7 | 5.8 | 5.9 | 6.0 |
| 0.7 | 5.6 | 5.7 | **5.8** | 5.9 | 5.9 | 6.0 | 6.1 | 6.3 | 6.4 | 6.5 | 6.6 |
| 0.8 | 6.0 | 6.1 | **6.2** | 6.3 | 6.4 | 6.5 | 6.6 | 6.7 | 6.8 | 7.0 | 7.1 |
| 0.9 | 6.3 | 6.4 | **6.5** | 6.6 | 6.7 | 6.9 | 7.0 | 7.1 | 7.3 | 7.4 | 7.6 |
| 1.0 | 6.6 | 6.7 | **6.8** | 7.0 | 7.1 | 7.2 | 7.3 | 7.5 | 7.6 | 7.8 | 8.0 |
| 1.1 | 6.9 | 7.0 | **7.1** | 7.2 | 7.4 | 7.5 | 7.6 | 7.8 | 8.0 | 8.1 | 8.3 |
| 1.2 | 7.1 | 7.3 | **7.4** | 7.5 | 7.6 | 7.8 | 7.9 | 8.1 | 8.2 | 8.4 | 8.6 |
| 1.3 | 7.4 | 7.5 | **7.6** | 7.7 | 7.8 | 8.0 | 8.1 | 8.3 | 8.5 | 8.7 | 8.9 |
| 1.4 | 7.5 | 7.7 | **7.8** | 7.9 | 8.1 | 8.2 | 8.4 | 8.5 | 8.7 | 8.9 | 9.1 |
| 1.5 | 7.7 | 7.8 | **8.0** | 8.1 | 8.2 | 8.4 | 8.6 | 8.7 | 8.9 | 9.1 | 9.4 |
| 1.6 | 7.9 | 8.0 | **8.1** | 8.3 | 8.4 | 8.6 | 8.7 | 8.9 | 9.1 | 9.3 | 9.6 |
| 2 | 8.4 | 8.5 | **8.6** | 8.8 | 8.9 | 9.1 | 9.3 | 9.5 | 9.7 | 9.9 | 10.2 |
| 3 | 9.1 | 9.2 | **9.4** | 9.6 | 9.7 | 9.9 | 10.1 | 10.3 | 10.5 | 10.8 | 11.0 |
| 4 | 9.5 | 9.6 | **9.8** | 10.0 | 10.1 | 10.3 | 10.5 | 10.7 | 11.0 | 11.2 | 11.5 |
| 5 | 9.7 | 9.9 | **10.0** | 10.2 | 10.4 | 10.6 | 10.8 | 11.0 | 11.2 | 11.5 | 11.8 |
| 6 | 9.9 | 10.0 | **10.2** | 10.4 | 10.5 | 10.7 | 10.9 | 11.2 | 11.4 | 11.7 | 11.9 |
| 7 | 10.0 | 10.1 | **10.3** | 10.5 | 10.7 | 10.8 | 11.1 | 11.3 | 11.5 | 11.8 | 12.1 |
| 8 | 10.1 | 10.2 | **10.4** | 10.5 | 10.7 | 10.9 | 11.1 | 11.4 | 11.6 | 11.9 | 12.1 |
| 9 | 10.1 | 10.3 | **10.4** | 10.6 | 10.8 | 11.0 | 11.2 | 11.4 | 11.7 | 11.9 | 12.2 |
| 10 | 10.2 | 10.3 | **10.5** | 10.7 | 10.8 | 11.0 | 11.2 | 11.5 | 11.7 | 12.0 | 12.3 |
| 15 | 10.3 | 10.4 | **10.6** | 10.8 | 11.0 | 11.2 | 11.4 | 11.6 | 11.8 | 12.1 | 12.4 |
| 20 | 10.4 | 10.5 | **10.7** | 10.8 | 11.0 | 11.2 | 11.4 | 11.7 | 11.9 | 12.2 | 12.4 |
| 40 | 10.4 | 10.6 | **10.7** | 10.9 | 11.1 | 11.3 | 11.5 | 11.7 | 12.0 | 12.2 | 12.5 |
| 100 | 10.5 | 10.6 | **10.8** | 11.0 | 11.1 | 11.3 | 11.5 | 11.8 | 12.0 | 12.3 | 12.6 |
| 1000 | 10.5 | 10.6 | **10.8** | 11.0 | 11.2 | 11.3 | 11.6 | 11.8 | 12.0 | 12.3 | 12.6 |

*In billions of light-years; flat universes in **bold**.*

## LOOKBACK DISTANCES*

HUBBLE CONSTANT= 65 OMEGA = 0.9

| REDSHIFT | LAMBDA 0.0 | **0.1** | 0.2 | 0.3 | 0.4 | 0.5 | 0.6 | 0.7 | 0.8 | 0.9 | 1.0 |
|---|---|---|---|---|---|---|---|---|---|---|---|
| 0.0 | 0.0 | **0.0** | 0.0 | 0.0 | 0.0 | 0.0 | 0.0 | 0.0 | 0.0 | 0.0 | 0.0 |
| 0.1 | 1.3 | **1.3** | 1.4 | 1.4 | 1.4 | 1.4 | 1.4 | 1.4 | 1.4 | 1.4 | 1.4 |
| 0.2 | 2.4 | **2.4** | 2.4 | 2.5 | 2.5 | 2.5 | 2.5 | 2.5 | 2.6 | 2.6 | 2.6 |
| 0.3 | 3.3 | **3.3** | 3.3 | 3.4 | 3.4 | 3.4 | 3.5 | 3.5 | 3.6 | 3.6 | 3.6 |
| 0.4 | 4.0 | **4.0** | 4.1 | 4.1 | 4.2 | 4.2 | 4.3 | 4.3 | 4.4 | 4.5 | 4.5 |
| 0.5 | 4.6 | **4.7** | 4.7 | 4.8 | 4.8 | 4.9 | 5.0 | 5.0 | 5.1 | 5.2 | 5.3 |
| 0.6 | 5.1 | **5.2** | 5.3 | 5.3 | 5.4 | 5.5 | 5.6 | 5.6 | 5.7 | 5.8 | 6.0 |
| 0.7 | 5.6 | **5.6** | 5.7 | 5.8 | 5.9 | 6.0 | 6.1 | 6.2 | 6.3 | 6.4 | 6.5 |
| 0.8 | 5.9 | **6.0** | 6.1 | 6.2 | 6.3 | 6.4 | 6.5 | 6.6 | 6.7 | 6.9 | 7.0 |
| 0.9 | 6.3 | **6.4** | 6.5 | 6.6 | 6.7 | 6.8 | 6.9 | 7.0 | 7.1 | 7.3 | 7.4 |
| 1.0 | 6.6 | **6.7** | 6.8 | 6.9 | 7.0 | 7.1 | 7.2 | 7.3 | 7.5 | 7.6 | 7.8 |
| 1.1 | 6.8 | **6.9** | 7.0 | 7.1 | 7.2 | 7.4 | 7.5 | 7.6 | 7.8 | 8.0 | 8.1 |
| 1.2 | 7.0 | **7.2** | 7.3 | 7.4 | 7.5 | 7.6 | 7.8 | 7.9 | 8.1 | 8.2 | 8.4 |
| 1.3 | 7.2 | **7.4** | 7.5 | 7.6 | 7.7 | 7.9 | 8.0 | 8.2 | 8.3 | 8.5 | 8.7 |
| 1.4 | 7.4 | **7.5** | 7.7 | 7.8 | 7.9 | 8.1 | 8.2 | 8.4 | 8.5 | 8.7 | 8.9 |
| 1.5 | 7.6 | **7.7** | 7.8 | 8.0 | 8.1 | 8.2 | 8.4 | 8.6 | 8.7 | 8.9 | 9.1 |
| 1.6 | 7.7 | **7.9** | 8.0 | 8.1 | 8.3 | 8.4 | 8.6 | 8.7 | 8.9 | 9.1 | 9.3 |
| 2 | 8.2 | **8.4** | 8.5 | 8.6 | 8.8 | 8.9 | 9.1 | 9.3 | 9.5 | 9.7 | 9.9 |
| 3 | 8.9 | **9.1** | 9.2 | 9.4 | 9.5 | 9.7 | 9.9 | 10.1 | 10.3 | 10.5 | 10.7 |
| 4 | 9.3 | **9.4** | 9.6 | 9.7 | 9.9 | 10.1 | 10.3 | 10.5 | 10.7 | 10.9 | 11.2 |
| 5 | 9.5 | **9.7** | 9.8 | 10.0 | 10.1 | 10.3 | 10.5 | 10.7 | 10.9 | 11.2 | 11.4 |
| 6 | 9.7 | **9.8** | 10.0 | 10.1 | 10.3 | 10.5 | 10.7 | 10.9 | 11.1 | 11.3 | 11.6 |
| 7 | 9.8 | **9.9** | 10.1 | 10.2 | 10.4 | 10.6 | 10.8 | 11.0 | 11.2 | 11.4 | 11.7 |
| 8 | 9.8 | **10.0** | 10.1 | 10.3 | 10.5 | 10.7 | 10.8 | 11.0 | 11.3 | 11.5 | 11.8 |
| 9 | 9.9 | **10.0** | 10.2 | 10.4 | 10.5 | 10.7 | 10.9 | 11.1 | 11.3 | 11.6 | 11.8 |
| 10 | 9.9 | **10.1** | 10.2 | 10.4 | 10.6 | 10.8 | 11.0 | 11.2 | 11.4 | 11.6 | 11.9 |
| 15 | 10.1 | **10.2** | 10.4 | 10.5 | 10.7 | 10.9 | 11.1 | 11.3 | 11.5 | 11.7 | 12.0 |
| 20 | 10.1 | **10.3** | 10.4 | 10.6 | 10.8 | 10.9 | 11.1 | 11.3 | 11.6 | 11.8 | 12.0 |
| 40 | 10.2 | **10.3** | 10.5 | 10.7 | 10.8 | 11.0 | 11.2 | 11.4 | 11.6 | 11.9 | 12.1 |
| 100 | 10.2 | **10.4** | 10.5 | 10.7 | 10.9 | 11.0 | 11.2 | 11.4 | 11.7 | 11.9 | 12.1 |
| 1000 | 10.2 | **10.4** | 10.5 | 10.7 | 10.9 | 11.1 | 11.2 | 11.4 | 11.7 | 11.9 | 12.2 |

**In billions of light-years; flat universes in* ***bold***.

## LOOKBACK DISTANCES *

HUBBLE CONSTANT= 65 OMEGA = 1.0

| | LAMBDA | | | | | | | | | | |
|---|---|---|---|---|---|---|---|---|---|---|---|
| REDSHIFT | **0.0** | 0.1 | 0.2 | 0.3 | 0.4 | 0.5 | 0.6 | 0.7 | 0.8 | 0.9 | 1.0 |
| 0.0 | **0.0** | 0.0 | 0.0 | 0.0 | 0.0 | 0.0 | 0.0 | 0.0 | 0.0 | 0.0 | 0.0 |
| 0.1 | **1.3** | 1.3 | 1.3 | 1.4 | 1.4 | 1.4 | 1.4 | 1.4 | 1.4 | 1.4 | 1.4 |
| 0.2 | **2.4** | 2.4 | 2.4 | 2.5 | 2.5 | 2.5 | 2.5 | 2.5 | 2.5 | 2.6 | 2.6 |
| 0.3 | **3.3** | 3.3 | 3.3 | 3.4 | 3.4 | 3.4 | 3.5 | 3.5 | 3.5 | 3.6 | 3.6 |
| 0.4 | **4.0** | 4.0 | 4.1 | 4.1 | 4.2 | 4.2 | 4.3 | 4.3 | 4.4 | 4.4 | 4.5 |
| 0.5 | **4.6** | 4.6 | 4.7 | 4.7 | 4.8 | 4.9 | 4.9 | 5.0 | 5.1 | 5.1 | 5.2 |
| 0.6 | **5.1** | 5.1 | 5.2 | 5.3 | 5.3 | 5.4 | 5.5 | 5.6 | 5.7 | 5.8 | 5.9 |
| 0.7 | **5.5** | 5.6 | 5.6 | 5.7 | 5.8 | 5.9 | 6.0 | 6.1 | 6.2 | 6.3 | 6.4 |
| 0.8 | **5.9** | 6.0 | 6.0 | 6.1 | 6.2 | 6.3 | 6.4 | 6.5 | 6.6 | 6.8 | 6.9 |
| 0.9 | **6.2** | 6.3 | 6.4 | 6.5 | 6.6 | 6.7 | 6.8 | 6.9 | 7.0 | 7.2 | 7.3 |
| 1.0 | **6.5** | 6.6 | 6.7 | 6.8 | 6.9 | 7.0 | 7.1 | 7.2 | 7.4 | 7.5 | 7.7 |
| 1.1 | **6.7** | 6.8 | 6.9 | 7.0 | 7.1 | 7.3 | 7.4 | 7.5 | 7.7 | 7.8 | 8.0 |
| 1.2 | **7.0** | 7.1 | 7.2 | 7.3 | 7.4 | 7.5 | 7.6 | 7.8 | 7.9 | 8.1 | 8.3 |
| 1.3 | **7.2** | 7.3 | 7.4 | 7.5 | 7.6 | 7.7 | 7.9 | 8.0 | 8.2 | 8.3 | 8.5 |
| 1.4 | **7.3** | 7.4 | 7.5 | 7.7 | 7.8 | 7.9 | 8.1 | 8.2 | 8.4 | 8.5 | 8.7 |
| 1.5 | **7.5** | 7.6 | 7.7 | 7.8 | 8.0 | 8.1 | 8.2 | 8.4 | 8.6 | 8.7 | 8.9 |
| 1.6 | **7.6** | 7.7 | 7.9 | 8.0 | 8.1 | 8.3 | 8.4 | 8.6 | 8.7 | 8.9 | 9.1 |
| 2 | **8.1** | 8.2 | 8.3 | 8.5 | 8.6 | 8.8 | 8.9 | 9.1 | 9.3 | 9.4 | 9.7 |
| 3 | **8.8** | 8.9 | 9.0 | 9.2 | 9.3 | 9.5 | 9.7 | 9.8 | 10.0 | 10.2 | 10.4 |
| 4 | **9.1** | 9.3 | 9.4 | 9.5 | 9.7 | 9.9 | 10.0 | 10.2 | 10.4 | 10.6 | 10.8 |
| 5 | **9.3** | 9.5 | 9.6 | 9.8 | 9.9 | 10.1 | 10.3 | 10.4 | 10.6 | 10.9 | 11.1 |
| 6 | **9.5** | 9.6 | 9.8 | 9.9 | 10.1 | 10.2 | 10.4 | 10.6 | 10.8 | 11.0 | 11.2 |
| 7 | **9.6** | 9.7 | 9.9 | 10.0 | 10.2 | 10.3 | 10.5 | 10.7 | 10.9 | 11.1 | 11.3 |
| 8 | **9.7** | 9.8 | 9.9 | 10.1 | 10.2 | 10.4 | 10.6 | 10.8 | 11.0 | 11.2 | 11.4 |
| 9 | **9.7** | 9.8 | 10.0 | 10.1 | 10.3 | 10.5 | 10.6 | 10.8 | 11.0 | 11.2 | 11.5 |
| 10 | **9.8** | 9.9 | 10.0 | 10.2 | 10.3 | 10.5 | 10.7 | 10.9 | 11.1 | 11.3 | 11.5 |
| 15 | **9.9** | 10.0 | 10.2 | 10.3 | 10.5 | 10.6 | 10.8 | 11.0 | 11.2 | 11.4 | 11.6 |
| 20 | **9.9** | 10.1 | 10.2 | 10.4 | 10.5 | 10.7 | 10.9 | 11.0 | 11.2 | 11.5 | 11.7 |
| 40 | **10.0** | 10.1 | 10.3 | 10.4 | 10.6 | 10.7 | 10.9 | 11.1 | 11.3 | 11.5 | 11.8 |
| 100 | **10.0** | 10.2 | 10.3 | 10.4 | 10.6 | 10.8 | 11.0 | 11.1 | 11.3 | 11.6 | 11.8 |
| 1000 | **10.0** | 10.2 | 10.3 | 10.5 | 10.6 | 10.8 | 11.0 | 11.2 | 11.4 | 11.6 | 11.8 |

**In billions of light-years; flat universes in* ***bold***.

## LOOKBACK DISTANCES*

HUBBLE CONSTANT= 65 OMEGA = 1.5

| | LAMBDA | | | | | | | | | | |
|---|---|---|---|---|---|---|---|---|---|---|---|
| REDSHIFT | 0.0 | 0.1 | 0.2 | 0.3 | 0.4 | 0.5 | 0.6 | 0.7 | 0.8 | 0.9 | 1.0 |
| 0.0 | 0.0 | 0.0 | 0.0 | 0.0 | 0.0 | 0.0 | 0.0 | 0.0 | 0.0 | 0.0 | 0.0 |
| 0.1 | 1.3 | 1.3 | 1.3 | 1.3 | 1.3 | 1.3 | 1.4 | 1.4 | 1.4 | 1.4 | 1.4 |
| 0.2 | 2.4 | 2.4 | 2.4 | 2.4 | 2.4 | 2.4 | 2.5 | 2.5 | 2.5 | 2.5 | 2.5 |
| 0.3 | 3.2 | 3.2 | 3.2 | 3.3 | 3.3 | 3.3 | 3.3 | 3.4 | 3.4 | 3.4 | 3.5 |
| 0.4 | 3.8 | 3.9 | 3.9 | 4.0 | 4.0 | 4.0 | 4.1 | 4.1 | 4.2 | 4.2 | 4.3 |
| 0.5 | 4.4 | 4.4 | 4.5 | 4.5 | 4.6 | 4.6 | 4.7 | 4.8 | 4.8 | 4.9 | 5.0 |
| 0.6 | 4.9 | 4.9 | 5.0 | 5.0 | 5.1 | 5.2 | 5.2 | 5.3 | 5.4 | 5.4 | 5.5 |
| 0.7 | 5.3 | 5.3 | 5.4 | 5.4 | 5.5 | 5.6 | 5.7 | 5.7 | 5.8 | 5.9 | 6.0 |
| 0.8 | 5.6 | 5.7 | 5.7 | 5.8 | 5.9 | 5.9 | 6.0 | 6.1 | 6.2 | 6.3 | 6.4 |
| 0.9 | 5.9 | 6.0 | 6.0 | 6.1 | 6.2 | 6.3 | 6.4 | 6.4 | 6.5 | 6.6 | 6.8 |
| 1.0 | 6.1 | 6.2 | 6.3 | 6.4 | 6.5 | 6.5 | 6.6 | 6.7 | 6.8 | 6.9 | 7.1 |
| 1.1 | 6.4 | 6.4 | 6.5 | 6.6 | 6.7 | 6.8 | 6.9 | 7.0 | 7.1 | 7.2 | 7.3 |
| 1.2 | 6.6 | 6.6 | 6.7 | 6.8 | 6.9 | 7.0 | 7.1 | 7.2 | 7.3 | 7.4 | 7.6 |
| 1.3 | 6.7 | 6.8 | 6.9 | 7.0 | 7.1 | 7.2 | 7.3 | 7.4 | 7.5 | 7.6 | 7.8 |
| 1.4 | 6.9 | 7.0 | 7.1 | 7.2 | 7.3 | 7.4 | 7.5 | 7.6 | 7.7 | 7.8 | 7.9 |
| 1.5 | 7.0 | 7.1 | 7.2 | 7.3 | 7.4 | 7.5 | 7.6 | 7.7 | 7.8 | 8.0 | 8.1 |
| 1.6 | 7.2 | 7.2 | 7.3 | 7.4 | 7.5 | 7.6 | 7.8 | 7.9 | 8.0 | 8.1 | 8.3 |
| 2 | 7.6 | 7.7 | 7.8 | 7.9 | 8.0 | 8.1 | 8.2 | 8.3 | 8.4 | 8.6 | 8.7 |
| 3 | 8.1 | 8.2 | 8.3 | 8.5 | 8.6 | 8.7 | 8.8 | 8.9 | 9.1 | 9.2 | 9.4 |
| 4 | 8.4 | 8.5 | 8.7 | 8.8 | 8.9 | 9.0 | 9.1 | 9.3 | 9.4 | 9.5 | 9.7 |
| 5 | 8.6 | 8.7 | 8.8 | 8.9 | 9.1 | 9.2 | 9.3 | 9.4 | 9.6 | 9.7 | 9.9 |
| 6 | 8.7 | 8.8 | 9.0 | 9.1 | 9.2 | 9.3 | 9.4 | 9.6 | 9.7 | 9.9 | 10.0 |
| 7 | 8.8 | 8.9 | 9.0 | 9.2 | 9.3 | 9.4 | 9.5 | 9.7 | 9.8 | 9.9 | 10.1 |
| 8 | 8.9 | 9.0 | 9.1 | 9.2 | 9.3 | 9.5 | 9.6 | 9.7 | 9.9 | 10.0 | 10.2 |
| 9 | 8.9 | 9.0 | 9.1 | 9.3 | 9.4 | 9.5 | 9.6 | 9.8 | 9.9 | 10.1 | 10.2 |
| 10 | 9.0 | 9.1 | 9.2 | 9.3 | 9.4 | 9.5 | 9.7 | 9.8 | 9.9 | 10.1 | 10.2 |
| 15 | 9.1 | 9.2 | 9.3 | 9.4 | 9.5 | 9.6 | 9.8 | 9.9 | 10.0 | 10.2 | 10.3 |
| 20 | 9.1 | 9.2 | 9.3 | 9.4 | 9.6 | 9.7 | 9.8 | 9.9 | 10.1 | 10.2 | 10.4 |
| 40 | 9.2 | 9.3 | 9.4 | 9.5 | 9.6 | 9.7 | 9.9 | 10.0 | 10.1 | 10.3 | 10.4 |
| 100 | 9.2 | 9.3 | 9.4 | 9.5 | 9.6 | 9.8 | 9.9 | 10.0 | 10.2 | 10.3 | 10.5 |
| 1000 | 9.2 | 9.3 | 9.4 | 9.5 | 9.6 | 9.8 | 9.9 | 10.0 | 10.2 | 10.3 | 10.5 |

***In billions of light-years.**

# LOOKBACK DISTANCES*

HUBBLE CONSTANT= 65 OMEGA = 2.0

| REDSHIFT | LAMBDA 0.0 | 0.1 | 0.2 | 0.3 | 0.4 | 0.5 | 0.6 | 0.7 | 0.8 | 0.9 | 1.0 |
|---|---|---|---|---|---|---|---|---|---|---|---|
| 0.0 | 0.0 | 0.0 | 0.0 | 0.0 | 0.0 | 0.0 | 0.0 | 0.0 | 0.0 | 0.0 | 0.0 |
| 0.1 | 1.3 | 1.3 | 1.3 | 1.3 | 1.3 | 1.3 | 1.3 | 1.3 | 1.4 | 1.4 | 1.4 |
| 0.2 | 2.3 | 2.3 | 2.3 | 2.4 | 2.4 | 2.4 | 2.4 | 2.4 | 2.4 | 2.5 | 2.5 |
| 0.3 | 3.1 | 3.1 | 3.1 | 3.2 | 3.2 | 3.2 | 3.3 | 3.3 | 3.3 | 3.3 | 3.4 |
| 0.4 | 3.7 | 3.8 | 3.8 | 3.8 | 3.9 | 3.9 | 3.9 | 4.0 | 4.0 | 4.1 | 4.1 |
| 0.5 | 4.3 | 4.3 | 4.3 | 4.4 | 4.4 | 4.5 | 4.5 | 4.6 | 4.6 | 4.7 | 4.7 |
| 0.6 | 4.7 | 4.7 | 4.8 | 4.8 | 4.9 | 4.9 | 5.0 | 5.0 | 5.1 | 5.2 | 5.2 |
| 0.7 | 5.0 | 5.1 | 5.2 | 5.2 | 5.3 | 5.3 | 5.4 | 5.5 | 5.5 | 5.6 | 5.7 |
| 0.8 | 5.4 | 5.4 | 5.5 | 5.5 | 5.6 | 5.7 | 5.7 | 5.8 | 5.9 | 5.9 | 6.0 |
| 0.9 | 5.6 | 5.7 | 5.7 | 5.8 | 5.9 | 5.9 | 6.0 | 6.1 | 6.2 | 6.3 | 6.3 |
| 1.0 | 5.9 | 5.9 | 6.0 | 6.1 | 6.1 | 6.2 | 6.3 | 6.4 | 6.4 | 6.5 | 6.6 |
| 1.1 | 6.1 | 6.1 | 6.2 | 6.3 | 6.3 | 6.4 | 6.5 | 6.6 | 6.7 | 6.7 | 6.8 |
| 1.2 | 6.2 | 6.3 | 6.4 | 6.5 | 6.5 | 6.6 | 6.7 | 6.8 | 6.9 | 7.0 | 7.0 |
| 1.3 | 6.4 | 6.5 | 6.5 | 6.6 | 6.7 | 6.8 | 6.9 | 6.9 | 7.0 | 7.1 | 7.2 |
| 1.4 | 6.5 | 6.6 | 6.7 | 6.8 | 6.8 | 6.9 | 7.0 | 7.1 | 7.2 | 7.3 | 7.4 |
| 1.5 | 6.7 | 6.7 | 6.8 | 6.9 | 7.0 | 7.1 | 7.1 | 7.2 | 7.3 | 7.4 | 7.5 |
| 1.6 | 6.8 | 6.9 | 6.9 | 7.0 | 7.1 | 7.2 | 7.3 | 7.4 | 7.5 | 7.6 | 7.7 |
| 2 | 7.1 | 7.2 | 7.3 | 7.4 | 7.5 | 7.6 | 7.6 | 7.7 | 7.8 | 8.0 | 8.1 |
| 3 | 7.7 | 7.7 | 7.8 | 7.9 | 8.0 | 8.1 | 8.2 | 8.3 | 8.4 | 8.5 | 8.6 |
| 4 | 7.9 | 8.0 | 8.1 | 8.2 | 8.3 | 8.4 | 8.5 | 8.6 | 8.7 | 8.8 | 8.9 |
| 5 | 8.1 | 8.2 | 8.3 | 8.3 | 8.4 | 8.5 | 8.6 | 8.7 | 8.8 | 9.0 | 9.1 |
| 6 | 8.2 | 8.3 | 8.4 | 8.5 | 8.5 | 8.6 | 8.7 | 8.8 | 9.0 | 9.1 | 9.2 |
| 7 | 8.3 | 8.4 | 8.4 | 8.5 | 8.6 | 8.7 | 8.8 | 8.9 | 9.0 | 9.1 | 9.3 |
| 8 | 8.3 | 8.4 | 8.5 | 8.6 | 8.7 | 8.8 | 8.9 | 9.0 | 9.1 | 9.2 | 9.3 |
| 9 | 8.4 | 8.4 | 8.5 | 8.6 | 8.7 | 8.8 | 8.9 | 9.0 | 9.1 | 9.2 | 9.4 |
| 10 | 8.4 | 8.5 | 8.6 | 8.6 | 8.7 | 8.8 | 8.9 | 9.0 | 9.2 | 9.3 | 9.4 |
| 15 | 8.5 | 8.6 | 8.6 | 8.7 | 8.8 | 8.9 | 9.0 | 9.1 | 9.2 | 9.4 | 9.5 |
| 20 | 8.5 | 8.6 | 8.7 | 8.8 | 8.9 | 9.0 | 9.1 | 9.2 | 9.3 | 9.4 | 9.5 |
| 40 | 8.6 | 8.6 | 8.7 | 8.8 | 8.9 | 9.0 | 9.1 | 9.2 | 9.3 | 9.4 | 9.6 |
| 100 | 8.6 | 8.7 | 8.8 | 8.8 | 8.9 | 9.0 | 9.1 | 9.2 | 9.3 | 9.5 | 9.6 |
| 1000 | 8.6 | 8.7 | 8.8 | 8.8 | 8.9 | 9.0 | 9.1 | 9.2 | 9.4 | 9.5 | 9.6 |

***In billions of light-years.**

# TABLE 5
# THE AGE OF THE UNIVERSE*

The universe's age depends on three cosmological parameters: the Hubble constant, the universe's present expansion rate; omega (Ω), the universe's matter density; and lambda (λ), the strength of the cosmological constant. The following tables give the universe's age for nearly 2,000 different combinations of these parameters. The numbers were calculated using Ned Wright's cosmological calculator (http://www.astro.ucla.edu/~wright/CosmoCalc.html), which includes the energy of the cosmic microwave background, even when omega is 0.

To use these tables:

1. Choose a value for the Hubble constant, in kilometers per second per megaparsec, and turn to the table headed by that number.
2. Choose a value for omega, and look in the far left-hand column for that value of omega.
3. Choose a value for lambda. Follow the row rightward until you reach the column headed by the value of lambda you want. There you will find the age of the universe, in billions of years.

For example, to compute the universe's age for a Hubble constant of 65, omega of 0.3, and lambda of 0.7, turn to the table that says "Hubble Constant = 65," then look down the column that says "Omega" until you find the row that starts with "0.3." Then move rightward along that row until you reach the column headed by a lambda of "0.7." There the table says "14.5," indicating that the age of such a universe is 14.5 billion years.

Flat universes are in **bold**. Open universes lie to the upper left of the bold diagonal, closed universes to the lower right. Open universes expand forever. Closed universes with lambda = 0 collapse, but most closed universes with nonzero lambda expand forever—see Table 3.

HUBBLE CONSTANT = 40

| | LAMBDA | | | | | | | | | | |
|---|---|---|---|---|---|---|---|---|---|---|---|
| OMEGA | 0.0 | 0.1 | 0.2 | 0.3 | 0.4 | 0.5 | 0.6 | 0.7 | 0.8 | 0.9 | 1.0 |
| 0.0 | 24.1 | 24.9 | 25.8 | 26.9 | 28.2 | 29.7 | 31.6 | 34.1 | 37.6 | 43.4 | **59.0** |
| 0.1 | 21.9 | 22.5 | 23.1 | 23.9 | 24.7 | 25.6 | 26.6 | 27.8 | 29.3 | **31.1** | 33.5 |
| 0.2 | 20.7 | 21.1 | 21.7 | 22.2 | 22.9 | 23.6 | 24.4 | 25.2 | **26.3** | 27.5 | 28.9 |
| 0.3 | 19.8 | 20.2 | 20.6 | 21.1 | 21.6 | 22.2 | 22.8 | **23.5** | 24.3 | 25.2 | 26.3 |
| 0.4 | 19.0 | 19.4 | 19.8 | 20.2 | 20.7 | 21.2 | **21.7** | 22.3 | 22.9 | 23.7 | 24.5 |
| 0.5 | 18.4 | 18.7 | 19.1 | 19.5 | 19.9 | **20.3** | 20.8 | 21.3 | 21.8 | 22.5 | 23.1 |
| 0.6 | 17.9 | 18.2 | 18.5 | 18.8 | **19.2** | 19.6 | 20.0 | 20.5 | 20.9 | 21.5 | 22.1 |
| 0.7 | 17.4 | 17.7 | 18.0 | **18.3** | 18.6 | 19.0 | 19.3 | 19.8 | 20.2 | 20.7 | 21.2 |
| 0.8 | 17.0 | 17.3 | **17.5** | 17.8 | 18.1 | 18.4 | 18.8 | 19.1 | 19.5 | 20.0 | 20.4 |
| 0.9 | 16.6 | **16.9** | 17.1 | 17.4 | 17.7 | 18.0 | 18.3 | 18.6 | 19.0 | 19.3 | 19.8 |
| 1.0 | **16.3** | 16.5 | 16.7 | 17.0 | 17.3 | 17.5 | 17.8 | 18.1 | 18.4 | 18.8 | 19.2 |
| 1.5 | 14.9 | 15.1 | 15.3 | 15.5 | 15.7 | 15.9 | 16.1 | 16.3 | 16.5 | 16.8 | 17.0 |
| 2.0 | 14.0 | 14.1 | 14.2 | 14.4 | 14.5 | 14.7 | 14.8 | 15.0 | 15.2 | 15.4 | 15.6 |

**In billions of years; flat universes in* ***bold***

## THE AGE OF THE UNIVERSE*

| | Hubble Constant = 45 | | | | | | | | | | |
|---|---|---|---|---|---|---|---|---|---|---|---|
| | Lambda | | | | | | | | | | |
| Omega | 0.0 | 0.1 | 0.2 | 0.3 | 0.4 | 0.5 | 0.6 | 0.7 | 0.8 | 0.9 | 1.0 |
| 0.0 | 21.4 | 22.2 | 23.0 | 24.0 | 25.1 | 26.5 | 28.2 | 30.4 | 33.6 | 38.9 | **53.7** |
| 0.1 | 19.5 | 20.0 | 20.6 | 21.2 | 21.9 | 22.7 | 23.7 | 24.8 | 26.1 | **27.7** | 29.8 |
| 0.2 | 18.4 | 18.8 | 19.3 | 19.8 | 20.3 | 21.0 | 21.7 | 22.4 | **23.4** | 24.4 | 25.7 |
| 0.3 | 17.6 | 17.9 | 18.3 | 18.8 | 19.2 | 19.7 | 20.3 | **20.9** | 21.6 | 22.4 | 23.4 |
| 0.4 | 16.9 | 17.2 | 17.6 | 18.0 | 18.4 | 18.8 | **19.3** | 19.8 | 20.4 | 21.0 | 21.8 |
| 0.5 | 16.4 | 16.7 | 17.0 | 17.3 | 17.7 | **18.0** | 18.5 | 18.9 | 19.4 | 20.0 | 20.6 |
| 0.6 | 15.9 | 16.2 | 16.4 | 16.7 | **17.1** | 17.4 | 17.8 | 18.2 | 18.6 | 19.1 | 19.6 |
| 0.7 | 15.5 | 15.7 | 16.0 | **16.3** | 16.6 | 16.9 | 17.2 | 17.6 | 17.9 | 18.4 | 18.8 |
| 0.8 | 15.1 | 15.3 | **15.6** | 15.8 | 16.1 | 16.4 | 16.7 | 17.0 | 17.4 | 17.7 | 18.1 |
| 0.9 | 14.8 | **15.0** | 15.2 | 15.5 | 15.7 | 16.0 | 16.2 | 16.5 | 16.9 | 17.2 | 17.6 |
| 1.0 | **14.5** | 14.7 | 14.9 | 15.1 | 15.3 | 15.6 | 15.8 | 16.1 | 16.4 | 16.7 | 17.0 |
| 1.5 | 13.3 | 13.4 | 13.6 | 13.8 | 13.9 | 14.1 | 14.3 | 14.5 | 14.7 | 14.9 | 15.1 |
| 2.0 | 12.4 | 12.5 | 12.6 | 12.8 | 12.9 | 13.1 | 13.2 | 13.4 | 13.5 | 13.7 | 13.8 |

| | Hubble Constant = 50 | | | | | | | | | | |
|---|---|---|---|---|---|---|---|---|---|---|---|
| | Lambda | | | | | | | | | | |
| Omega | 0.0 | 0.1 | 0.2 | 0.3 | 0.4 | 0.5 | 0.6 | 0.7 | 0.8 | 0.9 | 1.0 |
| 0.0 | 19.3 | 20.0 | 20.7 | 21.6 | 22.6 | 23.9 | 25.4 | 27.5 | 30.4 | 35.2 | **49.4** |
| 0.1 | 17.5 | 18.0 | 18.5 | 19.1 | 19.7 | 20.5 | 21.3 | 22.3 | 23.5 | **24.9** | 26.8 |
| 0.2 | 16.5 | 16.9 | 17.3 | 17.8 | 18.3 | 18.9 | 19.5 | 20.2 | **21.0** | 22.0 | 23.1 |
| 0.3 | 15.8 | 16.1 | 16.5 | 16.9 | 17.3 | 17.8 | 18.3 | **18.8** | 19.5 | 20.2 | 21.0 |
| 0.4 | 15.2 | 15.5 | 15.8 | 16.2 | 16.5 | 16.9 | **17.4** | 17.8 | 18.4 | 18.9 | 19.6 |
| 0.5 | 14.7 | 15.0 | 15.3 | 15.6 | 15.9 | **16.2** | 16.6 | 17.0 | 17.5 | 18.0 | 18.5 |
| 0.6 | 14.3 | 14.5 | 14.8 | 15.1 | **15.4** | 15.7 | 16.0 | 16.4 | 16.8 | 17.2 | 17.7 |
| 0.7 | 13.9 | 14.2 | 14.4 | **14.6** | 14.9 | 15.2 | 15.5 | 15.8 | 16.2 | 16.5 | 16.9 |
| 0.8 | 13.6 | 13.8 | **14.0** | 14.3 | 14.5 | 14.8 | 15.0 | 15.3 | 15.6 | 16.0 | 16.3 |
| 0.9 | 13.3 | **13.5** | 13.7 | 13.9 | 14.1 | 14.4 | 14.6 | 14.9 | 15.2 | 15.5 | 15.8 |
| 1.0 | **13.0** | 13.2 | 13.4 | 13.6 | 13.8 | 14.0 | 14.3 | 14.5 | 14.8 | 15.0 | 15.3 |
| 1.5 | 12.0 | 12.1 | 12.2 | 12.4 | 12.5 | 12.7 | 12.9 | 13.0 | 13.2 | 13.4 | 13.6 |
| 2.0 | 11.2 | 11.3 | 11.4 | 11.5 | 11.6 | 11.7 | 11.9 | 12.0 | 12.2 | 12.3 | 12.5 |

**In billions of years; flat universes in* ***bold***.

## THE AGE OF THE UNIVERSE*

HUBBLE CONSTANT= 55

| OMEGA \ LAMBDA | 0.0 | 0.1 | 0.2 | 0.3 | 0.4 | 0.5 | 0.6 | 0.7 | 0.8 | 0.9 | 1.0 |
|---|---|---|---|---|---|---|---|---|---|---|---|
| 0.0 | 17.6 | 18.2 | 18.9 | 19.7 | 20.6 | 21.8 | 23.2 | 25.0 | 27.7 | 32.2 | **45.7** |
| 0.1 | 15.9 | 16.4 | 16.8 | 17.4 | 18.0 | 18.6 | 19.4 | 20.3 | 21.3 | **22.7** | 24.4 |
| 0.2 | 15.0 | 15.4 | 15.8 | 16.2 | 16.6 | 17.2 | 17.7 | 18.4 | **19.1** | 20.0 | 21.0 |
| 0.3 | 14.4 | 14.7 | 15.0 | 15.3 | 15.7 | 16.2 | 16.6 | **17.1** | 17.7 | 18.4 | 19.1 |
| 0.4 | 13.8 | 14.1 | 14.4 | 14.7 | 15.0 | 15.4 | **15.8** | 16.2 | 16.7 | 17.2 | 17.8 |
| 0.5 | 13.4 | 13.6 | 13.9 | 14.2 | 14.5 | **14.8** | 15.1 | 15.5 | 15.9 | 16.3 | 16.8 |
| 0.6 | 13.0 | 13.2 | 13.5 | 13.7 | **14.0** | 14.2 | 14.6 | 14.9 | 15.2 | 15.6 | 16.1 |
| 0.7 | 12.7 | 12.9 | 13.1 | **13.3** | 13.5 | 13.8 | 14.1 | 14.4 | 14.7 | 15.0 | 15.4 |
| 0.8 | 12.4 | 12.6 | **12.8** | 13.0 | 13.2 | 13.4 | 13.7 | 13.9 | 14.2 | 14.5 | 14.8 |
| 0.9 | 12.1 | **12.3** | 12.5 | 12.6 | 12.8 | 13.1 | 13.3 | 13.5 | 13.8 | 14.1 | 14.4 |
| 1.0 | **11.8** | 12.0 | 12.2 | 12.4 | 12.5 | 12.7 | 13.0 | 13.2 | 13.4 | 13.7 | 13.9 |
| 1.5 | 10.9 | 11.0 | 11.1 | 11.3 | 11.4 | 11.5 | 11.7 | 11.8 | 12.0 | 12.2 | 12.4 |
| 2.0 | 10.1 | 10.2 | 10.3 | 10.5 | 10.6 | 10.7 | 10.8 | 10.9 | 11.1 | 11.2 | 11.3 |

HUBBLE CONSTANT = 60

| OMEGA \ LAMBDA | 0.0 | 0.1 | 0.2 | 0.3 | 0.4 | 0.5 | 0.6 | 0.7 | 0.8 | 0.9 | 1.0 |
|---|---|---|---|---|---|---|---|---|---|---|---|
| 0.0 | 16.1 | 16.7 | 17.3 | 18.1 | 18.9 | 20.0 | 21.3 | 23.0 | 25.5 | 29.7 | **42.6** |
| 0.1 | 14.6 | 15.0 | 15.4 | 15.9 | 16.5 | 17.1 | 17.8 | 18.6 | 19.6 | **20.8** | 22.4 |
| 0.2 | 13.8 | 14.1 | 14.5 | 14.8 | 15.3 | 15.7 | 16.2 | 16.8 | **17.5** | 18.3 | 19.3 |
| 0.3 | 13.2 | 13.5 | 13.7 | 14.1 | 14.4 | 14.8 | 15.2 | **15.7** | 16.2 | 16.8 | 17.5 |
| 0.4 | 12.7 | 12.9 | 13.2 | 13.5 | 13.8 | 14.1 | **14.5** | 14.9 | 15.3 | 15.8 | 16.3 |
| 0.5 | 12.3 | 12.5 | 12.7 | 13.0 | 13.2 | **13.5** | 13.9 | 14.2 | 14.6 | 15.0 | 15.4 |
| 0.6 | 11.9 | 12.1 | 12.3 | 12.6 | **12.8** | 13.1 | 13.3 | 13.6 | 14.0 | 14.3 | 14.7 |
| 0.7 | 11.6 | 11.8 | 12.0 | **12.2** | 12.4 | 12.7 | 12.9 | 13.2 | 13.5 | 13.8 | 14.1 |
| 0.8 | 11.3 | 11.5 | **11.7** | 11.9 | 12.1 | 12.3 | 12.5 | 12.8 | 13.0 | 13.3 | 13.6 |
| 0.9 | 11.1 | **11.2** | 11.4 | 11.6 | 11.8 | 12.0 | 12.2 | 12.4 | 12.6 | 12.9 | 13.2 |
| 1.0 | **10.9** | 11.0 | 11.2 | 11.3 | 11.5 | 11.7 | 11.9 | 12.1 | 12.3 | 12.5 | 12.8 |
| 1.5 | 10.0 | 10.1 | 10.2 | 10.3 | 10.4 | 10.6 | 10.7 | 10.9 | 11.0 | 11.2 | 11.3 |
| 2.0 | 9.3 | 9.4 | 9.5 | 9.6 | 9.7 | 9.8 | 9.9 | 10.0 | 10.1 | 10.3 | 10.4 |

**In billions of years; flat universes in* ***bold***.

## THE AGE OF THE UNIVERSE*

HUBBLE CONSTANT = 65

| OMEGA \ LAMBDA | 0.0 | 0.1 | 0.2 | 0.3 | 0.4 | 0.5 | 0.6 | 0.7 | 0.8 | 0.9 | 1.0 |
|---|---|---|---|---|---|---|---|---|---|---|---|
| 0.0 | 14.9 | 15.4 | 16.0 | 16.7 | 17.5 | 18.5 | 19.7 | 21.3 | 23.6 | 27.5 | **39.9** |
| 0.1 | 13.5 | 13.9 | 14.3 | 14.7 | 15.2 | 15.8 | 16.4 | 17.2 | 18.1 | **19.2** | 20.6 |
| 0.2 | 12.7 | 13.0 | 13.3 | 13.7 | 14.1 | 14.5 | 15.0 | 15.5 | **16.2** | 16.9 | 17.8 |
| 0.3 | 12.2 | 12.4 | 12.7 | 13.0 | 13.3 | 13.7 | 14.1 | **14.5** | 15.0 | 15.5 | 16.2 |
| 0.4 | 11.7 | 11.9 | 12.2 | 12.4 | 12.7 | 13.0 | **13.4** | 13.7 | 14.1 | 14.6 | 15.1 |
| 0.5 | 11.3 | 11.5 | 11.8 | 12.0 | 12.2 | **12.5** | 12.8 | 13.1 | 13.4 | 13.8 | 14.3 |
| 0.6 | 11.0 | 11.2 | 11.4 | 11.6 | **11.8** | 12.1 | 12.3 | 12.6 | 12.9 | 13.2 | 13.6 |
| 0.7 | 10.7 | 10.9 | 11.1 | **11.3** | 11.5 | 11.7 | 11.9 | 12.2 | 12.4 | 12.7 | 13.0 |
| 0.8 | 10.5 | 10.6 | **10.8** | 11.0 | 11.2 | 11.3 | 11.6 | 11.8 | 12.0 | 12.3 | 12.6 |
| 0.9 | 10.2 | **10.4** | 10.5 | 10.7 | 10.9 | 11.1 | 11.2 | 11.4 | 11.7 | 11.9 | 12.2 |
| 1.0 | **10.0** | 10.2 | 10.3 | 10.5 | 10.6 | 10.8 | 11.0 | 11.2 | 11.4 | 11.6 | 11.8 |
| 1.5 | 9.2 | 9.3 | 9.4 | 9.5 | 9.6 | 9.8 | 9.9 | 10.0 | 10.2 | 10.3 | 10.5 |
| 2.0 | 8.6 | 8.7 | 8.8 | 8.8 | 8.9 | 9.0 | 9.1 | 9.2 | 9.4 | 9.5 | 9.6 |

HUBBLE CONSTANT = 70

| OMEGA \ LAMBDA | 0.0 | 0.1 | 0.2 | 0.3 | 0.4 | 0.5 | 0.6 | 0.7 | 0.8 | 0.9 | 1.0 |
|---|---|---|---|---|---|---|---|---|---|---|---|
| 0.0 | 13.8 | 14.3 | 14.9 | 15.5 | 16.3 | 17.2 | 18.3 | 19.8 | 21.9 | 25.6 | **37.6** |
| 0.1 | 12.5 | 12.9 | 13.2 | 13.7 | 14.1 | 14.6 | 15.2 | 15.9 | 16.8 | **17.8** | 19.2 |
| 0.2 | 11.8 | 12.1 | 12.4 | 12.7 | 13.1 | 13.5 | 13.9 | 14.4 | **15.0** | 15.7 | 16.5 |
| 0.3 | 11.3 | 11.5 | 11.8 | 12.1 | 12.4 | 12.7 | 13.1 | **13.5** | 13.9 | 14.4 | 15.0 |
| 0.4 | 10.9 | 11.1 | 11.3 | 11.5 | 11.8 | 12.1 | **12.4** | 12.7 | 13.1 | 13.5 | 14.0 |
| 0.5 | 10.5 | 10.7 | 10.9 | 11.1 | 11.4 | **11.6** | 11.9 | 12.2 | 12.5 | 12.8 | 13.2 |
| 0.6 | 10.2 | 10.4 | 10.6 | 10.8 | **11.0** | 11.2 | 11.4 | 11.7 | 12.0 | 12.3 | 12.6 |
| 0.7 | 10.0 | 10.1 | 10.3 | **10.5** | 10.6 | 10.8 | 11.1 | 11.3 | 11.5 | 11.8 | 12.1 |
| 0.8 | 9.7 | 9.9 | **10.0** | 10.2 | 10.4 | 10.5 | 10.7 | 10.9 | 11.2 | 11.4 | 11.7 |
| 0.9 | 9.5 | **9.6** | 9.8 | 9.9 | 10.1 | 10.3 | 10.4 | 10.6 | 10.8 | 11.1 | 11.3 |
| 1.0 | **9.3** | 9.4 | 9.6 | 9.7 | 9.9 | 10.0 | 10.2 | 10.4 | 10.5 | 10.7 | 11.0 |
| 1.5 | 8.5 | 8.6 | 8.7 | 8.8 | 9.0 | 9.1 | 9.2 | 9.3 | 9.4 | 9.6 | 9.7 |
| 2.0 | 8.0 | 8.1 | 8.1 | 8.2 | 8.3 | 8.4 | 8.5 | 8.6 | 8.7 | 8.8 | 8.9 |

**In billions of years; flat universes in* ***bold.***

## THE AGE OF THE UNIVERSE*

**Hubble Constant = 75**

| Omega \ Lambda | 0.0 | 0.1 | 0.2 | 0.3 | 0.4 | 0.5 | 0.6 | 0.7 | 0.8 | 0.9 | 1.0 |
|---|---|---|---|---|---|---|---|---|---|---|---|
| 0.0 | 12.9 | 13.4 | 13.9 | 14.5 | 15.2 | 16.0 | 17.1 | 18.5 | 20.5 | 24.0 | **35.5** |
| 0.1 | 11.7 | 12.0 | 12.4 | 12.7 | 13.2 | 13.7 | 14.2 | 14.9 | 15.7 | **16.6** | 17.9 |
| 0.2 | 11.0 | 11.3 | 11.6 | 11.9 | 12.2 | 12.6 | 13.0 | 13.5 | **14.0** | 14.7 | 15.4 |
| 0.3 | 10.5 | 10.8 | 11.0 | 11.3 | 11.5 | 11.8 | 12.2 | **12.6** | 13.0 | 13.5 | 14.0 |
| 0.4 | 10.2 | 10.3 | 10.6 | 10.8 | 11.0 | 11.3 | **11.6** | 11.9 | 12.2 | 12.6 | 13.1 |
| 0.5 | 9.8 | 10.0 | 10.2 | 10.4 | 10.6 | **10.8** | 11.1 | 11.4 | 11.7 | 12.0 | 12.4 |
| 0.6 | 9.5 | 9.7 | 9.9 | 10.1 | **10.2** | 10.4 | 10.7 | 10.9 | 11.2 | 11.5 | 11.8 |
| 0.7 | 9.3 | 9.4 | 9.6 | **9.8** | 9.9 | 10.1 | 10.3 | 10.5 | 10.8 | 11.0 | 11.3 |
| 0.8 | 9.1 | 9.2 | **9.4** | 9.5 | 9.7 | 9.8 | 10.0 | 10.2 | 10.4 | 10.6 | 10.9 |
| 0.9 | 8.9 | **9.0** | 9.1 | 9.3 | 9.4 | 9.6 | 9.7 | 9.9 | 10.1 | 10.3 | 10.5 |
| 1.0 | **8.7** | 8.8 | 8.9 | 9.1 | 9.2 | 9.3 | 9.5 | 9.7 | 9.8 | 10.0 | 10.2 |
| 1.5 | 8.0 | 8.1 | 8.2 | 8.3 | 8.4 | 8.5 | 8.6 | 8.7 | 8.8 | 8.9 | 9.1 |
| 2.0 | 7.4 | 7.5 | 7.6 | 7.7 | 7.7 | 7.8 | 7.9 | 8.0 | 8.1 | 8.2 | 8.3 |

**Hubble Constant = 80**

| Omega \ Lambda | 0.0 | 0.1 | 0.2 | 0.3 | 0.4 | 0.5 | 0.6 | 0.7 | 0.8 | 0.9 | 1.0 |
|---|---|---|---|---|---|---|---|---|---|---|---|
| 0.0 | 12.1 | 12.5 | 13.0 | 13.6 | 14.2 | 15.0 | 16.0 | 17.4 | 19.3 | 22.5 | **33.7** |
| 0.1 | 11.0 | 11.3 | 11.6 | 11.9 | 12.4 | 12.8 | 13.3 | 14.0 | 14.7 | **15.6** | 16.8 |
| 0.2 | 10.3 | 10.6 | 10.8 | 11.1 | 11.4 | 11.8 | 12.2 | 12.6 | **13.1** | 13.7 | 14.5 |
| 0.3 | 9.9 | 10.1 | 10.3 | 10.6 | 10.8 | 11.1 | 11.4 | **11.8** | 12.2 | 12.6 | 13.2 |
| 0.4 | 9.5 | 9.7 | 9.9 | 10.1 | 10.3 | 10.6 | **10.9** | 11.1 | 11.5 | 11.8 | 12.3 |
| 0.5 | 9.2 | 9.4 | 9.5 | 9.7 | 9.9 | **10.2** | 10.4 | 10.6 | 10.9 | 11.2 | 11.6 |
| 0.6 | 8.9 | 9.1 | 9.3 | 9.4 | **9.6** | 9.8 | 10.0 | 10.2 | 10.5 | 10.7 | 11.0 |
| 0.7 | 8.7 | 8.8 | 9.0 | **9.2** | 9.3 | 9.5 | 9.7 | 9.9 | 10.1 | 10.3 | 10.6 |
| 0.8 | 8.5 | 8.6 | **8.8** | 8.9 | 9.1 | 9.2 | 9.4 | 9.6 | 9.8 | 10.0 | 10.2 |
| 0.9 | 8.3 | **8.4** | 8.6 | 8.7 | 8.8 | 9.0 | 9.1 | 9.3 | 9.5 | 9.7 | 9.9 |
| 1.0 | **8.1** | 8.3 | 8.4 | 8.5 | 8.6 | 8.8 | 8.9 | 9.1 | 9.2 | 9.4 | 9.6 |
| 1.5 | 7.5 | 7.6 | 7.6 | 7.7 | 7.8 | 7.9 | 8.0 | 8.1 | 8.3 | 8.4 | 8.5 |
| 2.0 | 7.0 | 7.0 | 7.1 | 7.2 | 7.3 | 7.3 | 7.4 | 7.5 | 7.6 | 7.7 | 7.8 |

**In billions of years; flat universes in **bold**.*

## THE AGE OF THE UNIVERSE*

HUBBLE CONSTANT = 85

| OMEGA \ LAMBDA | 0.0 | 0.1 | 0.2 | 0.3 | 0.4 | 0.5 | 0.6 | 0.7 | 0.8 | 0.9 | 1.0 |
|---|---|---|---|---|---|---|---|---|---|---|---|
| 0.0 | 11.4 | 11.8 | 12.3 | 12.8 | 13.4 | 14.2 | 15.1 | 16.4 | 18.1 | 21.2 | **32.1** |
| 0.1 | 10.3 | 10.6 | 10.9 | 11.2 | 11.6 | 12.1 | 12.6 | 13.1 | 13.8 | **14.7** | 15.8 |
| 0.2 | 9.7 | 10.0 | 10.2 | 10.5 | 10.8 | 11.1 | 11.5 | 11.9 | **12.4** | 12.9 | 13.6 |
| 0.3 | 9.3 | 9.5 | 9.7 | 9.9 | 10.2 | 10.5 | 10.8 | **11.1** | 11.5 | 11.9 | 12.4 |
| 0.4 | 9.0 | 9.1 | 9.3 | 9.5 | 9.7 | 10.0 | **10.2** | 10.5 | 10.8 | 11.1 | 11.5 |
| 0.5 | 8.7 | 8.8 | 9.0 | 9.2 | 9.4 | **9.6** | 9.8 | 10.0 | 10.3 | 10.6 | 10.9 |
| 0.6 | 8.4 | 8.6 | 8.7 | 8.9 | **9.0** | 9.2 | 9.4 | 9.6 | 9.9 | 10.1 | 10.4 |
| 0.7 | 8.2 | 8.3 | 8.5 | **8.6** | 8.8 | 8.9 | 9.1 | 9.3 | 9.5 | 9.7 | 10.0 |
| 0.8 | 8.0 | 8.1 | **8.3** | 8.4 | 8.5 | 8.7 | 8.8 | 9.0 | 9.2 | 9.4 | 9.6 |
| 0.9 | 7.8 | **7.9** | 8.1 | 8.2 | 8.3 | 8.5 | 8.6 | 8.8 | 8.9 | 9.1 | 9.3 |
| 1.0 | **7.7** | 7.8 | 7.9 | 8.0 | 8.1 | 8.2 | 8.4 | 8.5 | 8.7 | 8.8 | 9.0 |
| 1.5 | 7.0 | 7.1 | 7.2 | 7.3 | 7.4 | 7.5 | 7.6 | 7.7 | 7.8 | 7.9 | 8.0 |
| 2.0 | 6.6 | 6.6 | 6.7 | 6.8 | 6.8 | 6.9 | 7.0 | 7.1 | 7.2 | 7.2 | 7.3 |

HUBBLE CONSTANT = 90

| OMEGA \ LAMBDA | 0.0 | 0.1 | 0.2 | 0.3 | 0.4 | 0.5 | 0.6 | 0.7 | 0.8 | 0.9 | 1.0 |
|---|---|---|---|---|---|---|---|---|---|---|---|
| 0.0 | 10.8 | 11.2 | 11.6 | 12.1 | 12.7 | 13.4 | 14.3 | 15.5 | 17.2 | 20.1 | **30.6** |
| 0.1 | 9.8 | 10.0 | 10.3 | 10.6 | 11.0 | 11.4 | 11.9 | 12.4 | 13.1 | **13.9** | 14.9 |
| 0.2 | 9.2 | 9.4 | 9.6 | 9.9 | 10.2 | 10.5 | 10.8 | 11.2 | **11.7** | 12.2 | 12.9 |
| 0.3 | 8.8 | 9.0 | 9.2 | 9.4 | 9.6 | 9.9 | 10.2 | **10.5** | 10.8 | 11.2 | 11.7 |
| 0.4 | 8.5 | 8.6 | 8.8 | 9.0 | 9.2 | 9.4 | **9.6** | 9.9 | 10.2 | 10.5 | 10.9 |
| 0.5 | 8.2 | 8.3 | 8.5 | 8.7 | 8.8 | **9.0** | 9.2 | 9.5 | 9.7 | 10.0 | 10.3 |
| 0.6 | 8.0 | 8.1 | 8.2 | 8.4 | **8.5** | 8.7 | 8.9 | 9.1 | 9.3 | 9.6 | 9.8 |
| 0.7 | 7.7 | 7.9 | 8.0 | **8.1** | 8.3 | 8.4 | 8.6 | 8.8 | 9.0 | 9.2 | 9.4 |
| 0.8 | 7.6 | 7.7 | **7.8** | 7.9 | 8.1 | 8.2 | 8.3 | 8.5 | 8.7 | 8.9 | 9.1 |
| 0.9 | 7.4 | **7.5** | 7.6 | 7.7 | 7.9 | 8.0 | 8.1 | 8.3 | 8.4 | 8.6 | 8.8 |
| 1.0 | **7.2** | 7.3 | 7.4 | 7.6 | 7.7 | 7.8 | 7.9 | 8.1 | 8.2 | 8.4 | 8.5 |
| 1.5 | 6.6 | 6.7 | 6.8 | 6.9 | 7.0 | 7.0 | 7.1 | 7.2 | 7.3 | 7.5 | 7.6 |
| 2.0 | 6.2 | 6.3 | 6.3 | 6.4 | 6.5 | 6.5 | 6.6 | 6.7 | 6.8 | 6.8 | 6.9 |

**In billions of years; flat universes in* ***bold.***

## THE AGE OF THE UNIVERSE*

HUBBLE CONSTANT= 95

| | LAMBDA | | | | | | | | | | |
|---|---|---|---|---|---|---|---|---|---|---|---|
| OMEGA | 0.0 | 0.1 | 0.2 | 0.3 | 0.4 | 0.5 | 0.6 | 0.7 | 0.8 | 0.9 | 1.0 |
| 0.0 | 10.2 | 10.6 | 11.0 | 11.5 | 12.0 | 12.7 | 13.5 | 14.7 | 16.3 | 19.1 | **29.3** |
| 0.1 | 9.2 | 9.5 | 9.8 | 10.1 | 10.4 | 10.8 | 11.2 | 11.8 | 12.4 | **13.1** | 14.1 |
| 0.2 | 8.7 | 8.9 | 9.1 | 9.4 | 9.6 | 9.9 | 10.3 | 10.6 | **11.1** | 11.6 | 12.2 |
| 0.3 | 8.3 | 8.5 | 8.7 | 8.9 | 9.1 | 9.4 | 9.6 | **9.9** | 10.3 | 10.6 | 11.1 |
| 0.4 | 8.0 | 8.2 | 8.3 | 8.5 | 8.7 | 8.9 | **9.1** | 9.4 | 9.7 | 10.0 | 10.3 |
| 0.5 | 7.8 | 7.9 | 8.0 | 8.2 | 8.4 | **8.6** | 8.8 | 9.0 | 9.2 | 9.5 | 9.8 |
| 0.6 | 7.5 | 7.7 | 7.8 | 7.9 | **8.1** | 8.3 | 8.4 | 8.6 | 8.8 | 9.0 | 9.3 |
| 0.7 | 7.3 | 7.5 | 7.6 | **7.7** | 7.8 | 8.0 | 8.2 | 8.3 | 8.5 | 8.7 | 8.9 |
| 0.8 | 7.2 | 7.3 | **7.4** | 7.5 | 7.6 | 7.8 | 7.9 | 8.1 | 8.2 | 8.4 | 8.6 |
| 0.9 | 7.0 | **7.1** | 7.2 | 7.3 | 7.4 | 7.6 | 7.7 | 7.8 | 8.0 | 8.1 | 8.3 |
| 1.0 | **6.9** | 7.0 | 7.1 | 7.2 | 7.3 | 7.4 | 7.5 | 7.6 | 7.8 | 7.9 | 8.1 |
| 1.5 | 6.3 | 6.4 | 6.4 | 6.5 | 6.6 | 6.7 | 6.8 | 6.9 | 7.0 | 7.1 | 7.2 |
| 2.0 | 5.9 | 5.9 | 6.0 | 6.1 | 6.1 | 6.2 | 6.3 | 6.3 | 6.4 | 6.5 | 6.6 |

HUBBLE CONSTANT= 100

| | LAMBDA | | | | | | | | | | |
|---|---|---|---|---|---|---|---|---|---|---|---|
| OMEGA | 0.0 | 0.1 | 0.2 | 0.3 | 0.4 | 0.5 | 0.6 | 0.7 | 0.8 | 0.9 | 1.0 |
| 0.0 | 9.7 | 10.1 | 10.4 | 10.9 | 11.4 | 12.1 | 12.9 | 13.9 | 15.5 | 18.1 | **28.1** |
| 0.1 | 8.8 | 9.0 | 9.3 | 9.6 | 9.9 | 10.2 | 10.7 | 11.2 | 11.8 | **12.5** | 13.4 |
| 0.2 | 8.3 | 8.5 | 8.7 | 8.9 | 9.2 | 9.4 | 9.8 | 10.1 | **10.5** | 11.0 | 11.6 |
| 0.3 | 7.9 | 8.1 | 8.3 | 8.4 | 8.7 | 8.9 | 9.1 | **9.4** | 9.7 | 10.1 | 10.5 |
| 0.4 | 7.6 | 7.8 | 7.9 | 8.1 | 8.3 | 8.5 | **8.7** | 8.9 | 9.2 | 9.5 | 9.8 |
| 0.5 | 7.4 | 7.5 | 7.6 | 7.8 | 8.0 | **8.1** | 8.3 | 8.5 | 8.7 | 9.0 | 9.3 |
| 0.6 | 7.2 | 7.3 | 7.4 | 7.5 | **7.7** | 7.8 | 8.0 | 8.2 | 8.4 | 8.6 | 8.8 |
| 0.7 | 7.0 | 7.1 | 7.2 | **7.3** | 7.5 | 7.6 | 7.7 | 7.9 | 8.1 | 8.3 | 8.5 |
| 0.8 | 6.8 | 6.9 | **7.0** | 7.1 | 7.2 | 7.4 | 7.5 | 7.7 | 7.8 | 8.0 | 8.2 |
| 0.9 | 6.7 | **6.8** | 6.8 | 7.0 | 7.1 | 7.2 | 7.3 | 7.4 | 7.6 | 7.7 | 7.9 |
| 1.0 | **6.5** | 6.6 | 6.7 | 6.8 | 6.9 | 7.0 | 7.1 | 7.3 | 7.4 | 7.5 | 7.7 |
| 1.5 | 6.0 | 6.0 | 6.1 | 6.2 | 6.3 | 6.3 | 6.4 | 6.5 | 6.6 | 6.7 | 6.8 |
| 2.0 | 5.6 | 5.6 | 5.7 | 5.8 | 5.8 | 5.9 | 5.9 | 6.0 | 6.1 | 6.2 | 6.2 |

***In billions of years; flat universes in bold.**

# GLOSSARY

*I don't pretend to understand the universe—it's a great deal bigger than I am.*

—THOMAS CARLYLE, RECORDED BY WILLIAM ALLINGHAM

**Absolute Magnitude** A measure of intrinsic brightness, the amount of light an object emits, as opposed to apparent magnitude, which merely indicates how bright it looks. Absolute magnitude is the apparent magnitude an object would have if it were viewed from a distance of 32.6 light-years (10 parsecs). The Sun's absolute magnitude is +4.83. More powerful stars have *lower* absolute magnitudes—for example, Sirius has an absolute magnitude of +1.45, and the even more powerful Canopus has a *negative* absolute magnitude, –5.5. A difference of 5 in absolute magnitude equals a luminosity difference of 100. Thus, a galaxy with absolute magnitude –21, such as Andromeda, emits 100 times more light than one with absolute magnitude –16, such as the Small Magellanic Cloud. See also **Apparent Magnitude**.

**Andromeda Galaxy** The closest giant galaxy to our own and the largest member of the Local Group, the Andromeda Galaxy is a spiral somewhat larger and brighter than the Milky Way. It is 2.4 million light-years from Earth, has an absolute visual magnitude of –21.1, and shines with the strength of 25 billion solar luminosities. Several Local Group galaxies orbit Andromeda, the largest being M33. Andromeda is moving toward the Milky Way and in several billion years may collide with it; if so, the collision will likely transform the two spirals into a single giant elliptical galaxy.

**Angstrom** A unit of wavelength used especially by optical astronomers. One angstrom equals 0.00000001 centimeter. Violet light has a wavelength of about 4,000 angstroms, blue light 4,400 angstroms, yellow light 5,500 angstroms, red light 6,500 angstroms, and infrared light longer than about 7,000 angstroms. A wavelength of 10,000 angstroms equals 1 micron, the unit used by most infrared astronomers.

**Antimatter** A mirror of normal matter, antimatter really exists. In ordinary matter, protons are positively charged and electrons negatively charged; but in antimatter, the protons are negative and the electrons positive. When matter meets antimatter, they annihilate in a burst of energy. The absence of such annihilation indicates that all galaxies in the observable universe consist primarily of matter. This matter-antimatter imbalance arose shortly after the big bang, when matter particles slightly outnumbered their antimatter peers: for every billion antimatter particles, there were a billion and one matter particles. The billion matter particles annihilated the billion antimatter particles, leaving only one matter particle in a billion to survive. You are made of those matter particles now—slender indeed was the thread on which our existence hung.

**Apparent Magnitude** A measure of how bright a celestial object looks: the brighter, the *less* is the apparent magnitude. Thus Sirius (apparent magnitude −1.44) looks brighter than Procyon (apparent magnitude +0.40), which looks brighter than Aldebaran (apparent magnitude +0.87). An object's apparent magnitude depends on its luminosity—its output of light—as well as on its distance and the amount of light-absorbing dust in front of it. A nearby star can look brighter than a more luminous one. For example, Sirius looks brighter than Aldebaran only because Sirius is closer; Aldebaran actually outshines it, so if the two were equidistant from Earth, Aldebaran would look brighter. See also **Absolute Magnitude**.

* * *

**Barred Spiral Galaxy** A spiral galaxy whose central region, or bulge, is oval rather than round. Some barred spirals look like the letter S. The Milky Way is a barred spiral galaxy.

**Baryonic** Matter composed of ordinary particles—protons, neutrons, and electrons. The greater the universe's density of baryonic material, the faster the nuclear reactions after the big bang created hydrogen, helium, and lithium. Hence, the primordial abundances of these three elements reveal the density of baryonic matter in the universe: about 5 percent of what is needed to close the universe, if the Hubble constant is 65. Since the universe's total density of matter far exceeds this, most matter in the universe must be *non*baryonic.

**Bias** The concept that galaxies form, or at least shine, only in the densest regions of the universe—in galaxy clusters and groups—even though matter is spread throughout space. Bias reconciles predictions that omega be 1 with observations of galaxy clusters suggesting a much lower omega. However, if recent observations are correct, the universe has little bias; the voids are as empty as they look.

**Big Bang Theory** The proposal that the universe began in an extremely dense state some 15 billion years ago and has expanded ever since. Key observational evidence for the big bang theory includes the expansion of the universe (which resulted from the big bang's initial kick), the high abundance of helium (most of which was manufactured just after the big bang), the uniform ages of giant elliptical galaxies (which formed shortly after the big bang), the existence of radio galaxies and quasars at large redshifts (which indicate the universe has evolved during its life), the cosmic microwave background (which is the big bang's afterglow), and the concordance between the universe's age, as derived from the Hubble constant, omega, and lambda, and the age of its oldest stars.

**Blackbody** The spectrum emitted by an object in thermodynamic equilibrium: it shines most intensely at one wavelength, and its intensity falls off slowly at longer wavelengths, sharply at shorter ones. The hotter a blackbody, the shorter is the wavelength at which its radiation peaks. Stars are approximate blackbodies, and the universe is a near-perfect blackbody. Contrary to their misleading name, blackbodies are *not* black: they radiate profusely, and have nothing to do with black holes.

**Black Hole** An object with such great gravity that nothing, not even light, can escape. Giant black holes reside at the centers of most large galaxies, including our own.

**Blueshift** The scrunching up of light waves, normally caused by an object's motion toward Earth.

**Blue Supergiant** A luminous blue star, like Rigel, that has evolved off the main sequence. Blue supergiants dot spiral arms and shine so brightly that they can be seen in other galaxies.

**Bottom-Up Theory** The proposal that galaxies formed first, then attracted one another through their gravity to form clusters and finally superclusters. The bottom-up theory became wedded to cold dark matter, slow-moving particles that gather into even small galaxies. See also **Top-Down Theory**.

**Brown Dwarf** A star born with less than 8 percent of the Sun's mass, too little to sustain the hydrogen-to-helium nuclear reaction that powers the Sun and other main-sequence stars. When young, a brown dwarf shines red, by converting gravitational energy into heat; then it cools, fades, and turns black.

* * *

**Cepheid** A yellow supergiant star that expands and contracts, typically every few days, weeks, or months. During this pulsation cycle, the star brightens fast, then fades slowly. The variability of the first Cepheids, Eta Aquilae and Delta Cephei, was discovered in 1784 by Edward Pigott and John Goodricke, respectively. In 1907, Henrietta Leavitt discovered the period-luminosity relation that makes these stars excellent yardsticks: the longer the pulsation period, the greater is the star's intrinsic brightness. Measuring the star's period therefore yields its intrinsic brightness; comparing the latter with the star's apparent brightness then reveals the distance to the star and its host galaxy.

**Closed Universe** A universe in which omega plus lambda exceeds 1. *If* lambda is 0, then a closed universe eventually collapses; however, if lambda is sufficiently large, a closed universe can expand forever, as an open universe does. Nevertheless, a closed universe is finite, parallel lines converge, and a triangle's angles add up to more than 180 degrees.

**Cluster** 1. A collection of stars. See **Globular Cluster**; **Open Cluster**. 2. A collection of hundreds or thousands of galaxies, such as the Virgo cluster, whose gravity corrals the members.

**COBE** Pronounced so as to rhyme with "Moby," the COBE (COsmic Background Explorer) satellite established the blackbody spectrum of the cosmic microwave background and detected its ripples, the seeds of the superclusters and voids that crisscross the cosmos today.

**Cold Dark Matter** Slow-moving nonbaryonic particles that may constitute most of the universe's matter. WIMPs are cold dark matter.

**Coma Cluster** A rich cluster of galaxies in the constellation Coma Berenices, nearly due north of the plane of the Galaxy. If the Hubble constant is 65, the Coma cluster is 350 million light-years from Earth, in the same direction as the Virgo cluster but nearly 300 million light-years beyond. The Coma cluster is much richer than Virgo. In 1933, Fritz Zwicky studied the Coma cluster and reported the first strong evidence for dark matter: the Coma galaxies moved fast, so the cluster must possess unseen material whose gravity binds the member galaxies to the cluster.

**Cosmic Microwave Background** The big bang's afterglow, the redshifted remains of the light that broke free of matter some 300,000 years after the big bang and has streamed toward Earth nearly unimpeded ever since.

**Cosmological Constant** Designated lambda ($\lambda$), the cosmological constant pervades space, exerts a repulsive force, and hastens the universe's expansion. Along with omega, the cosmological constant helps to flatten the universe. A lambda-dominated universe expands forever, forever faster.

**Cosmological Principle** The assumption that the universe looks much the same to observers in all galaxies and in particular that galaxies speckle space rather uniformly. Strict adherence to the cosmological principle prevented many astronomers from accepting the reality of enormous cosmic structures—the superclusters and voids. See also **Perfect Cosmological Principle.**

**Cosmological Redshift** The shift to longer wavelengths of an object's spectrum caused by the expansion of space between the object and Earth. The farther the object, the more space the light has traveled through to reach Earth, so the more that space has stretched the light waves; hence, as Edwin Hubble found, the farther a galaxy, the greater is its redshift. If the galaxy is also moving through space, then that motion will produce a Doppler shift that will add to or subtract from the cosmological redshift.

**Cosmology** The study of the overall universe—in particular, its origin, evolution, and fate.

**Critical Density** The minimum density of matter that the universe must have in order for gravity to halt the universe's expansion and cause the cosmos to collapse. The critical density is approximately $10^{-29}$ grams per cubic centimeter, more tenuous than the lunar atmosphere; the exact number depends on the Hubble constant (see Table 2 in the Appendix). Omega is the universe's actual density of matter divided by the critical density. Observations suggest that the universe's matter density falls far shy of the critical density, so the universe will expand forever.

* * *

**Dark Halo** The enormous envelope of dark material that surrounds and outweighs the luminous part of a galaxy. Most of the Milky Way's dark halo lies beyond the Sun's orbit around the Galactic center.

**Dark Matter** Material that emits little or no light. Still, dark matter tugs other matter via gravity, allowing astronomers to detect it. Most matter is dark, some baryonic, most nonbaryonic. In particular, giant galaxies are embedded in enormous halos of dark matter, and galaxy clusters bear large quantities, too. Nevertheless, the universe hasn't enough dark matter to halt its expansion.

**Dark Night Sky** See **Olbers' Paradox**.

**Deuterium** Hydrogen-2, or heavy hydrogen, the rare isotope of hydrogen possessing both a proton and a neutron. Deuterium was manufactured moments after the big bang, but most got transformed into helium. The greater the universe's density of baryonic material, the faster the primordial nuclear reactions destroyed deuterium, so the less that should exist.

**Doppler Shift** The shift of an object's spectrum caused by its movement through space. Movement toward Earth scrunches up the light waves, producing a blueshift, so called because blue light has a short wavelength; movement away from Earth stretches the light waves, causing a redshift, because red light has a long wavelength. In the conventional cosmological interpretation, the redshift that the typical galaxy exhibits is *not* a Doppler shift, because it is caused not by movement through space but by the expansion of space itself—see **Cosmological Redshift**.

**Dwarf Spheroidal Galaxy** A faint, diffuse elliptical galaxy with stars spread out from one another. Dwarf spheroidals outnumber all other galactic types, and at least nine orbit the Milky Way. In 1982, Marc Aaronson discovered that one of them, Draco, bore large quantities of dark matter, indicating that dark matter can't be neutrinos, because a small galaxy can't hold on to them.

* * *

**Eclipsing Binary** Two stars that revolve around each other so that one periodically blocks the other's light. By observing the stars' brightness, temperature, and orbital velocity, astronomers can deduce their distance—without knowing the parallax.

**Electron** The negatively charged subatomic particle that appears in every neutral atom and weighs only 1/1,836 as much as a proton.

**Elliptical Galaxy** A round or ellipsoidal galaxy that typically lacks gas, dust, and new stars. Elliptical galaxies range from giants like M87 in the Virgo cluster to lesser galaxies like M32, a satellite of Andromeda, to minuscule dwarf spheroidals like Draco, a satellite of the Milky Way. Because elliptical galaxies lack youthful stars, they don't sport the Cepheids that could mark their distances; they do, however, bear fainter yardsticks, such as RR Lyrae stars.

* * *

**Flatness Problem** A problem that inflation theory solves: if omega, the universe's matter density, is not exactly 1, then it deviates from 1 as the universe expands, making it unlikely that it would still be close to 1 all these many years after the big bang. According to inflation theory, the hyperexpansion right after the big bang flattened the universe, driving omega (or omega plus lambda) to 1.

**Flat Universe** As predicted by inflation theory, a universe in which omega plus lambda equals 1. Such a universe is infinite and expands forever; parallel lines stay parallel and triangles' angles add up to 180 degrees.

**Fornax Cluster** A compact galaxy cluster in the dim constellation Fornax, appearing in the south during the northern hemisphere's autumn. Although the Fornax cluster boasts the stunning barred spiral NGC 1365, most other prominent members are elliptical galaxies. Fornax is about as far from Earth as the Virgo cluster is but lies in the opposite direction, south of the Galactic plane.

* * *

**Galaxy** An enormous collection of millions, billions, or sometimes trillions of stars. The Milky Way is our Galaxy; when capitalized, "Galaxy" and "Galactic" refer to the Milky Way.

**Galaxy Cluster** Hundreds or thousands of galaxies held to one another by gravity. Examples include the Virgo, Coma, and Fornax clusters. Cluster galaxies move so fast relative to one another that they would escape the cluster were it not for the gravitational attraction of dark matter.

**Galaxy Group** A gathering of galaxies—usually a few bright ones and a few dozen faint ones—like the Local Group, to which the Milky Way belongs. Other nearby groups include the Sculptor, Maffei, M81, and Centaurus groups.

**Gamma Rays** The most energetic form of electromagnetic radiation, even more powerful than x-rays.

**Giant Star** A large, bright star. Most giant stars are roughly a hundred times more luminous than the Sun. The five brightest giants in Earth's sky are Arcturus, Capella, Beta Centauri, Aldebaran, and Pollux. In a few billion years, after the Sun's center depletes its hydrogen fuel, the Sun will swell into a giant.

**Globular Cluster** A compact star cluster, typically housing hundreds of thousands of stars. In the Milky Way, globular clusters are among the most ancient objects known, so they constrain the age of the universe.

**Gravitational Lensing** Gravity's bending of light, predicted by both Newtonian and Einsteinian theory. Gravitational lensing can produce multiple images of a single object and make it look brighter than it really is. "Lensing" usually refers to instances when an entire galaxy bends light; "microlensing" refers to cases when an individual object within a galaxy does so. Lensing of distant quasars provides one measure of the Hubble constant; microlensing of stars in the Magellanic Clouds yields information about the Milky Way's dark halo.

**Great Attractor** A cluster of superclusters centered in the southern constellation Norma, 250 million light-years away, that tries to pull the Milky Way toward it. Its great rival is the Pisces-Perseus supercluster, which sits on the other side of the sky and also tries to pull us toward it. Because of the expansion of space, however, we are actually moving away from both objects.

**Group** A small gathering of galaxies, held to one another by gravity. The

Local Group has thirty-six known galaxies, including Andromeda, the Milky Way, M33, and the Magellanic Clouds.

* * *

**H II Region** A region of hydrogen gas ionized and set aglow by hot stars. Most H II regions mark birthplaces of stars; the Orion Nebula is the best example.

**Halo** A structure embedding part of a galaxy. The Milky Way has two halos: the stellar halo, most of whose stars reside closer to the Galactic center than does the Sun; and the dark halo, most of which resides farther from the Galactic center than does the Sun. See **Dark Halo; Stellar Halo.**

**Helium** Rare on Earth, helium is the second lightest and second most common chemical element in the universe. About one atom in eleven is helium—primarily helium-4, whose nucleus has two protons and two neutrons. Most helium in the universe arose just after the big bang, so the primordial helium abundance constrains big bang models; however, terrestrial helium comes from the decay of radioactive elements like uranium.

**Horizon** The edge of the observable universe—the boundary set by the finite speed of light. For example, if the universe is 14.5 billion years old, then the horizon has a lookback distance of 14.5 billion light-years, because light from more distant objects hasn't had time to reach the Earth. Observers in other galaxies have different horizons and see different parts of the universe, just as people in different cities have different horizons and see different parts of the world.

**Horizon Problem** A problem solved by inflation theory: the universe looks similar in opposite directions, yet these regions are so far apart that they couldn't have smoothed themselves out during its lifetime. According to inflation theory, these disparate regions were actually much closer together, so that they did smooth themselves out; inflation then separated them to enormous distances.

**Hot Dark Matter** Swift nonbaryonic particles, like neutrinos, traveling near the speed of light. Hot dark matter probably constitutes only a tiny fraction of the universe's total mass.

**Hubble Constant** The universe's present expansion rate. For example, if the Hubble constant is 65 kilometers per second per megaparsec, then a galaxy one megaparsec farther than another will typically have a redshift 65 kilometers per second greater (one megaparsec is 3.26 million light-years). The Hubble constant converts galaxy redshifts into distances. It also indicates the approximate age of the universe: the higher the Hubble constant, the faster the universe has expanded since the big bang, so the sooner it attained its present size and the younger it must be.

**Hubble Time** The age of the universe if the expansion rate has neither accelerated nor decelerated—that is, if omega and lambda are both 0. Mathematically, the Hubble time is simply the inverse of the Hubble constant, so the greater the Hubble constant, the less is the Hubble time. For example, if the Hubble constant is 65, the Hubble time is 15 billion years.

**Hydrogen** The lightest and most common chemical element in the cosmos, hydrogen accounts for 91 percent of all atoms. Most is hydrogen-1, whose nucleus has one proton and no neutrons; some is hydrogen-2, or deuterium, with one proton and one neutron; hydrogen-3, or tritium, is radioactive. Most hydrogen formed in the big bang. Neutral hydrogen, called H I, has one electron and emits radio waves that are 21 centimeters long. If the electron escapes, the hydrogen is said to be ionized and is designated H II. Regions of ionized hydrogen (H II regions), such as the Orion Nebula, glow red or magenta, because after the electron rejoins the atom, it can emit red light. H II regions usually mark sites of star formation. Two hydrogen atoms can join to form molecular hydrogen ($H_2$), the most common interstellar molecule.

* * *

**IC** Designation for objects in the *Index Catalogue*, which lists star clusters, nebulae, and galaxies. See also **NGC**.

**IC 4182** A small galaxy in the constellation Canes Venatici, 16 million light-years from Earth, where a type Ia supernova exploded in 1937.

**Inflation** The hypothetical hyperexpansion a split second after the big bang that flattened the universe. Inflation theory predicts that the universe is flat and solves the flatness and horizon problems.

**Infrared** Electromagnetic radiation whose waves are somewhat longer than those of visible light. Infrared radiation penetrates gas and dust much more readily than light does.

**Intrinsic Brightness** The amount of light a star radiates, as opposed to how bright it looks. See **Absolute Magnitude**.

**Ionized Hydrogen Region** Also called H II region, hydrogen gas that glows red or magenta because the radiation of hot stars has stripped electrons off the hydrogen atoms. Ionized hydrogen regions, such as the Orion Nebula, usually give birth to new stars.

**Irregular Galaxy** An amorphous galaxy, neither spiral nor elliptical. Most irregular galaxies are small but sport gas clouds that create new stars. The two nearest irregular galaxies are the Magellanic Clouds.

* * *

**Kelvin** A temperature scale on which absolute zero is 0, the freezing point of water is 273, and the boiling point 373. The universe has a temperature of 2.73 Kelvin, or –454.76 degrees Fahrenheit.

**Kilometer Per Second** A unit of speed. One kilometer per second is 2,237 miles per hour; a plane this fast could cross the United States from east to west in about an hour.

* * *

**Lambda (λ)** The numerical strength of the cosmological constant, the repulsive "antigravity" force that pervades space and accelerates the universe's expansion. Lambda increases age estimates of the universe.

**Large Magellanic Cloud** The biggest and brightest of the many satellites that revolve around the Milky Way, the Large Magellanic Cloud is an irregular galaxy giving birth to new stars. It is 160,000 light-years from Earth, has an absolute visual magnitude of –18.1, and emits one tenth as much light as the Milky Way. Although visible to the unaided eye, it lies so far south that most northern observers can't see it.

**Lensing** See **Gravitational Lensing**.

**Light Elements** The elements—hydrogen, helium, and lithium—created moments after the big bang. Because their production depended on the density of baryonic material, their primordial abundances yield the amount of baryonic matter in the universe.

**Light-Year** The amount of distance that light speeds through in a year—5.88 trillion miles, or 9.46 trillion kilometers. The nearest star system beyond the Sun, Alpha Centauri, is just over 4 light-years from Earth, and the nearest giant galaxy, Andromeda, is 2.4 million light-years distant. By a lucky coincidence, the number of Earth-Sun distances in a light-year equals the number of inches in a mile, to an accuracy of much better than 1 percent; so if the distance between the Sun and the Earth were one inch, then on the same scale one light-year would equal one mile.

**Lithium** The third lightest chemical element, after hydrogen and helium. The lithium manufactured moments after the big bang provides clues about the universe's density of baryonic matter.

**Local Group** The collection of thirty-six nearby galaxies that is held together by gravity. The Local Group's largest members, by far, are the Andromeda Galaxy and the Milky Way. Most Local Group galaxies are satellites of Andromeda, such as M33, M32, and NGC 205, or of the Milky Way, such as the Magellanic Clouds and the Ursa Minor and Draco dwarfs. Most Local Group galaxies reside within 3 million light-years of the Milky Way. The three largest—Andromeda, the Milky Way, and M33—are spirals.

**Local Supercluster** The vast assemblage of thousands of galaxies to which the Milky Way belongs. The Milky Way resides on its outskirts; the Virgo cluster marks its center.

**Lookback Distance** When astronomers give the distance of a galaxy or quasar, they traditionally mean its lookback distance: how far the light traveled to reach the Earth. Thus, we see a galaxy with a lookback distance of 8 billion light-years as it was 8 billion years ago. However, when the galaxy sent out that light, it was closer than the lookback distance, because the universe is expanding, and today it is farther. Unfortunately, the ambiguity allows an unscrupulous scientist to garner headlines by announcing, for example, the discovery of a quasar 26 billion light-years away—when in fact its lookback distance is only 13 billion light-years.

**Luminosity** Intrinsic brightness—how much light an object sends into space. A luminous star can *look* faint if it's distant; a weak star can *look* bright if it's nearby. See also **Absolute Magnitude; Apparent Magnitude.**

* * *

**M** Designation for the 103 objects in French comet hunter Charles Messier's catalogue, which he completed in 1781. Most famous galaxies—such as Andromeda (M31), the Whirlpool (M51), and the brightest members of the Virgo cluster—bear M numbers. See also **IC** and **NGC.**

**M31** The closest giant galaxy to ours. See **Andromeda Galaxy.**

**M32** A small elliptical galaxy that orbits the Andromeda Galaxy.

**M33** A delicate spiral galaxy in the constellation Triangulum, M33 is the Local Group's third largest member, after the Andromeda Galaxy and the Milky Way. It is 2.6 million light-years away and has an absolute visual magnitude of –18.9, giving it a luminosity one fifth of the Milky Way's. M33 is probably one of Andromeda's satellites.

**M51** An incredible spiral galaxy. See **Whirlpool Galaxy.**

**M81** A large spiral galaxy near the Big Dipper. M81 is 11 million light-years from Earth. Orbiting M81 is the irregular galaxy M82. The spiral NGC 2403, a twin of M33, also belongs to the M81 group.

**M87** A giant elliptical galaxy that may mark the center of the Virgo cluster and thus of the entire Local Supercluster. Its distance is in dispute, with different astronomers claiming numbers between 55 and 75 million light-years.

**M100** An attractive spiral galaxy in the Virgo cluster.

**M101** A striking spiral galaxy near the Big Dipper, 24 million light-years from Earth.

**MACHO** MAssive Compact Halo Object—a dark star, planet, or asteroid that may populate a galaxy's dark halo.

**Magellanic Clouds** A pair of irregular galaxies and the brightest of the Milky Way's many satellites. See **Large Magellanic Cloud; Small Magellanic Cloud.**

**Magnitude** A measure of brightness. The brighter the object, the *lower* is the numerical value of its magnitude—a magnitude 1 star is brighter than a magnitude 2 star, just as a number 1 bestseller outsells a number 2 bestseller. Apparent magnitude says how bright the object looks, absolute magnitude how powerful it is. See **Absolute Magnitude; Apparent Magnitude.**

**Main-Sequence Star** A star, like the Sun, whose center converts hydrogen into helium. Most of the known nearby stars are main-sequence stars.

**Malmquist Bias** A statistical effect that can cause astronomers to underestimate distances and overestimate the Hubble constant. The greater the scatter in a standard candle, the worse is the Malmquist bias. However, if the scatter is known, the Malmquist bias can be corrected.

**Mass-to-Light Ratio** A measure of how bright or dark material is. The mass-to-light ratio is expressed in solar units, so the Sun's mass-to-light ratio is 1. More luminous stars have lower mass-to-light ratios, since they usually have somewhat more mass but give off much more light. Less luminous stars, the vast majority, have higher mass-to-light ratios. Dark matter, of course, has an extremely high mass-to-light ratio.

**Megaparsec** A unit of distance equal to a million parsecs—3.26 million light-years. The nearest giant galaxy, Andromeda, is a bit less than a megaparsec from the Milky Way.

**Metallicity** The abundance of elements heavier than helium. Halo stars have low metallicities, because they formed before stars had substantially enriched the Galaxy with heavy elements.

**Microlensing** See **Gravitational Lensing**.

**Microwaves** Electromagnetic radiation with longer wavelengths than those of infrared radiation. The light that broke free of matter some 300,000 years after the big bang has been so redshifted that it appears as microwaves.

**Milky Way** Our Galaxy, a giant spiral whose luminous disk is 120,000 light-years across and 2,000 light-years thick, but whose dark halo extends far beyond. The Sun and Earth hover a few dozen light-years north of the Galaxy's plane, 27,000 light-years from the Galaxy's center. The Milky Way is the Local Group's second largest galaxy, after Andromeda.

* * *

**Nebula** A cloud of gas and dust in space. In olden times, the term also referred to galaxies, which from afar look like clouds.

**Neutrino** An uncharged nonbaryonic particle with little mass, zipping through space—and our bodies—at near light speed. Neutrinos are a form of hot dark matter.

**Neutron** An uncharged subatomic baryonic particle that appears in every atomic nucleus except hydrogen-1 and weighs slightly more than a proton.

**Neutron Star** A tiny star that results when a big star explodes and collapses, crushing the protons and electrons into neutrons. It measures only about 10 miles across but weighs more than the Sun.

**NGC** Designation for the 7,840 star clusters, nebulae, and galaxies in the *New General Catalogue*, which appeared in 1888. See also **IC**.

**NGC 2403** A small spiral galaxy, 11 million light-years away, that resembles the Local Group's M33. It belongs to the M81 group.

**Nonbaryonic** Consisting of exotic material unlike the protons, neutrons, and electrons that make up the Earth and our bodies. Most material in the universe seems to be nonbaryonic. Cold dark matter and hot dark matter are both nonbaryonic.

**Nova** A stellar explosion far weaker than a supernova that does not destroy the star.

**Nucleosynthesis** The creation of new elements. Nucleosynthesis right after the big bang forged hydrogen, helium, and lithium. Today nucleosynthesis occurs chiefly in stars, making helium, carbon, nitrogen, oxygen, iron, and most other elements. See also **Primordial Nucleosynthesis**.

* * *

**Observable Universe** The part of the universe that telescopes could see if all space were transparent. If the universe is 14.5 billion years old, then astronomers can receive light only within a lookback distance of 14.5 billion light-years; this, then, is the observable universe. The edge of the observable universe, located 14.5 billion light-years in all directions from Earth, is called the horizon. Observers in other galaxies have different observable universes, just as people in different cities have different views out their windows.

**Olbers' Paradox** The puzzle of the dark night sky. If the universe is infinite, filled with infinitely many stars and galaxies, then an infinite amount of light should stream toward Earth and brighten the night, contrary to observation. Olbers' paradox is resolved by the universe's finite age, which means the Earth can receive light from only a finite volume of space, and by the scarcity of stars, which fail to flood all space with light. Contrary to popular belief, the universe's expansion does *not* resolve Olbers' paradox.

**Omega (Ω)** The cosmological parameter that denotes the universe's density of matter, which slows its expansion. Mathematically, omega is the actual matter density divided by the critical density. Thus, if the matter density exceeds the critical value—so that the universe will someday collapse—then omega is greater than 1. If the density is less than the critical density—so that the universe will expand forever—then omega is less than 1. Omega is probably well under 1, which means the universe will expand forever. Some astronomers use "omega" for the matter density *plus* lambda. This book, however, restricts the definition of omega to the matter density, since lambda affects the universe's future in the opposite way—lambda speeds up the universe's expansion; matter slows it down. See also **Critical Density**.

**Open Cluster** A loose collection of stars, like the Hyades and Pleiades, held together by the stars' own gravity. Most open clusters are young, because they usually get torn apart soon after birth.

**Open Universe** A universe in which omega plus lambda is less than 1: parallel lines diverge and the angles of a triangle add up to less than 180 degrees. An open universe is infinite and expands forever.

* * *

**Parallax** The tiny shift in a star's apparent position that results because astronomers view it from slightly different perspectives as the Earth circles the Sun. The closer the star, the greater is this shift, allowing astronomers to deduce the star's distance by measuring the size of its parallax.

**Parsec** A unit of distance equal to 3.26 light-years. The nearest star system to the Sun, Alpha Centauri, is just over a parsec away.

**Perfect Cosmological Principle** A cornerstone of the steady state theory, the perfect cosmological principle contends that the universe is uniform not only in space (the cosmological principle) but also in time. Thus the big bang theory violates the perfect cosmological principle, because the initial explosion marked a special time. See also **Cosmological Principle**.

**Period-Luminosity Relation** Discovered by Henrietta Leavitt in 1907: the bigger and brighter a Cepheid, the longer the star takes to pulsate. If astronomers know which luminosity goes with which period, they can observe the pulsation periods and apparent brightnesses of Cepheids in other galaxies and thereby ascertain their distances. See also **Cepheid**.

**Pisces-Perseus Supercluster** A large supercluster, 270 million light-years away, that winds through the northern autumn sky and sits on the opposite side of the sky from the Great Attractor. A large void sits in front of it.

**Planetary Nebula** A gaseous shell puffed off by a dying red giant star. A planetary nebula marks the transition from red giant to white dwarf and often looks like a colorful smoke ring. The Sun will one day form a planetary nebula. Prominent planetaries include the Ring Nebula, the Dumbbell Nebula, and the Helix Nebula.

**Population** A galaxy-wide assemblage of stars sharing similar ages, locations, orbital properties, and metallicities. The Milky Way has four stellar populations: the thin disk, the most luminous component, to which the Sun belongs; the thick disk, extending above and below the thin disk but bearing only old stars; the bulge, old metal-rich stars near the Galaxy's center; and the stellar halo, a diffuse collection of old stars that huddles around the Galactic bulge but also extends beyond the disk's edge.

**Primordial Black Hole** A hypothetical black hole that formed shortly after the big bang. Such objects might make up the dark halos surrounding galaxies like ours.

**Primordial Nucleosynthesis** The creation of the light elements—hydrogen, helium, and lithium—moments after the big bang. The rate at which these elements formed depended on the density of baryonic matter; thus, their primordial abundances reveal the density of baryonic matter in the universe.

**Proper Motion** The apparent motion of an object across the sky, year after year. If the object's distance is known, then the true velocity across our line of sight can be deduced.

**Proton** The positively charged subatomic baryonic particle that appears in every atomic nucleus. A proton has slightly less mass than a neutron but 1,836 times more mass than an electron.

* * *

**Quasar** The extraordinarily luminous center of a galaxy, billions of light-years away, emitting hundreds of times more light than the entire Milky Way.

**Quintessence** A hypothetical entity that pervades space and behaves like a cosmological constant that isn't constant. Quintessence can explain why omega and lambda are roughly equal; without quintessence, omega falls and lambda rises as the universe expands.

* * *

**Radial Velocity** The component of an object's motion toward or away from us. Astronomers can measure this from the object's blueshift or redshift.

**Radio** Electromagnetic radiation with the longest wavelength and lowest energy.

**Radio Galaxy** A galaxy, usually disturbed, that emits copious radio waves.

**Recombination** The era roughly 300,000 years after the big bang when the universe cooled enough so that protons joined electrons to form neutral hydrogen atoms. Before recombination, free electrons had scattered light; after recombination, they were gone, so the light could stream freely through space. This light, redshifted by a factor of 1,100, now appears as the cosmic microwave background.

**Red Clump Star** A yellow or orange giant star whose core burns helium into carbon and oxygen. The nearest to Earth is the cooler of the two yellow giants in the Capella system. Some astronomers use red clump stars to measure distances to nearby galaxies.

**Reddening** The scattering away of blue light by dust, rendering stars and galaxies redder than they really are. Reddening has nothing to do with redshifts.

**Red Dwarf** Fainter, cooler, and smaller than the Sun, a red dwarf nevertheless powers itself the same way, by converting hydrogen into helium at its core. Red dwarfs range from 8 to 60 percent of the Sun's mass. Because of their small mass, they burn their hydrogen slowly—which accounts for their faintness—and can live for trillions of years. Red dwarfs outnumber all other main-sequence stars put together.

**Red Giant** A big, bright, cool star, like Mira or Gamma Crucis, that emits roughly 100 times more light than the Sun. Billions of years from now, the Sun will swell into a red giant. Red giants cast off planetary nebulae and shrivel into white dwarf stars.

**Redshift** The shift to longer wavelengths of an object's spectrum. A redshift results if an object moves away from Earth (Doppler shift), if space expands between the object and Earth (cosmological redshift), or if light climbs out of a gravitational field (gravitational redshift).

**Red Supergiant** A gargantuan cool star, such as Antares or Betelgeuse, that outshines the Sun some 10,000 times. If put in place of the Sun, a red supergiant would swallow all the planets from Mercury to Mars; the largest would also swallow Jupiter and Saturn. Red supergiants explode as type Ib, Ic, or II supernovae, leaving behind neutron stars or black holes.

**Rotation Curve** How a galaxy's spin speed varies with distance from its center. A typical spiral galaxy has a flat rotation curve, which means its inner stars revolve around its center no faster than do its outer stars. A flat rotation curve implies that the amount of mass within some distance of the galactic center is proportional to that distance: if 100 billion solar masses lie within a certain distance, then twice that mass lies within twice that distance, thrice that mass within thrice that distance, and so on.

**RR Lyrae Star** An old white or yellow-white pulsating star that reveals the distances of elliptical galaxies and the older stellar populations of spiral and irregular galaxies. Because of their faintness, however, RR Lyrae stars cannot be seen much beyond the Local Group. RR Lyrae stars are common in globular clusters and the Milky Way's stellar halo and thick disk.

* * *

**Satellite Galaxy** A smaller galaxy that orbits a larger one. Because satellites feel their master's gravity, their motions reveal its mass. The brightest of the Milky Way's many satellite galaxies are the Magellanic Clouds.

**Sculptor Group** A nearby galaxy group due south of the Milky Way's disk. It houses several spirals, including edge-on NGC 253 and face-on NGC 300, which is 7 million light-years from Earth and resembles the Local Group's M33.

**Small Magellanic Cloud** An irregular southern galaxy that orbits the Milky Way, about 190,000 light-years from the Earth. Its absolute visual magnitude of –16.2 means that it emits about 2 percent as much light as the Milky Way.

**Spectrum** A breakdown of light, visible or invisible, by wavelength. A spectrum can reveal metallicity, because different atoms absorb different wavelengths, and speed, because the blueshift or redshift reveals how fast the object is moving toward or away from Earth.

**Spiral Galaxy** The most beautiful type of galaxy, spirals resemble celestial hurricanes, their arms lit by blue supergiants and dotted with pink regions of ionized hydrogen. Spirals bear gas and dust to create new stars, so the stars in spirals range from young to old. Spiral galaxies have Cepheids that reveal their distances. The Milky Way, Andromeda, M33, M81, M101, and the Whirlpool Galaxy are all spirals.

**Standard Candle** A celestial object of known intrinsic brightness, such as a Cepheid or type Ia supernova. By observing the standard candle's apparent brightness, astronomers can deduce its distance.

**Star Cluster** A gathering of stars held to one another by their gravitational attraction. Star clusters come in two types: the loose open clusters and the tightly packed globular clusters. See **Globular Cluster**; **Open Cluster**.

**Steady State Theory** The proposal, put forth by Hermann Bondi, Thomas Gold, and Fred Hoyle in 1948, that the universe has existed forever. Because the expansion of space would otherwise reduce the universe's density to zero, the steady state theory posited the continuous creation of matter.

**Stellar Halo** The population of old stars that engulfs the Milky Way's disk and bears low abundances of heavy elements such as oxygen and iron. Most stars in the stellar halo lie inside the Sun's orbit around the Galaxy, but some reside beyond the edge of the disk. The most prominent components of the stellar halo are metal-poor globular clusters.

**Stellar Population** A galaxy-wide assemblage of stars sharing similar ages, locations, orbital properties, and metallicities. The Milky Way has four stellar populations: the thin disk, the most luminous component, to which the Sun belongs; the thick disk, extending above and below the thin disk but bearing only old stars; the bulge, old metal-rich stars near the Galaxy's center; and the stellar halo, a diffuse collection of old stars that huddles around the Galactic center but also extends beyond the disk's edge.

**Sunyaev-Zel'dovich Effect** The boost to higher energies that the cold photons from the cosmic microwave background receive when they pass through the hot gas of a galaxy cluster. The Sunyaev-Zel'dovich effect provides a way to measure the Hubble constant.

**Supercluster** A huge assemblage of galaxies, often spanning over 100 million light-years. The Milky Way resides on the outskirts of the Local Supercluster.

**Supergiant Star** An extremely luminous star that has evolved off the main sequence. In Earth's sky, the five brightest supergiants are Canopus, Rigel, Betelgeuse, Antares, and Deneb.

**Supernova** A tremendous explosion that destroys a star. A single supernova can outshine all the billions of stars in a modest galaxy like M33. To cosmologists, the most useful of these explosions are type Ia supernovae. Such a supernova results when a star dumps material onto a white dwarf star, which then annihilates itself. All white dwarfs explode at nearly the same mass—1.4 Suns—so all type Ia supernovae have nearly the same intrinsic brightness. Thus, the supernova's apparent brightness reveals the distance to its host galaxy, making type Ia supernovae useful to astronomers who seek the numerical values of the Hubble constant, omega, and lambda.

Supernovae also occur when massive stars—born with over eight times the Sun's mass—explode. These supernovae are designated type Ib (if they lack hydrogen), type Ic (if they lack hydrogen and helium), or type II (if they have hydrogen). Unlike type Ia supernovae, massive-star supernovae arise from a variety of stars and therefore peak at different brightnesses, limiting their cosmological utility.

**Surface Brightness** A measure of how concentrated or diffuse an object's light is. In a typical galaxy, the surface brightness peaks at the center—where stars are most crowded—and drops toward the periphery.

* * *

**Timing Argument** Although ignored when it was published in 1959, the timing argument supports the existence of dark halos around the Milky Way and Andromeda. Assume that the two galaxies initially set off from each other because of the universe's expansion, but their gravitational attraction turned them around and made them move toward each other. To achieve such a turnaround during the universe's lifetime, each galaxy must have far more mass than exists in its visible stars.

**Tired Light Theory** A discredited idea which proposes that the universe is not expanding. Instead, says this theory, light loses energy as it travels through space and as a result becomes redshifted. The longer the light travels, the more "tired" it becomes; thus, as Hubble discovered, the farther a galaxy from Earth, the greater is the galaxy's redshift. However, two observations contradict the tired light theory. First, distant supernovae appear to wax and wane more slowly than those nearby, an effect caused by the expansion of space. Second, the tired light theory predicts that the cosmic microwave background should not have a blackbody spectrum, when in fact it does.

**Top-Down Theory** The proposal that superclusters of galaxies formed before individual galaxies did. The top-down theory became wedded to hot dark matter, which speeds through space fast and smooths out small structures, leaving large ones. See also **Bottom-Up Theory.**

**Tully-Fisher Relation** Named for R. Brent Tully and J. Richard Fisher, this relation applies to spiral galaxies: the faster they spin, the brighter they shine. Thus, by measuring a spiral's spin speed, astronomers can deduce its luminosity—and hence its distance.

* * *

**Ultraviolet** Radiation with a slightly shorter wavelength, and a slightly higher energy, than visible light.

* * *

**Velocity Dispersion** The spread of velocities among a group of objects. The more the velocities differ from one another, the greater is the velocity dispersion.

**Virgo Cluster** The closest galaxy cluster to Earth and the center of the Local Supercluster, which houses the Milky Way and other Local Group galaxies. The Virgo cluster's thousands of galaxies include ellipticals such as M87 and spirals such as M100. The distance to the Virgo cluster's center is in dispute, with estimates ranging between 55 and 75 million light-years.

**Void** An enormous region of space that harbors few bright galaxies.

* * *

**Whirlpool Galaxy** A breathtaking face-on spiral galaxy in the constellation Canes Venatici. Here, in 1845, Lord Rosse first discerned spiral structure.

**White Dwarf** A dim, dense, dying star that once shone as splendidly as the Sun but has used up its nuclear fuel. The typical white dwarf is 60 percent as massive as the Sun but little larger than Earth; a spoonful of white dwarf matter weighs tons. White dwarfs account for 5 percent of all stars, but they are so dim that the unaided eye sees none. The two nearest orbit the bright stars Sirius and Procyon, 8.6 and 11.4 light-years from Earth. If another star dumps material onto a white dwarf and raises its mass above 1.4 Suns, the white dwarf explodes as a type Ia supernova.

**WIMP** Weakly Interacting Massive Particle—a hypothetical nonbaryonic particle that may constitute most matter in the universe, including the dark halos that surround the Milky Way and other giant galaxies. WIMPs are cold dark matter.

* * *

**X-Rays** High-energy radiation, emitted by extremely hot objects.

* * *

**Yellow Supergiant** A star that resembles the Sun in color and temperature but outshines it thousands of times. Cepheids are yellow supergiants.

* * *

**Zone of Avoidance** The swath of sky that the Milky Way's gas and dust obscure, preventing easy observation of galaxies beyond. The zone of avoidance covers about a fifth of the sky. Fortunately, the Andromeda Galaxy, the Virgo cluster, and the Coma cluster lie outside the zone of avoidance; unfortunately, the center of the Great Attractor lies within it.

# NOTES

## INTRODUCTION: A GREAT ATTRACTION

All quotes are from Kraan-Korteweg (2000). Kraan-Korteweg and Lahav (2000) review searches for galaxies behind the Milky Way; Kraan-Korteweg and Lahav (1998) do so at a more popular level. Kraan-Korteweg et al. (1994) discovered a barred spiral galaxy in Cassiopeia, Dwingeloo 1, a member of the nearby Maffei galaxy group (Buta and McCall 1999). Ibata, Gilmore, and Irwin (1994) discovered the Sagittarius dwarf. Kraan-Korteweg et al. (1996) discovered the Great Attractor's core, which Abell, Corwin, and Olowin (1989) had previously catalogued as Abell 3627. The constellation Norma was invented by Nicolas Louis de Lacaille, whose life is described by Evans (1980).

## 1 HIGH MIDNIGHT

Harrison (2000) provides an introduction to cosmology. Peacock (1999) and Peebles (1993) are much more advanced. Harrison (1998) compares cosmology to comedy. Disney (2000) is much more pessimistic: "Cosmology must be the slowest moving branch of science. The number of practitioners per relevant observation is ridiculous. Consequently the same old things have to be said by the same old people (and by new ones) over and over and over again. For instance 'Cold Dark Matter' now sounds to me like a religious liturgy which its adherents chant like a mantra in the mindless hope that it will spring into existence."

## 2 IN THE DARK

Harrison (1987)'s book discusses Olbers' paradox and most of the protagonists. The first Harrison quote is from the preface of Harrison (1987); the other Harrison quotes in the opening are from Harrison (1999).

Harrison (1987) and Jaki (1969, 1991) discuss Olbers himself. Johnson (1937, 1957) describes Digges, as does Harrison (1987); the Digges quotation appears on page 165 of Johnson (1937), page 188 of Johnson (1957), and pages 37 and 213 of Harrison (1987). The introduction to Croswell (1997) describes Bruno. Bruno's words are from page 246 of Singer (1950). Harrison (1987) discusses Johannes Kepler.

Christianson (1984) and Westfall (1980) are biographies of Newton. Hoskin (1977, 1985) and Harrison (1986, 1987) discuss Newton's correspondence with Bentley. Newton's four letters to Bentley appear on pages 211 to 219 of Munitz (1957). Halley (1721a,b)'s papers are on pages 249 to 252 of Jaki (1969) and pages 218 to 220 of Harrison (1987). Harrison (1987) and Jaki (1969) discuss Chéseaux.

The Harrison quotes about Poe are from Harrison (1999) and page 258 of Harrison (1981). The Poe quotation "I have no desire to live . . ." is from page i of Poe (1973), a reprint of *Eureka* (Poe 1848). Pollin (1975) discusses reviews of Poe's *Eureka*, as does Tipler (1988a,b). Poe's quotation about the universe as a poem appears on page 130 of *Eureka*; his statement about what is now called Olbers' paradox appears on page 100. Tipler (1988a,b) discusses Mädler and includes his quotation about Olbers' paradox. Harrison (1987) discusses Kelvin. Harrison (1999) describes the influence of religion on Olbers' paradox.

Chapter 3 of Bondi (1952a) coined the term *Olbers' paradox*. The Hoyle quotation is from page 278 of Hoyle (1955). The Harrison quote about the popularity of the redshift solution is on page 192 of Harrison (1987). Wesson, Valle, and Stabell (1987) found that only three out of ten astronomy textbooks got Olbers' paradox right. The Harrison quote about Jaki is from page 229 of Harrison (1987); Jaki's quotation is from pages 46 to 47 of Jaki (1991); Olbers' quotation is from page 68 of Jaki (1991).

Harrison (1964, 1965) proposed the energy solution for Olbers' paradox, as further discussed by Harrison (1974, 1987, 1990a,b); Wesson, Valle, and Stabell (1987); and Wesson (1999). The first Harrison quote in the closing of the chapter is from page 194 of Harrison (1987); the last quote is from Harrison (1999).

## 3 GALAXIES BEYOND

Berendzen, Hart, and Seeley (1976) and Smith (1982) discuss most of the developments in this chapter.

Fernie (1970), Gingerich (1987), and Russell (1999) describe the history of thought about the nebulae. As discussed by Whitrow (1967), Wren in 1657 and Swedenborg in 1734 proposed the existence of other galaxies—see page 205 of Wren and Wren (1750) and page 161 of Swedenborg (1734). The final paragraphs of Wright (1750) state this idea more forcefully, as Hoskin (1970) and Hoskin and Rochester (1992) note. Calinger

(1979), Campbell (1914), Charlier (1925), and Glyn Jones (1971) discuss Kant (1755)'s work.

Messier's work is discussed by Gingerich (1953, 1960), Glyn Jones (1991), and Williams (1969); the quotation about Messier's catalogue is from page 63 of Whitney (1971). Gingerich (1988) and Skiff (1993) discuss catalogues after Messier's.

Dewhirst and Hoskin (1991) and Hoskin (1982b, 1990) describe Rosse's work; C. (1869) gives Rosse's obituary.

Anonymous (1864) and Huggins (1864) report spectroscopic observations of nebulae; Becker (2001) describes Huggins' work. Following up on work by Abbe (1867), Proctor (1869) and, later, Sanford (1917) describe the distribution of nebulae on the sky. De Vaucouleurs (1985), de Vaucouleurs and Corwin (1985), and Glyn Jones (1976) discuss the 1885 supernova in Andromeda; Pickering (1895) reported Williamina Fleming's discovery of the 1895 supernova in NGC 5253. Roberts (1888) detected Andromeda's spiral structure, as de Vaucouleurs (1987) describes. Clerke's quotation is from page 368 of Clerke (1890).

Woolf (1959) describes how the transits of Venus established the Earth-Sun distance. Bessel (1838) first detected the parallax of a star other than the Sun; Olbers' words are from page 25 of Jaki (1991). Bohlin (1907) claimed to detect Andromeda's parallax.

The "bad boy" quotation about Lowell Observatory is from page 131 of Tombaugh and Moore (1980); Hoyt (1976) describes Percival Lowell and Mars; Lowell's own quote about life on Mars is from page 376 of Lowell (1906). Hoyt (1980) and Parker (1986) describe Slipher's work on the spirals. Slipher (1913) discovered the large blueshift of Andromeda, Slipher (1914) the rotation of the Sombrero Galaxy, and Slipher (1915, 1917) the large velocities of other galaxies. Hall (1970) describes Slipher's life.

Hoskin (1976) describes the novae discovered in spirals during 1917 (Curtis 1917a,b,c; Pease 1917; Ritchey 1917a,b; Shapley 1917a,b); Curtis's quotation is from page 109 of Curtis (1917c).

Shapley's remarks about the Perseid meteors are from page 22 of Shapley (1969); about archeology and astronomy, from page 17 of Shapley (1969); about man's place in the universe, from page 60 of Shapley (1969). The quotation about Shapley's lack of humility is from page 218 of Whitney (1971). Although Shapley (1916) believed the globular clusters lay outside the Milky Way, Shapley (1918a,b,c,d, 1919) used the globulars to delineate the Milky Way's size.

Shapley's remarks about van Maanen are from pages 56 to 57 of Shapley (1969). Berendzen and Hart (1973), Brashear and Hetherington (1991), Hetherington (1974a,b, 1975), and Hetherington and Brashear (1992) describe the spurious proper motions that van Maanen (1916, 1921a,b,c, 1922a,b, 1923a,b) found.

Gingerich and Welther (1985) and Hoskin (1979) describe the use of Cepheids; Gilman (1978), Goodricke (1912), and Joy (1937) describe Goodricke's life. Leavitt (1907) first reported a correlation between a Cepheid's pulsation period and its luminosity; Leavitt (1912) strengthened the Cepheid period-luminosity relation; and Hertzsprung (1913) and Shapley (1918a) attempted to calibrate it. Fernie (1969) describes efforts to calibrate the Cepheid period-luminosity relation.

Christianson (1995), Mayall (1954), and Smith (1990) describe Hubble's life and work; Berendzen and Hoskin (1971), Hetherington (1990, 1996), and Hoskin (1982a) discuss Hubble's discovery of Cepheids in spirals (Hubble 1925, 1926a, 1929a).

Einstein (1917) and de Sitter (1917) attempted to apply general relativity to the universe, the latter predicting a distance-redshift relation. (For an English translation of Einstein's paper, see Engel and Schucking 1997.) The first sentence of Chapter 11 of Eddington (1939) declares the number of protons and electrons in the universe.

Hubble (1929b) discovered the correlation between distance and redshift; Hubble and Humason (1931) extended it. Hetherington (1971, 1990, 1996), Parker (1986), and Smith (1979) discuss the origins of the distance-redshift relation. An English translation of Lemaître's 1927 paper appears as Lemaître (1931a); see also Lemaître (1950). Earlier, Friedmann (1922, 1924) had proposed expanding universes. Ellis (1990) describes the shift to the idea of an expanding universe. Poe's quotation is from pages 42 to 43 of Poe (1848). The retort of Einstein's wife is on page 434 of Clark (1971).

## 4 BIG BANG BATTLES

Kragh (1996) describes most of the developments in this chapter.

Deprit (1984) and Kragh (1987) discuss Lemaître. His remark about finding the cause of the expansion appears in the last paragraph of Lemaître (1931a). Einstein's "abominable" remark about Lemaître appears in Smith (1990), a loose translation from Lemaître (1958). Einstein's remark about Friedmann's paper appears on page 244 of Stachel (1986).

Chapter 2, section 6, of Eddington (1933) says that the beginning should not be too unaesthetically abrupt; page 85 of Eddington (1928) narrowly missed coining "big bang." Eddington (1928) also discusses the running down of the universe.

Lemaître (1931b) originated the big bang theory; Anonymous (1931) reports the find for *The New York Times*. Page 73 of Godart and Heller (1985) gives Lemaître's deleted statement. Lemaître (1931c) refers to the primeval atom and expresses the wish that the all-knowing oracle not answer the question. Page 78 of Lemaître (1950) compares the evolution of the world to a display of fireworks; the original, in French, appears on page 408 of Lemaître (1931d).

Gamow (1970), page 15, recounts his study of a communion bread crumb; page 81, the determination of God's distance from Earth. All quotes from Alpher are from Alpher (1999); Alpher and Herman (1972a,b, 1988, 1990, 2001) describe their work with Gamow on the big bang; d'Agnese (1999) profiles Alpher's life. The conclusion of Gamow (1940) mentions the "'prehistoric' kitchen of the universe." Alpher, Bethe, and Gamow (1948) described the origin of the elements in the big bang; Alpher and Herman (1948) predicted the universe's temperature, as did the end of Chapter 2 of Gamow (1952a).

Hubble (1936) reported a Hubble constant of 530 kilometers per second per megaparsec, implying an extremely young universe.

Bondi and Gold (1948) and Hoyle (1948) proposed the steady state theory. Bondi (1982, 1988, 1990a,b, 1999), Gold (1982, 1999), Hoyle (1981, 1999), and Bondi, Gold, and Hoyle (1995) all described the origin of the steady state theory. All quotes from Bondi are from Bondi (1999). Hoyle's quote about the first priority among scientists is from page 328 of Hoyle (1994); his story of the Roman numerals is on page 45 of Hoyle (1994); his story about the six-petaled flower is on page 48 of Hoyle (1994). Chapter 3 of Bondi (1952a) coins "Olbers' paradox"; Gold (1968) proposed that pulsars are rapidly rotating neutron stars.

Gold's quote about religion is from Gold (1999); on page 67, Bondi (1990a) describes his marriage; page 7 of Hoyle (1977) proposes to solve the conflict in Ireland, page 421 of Hoyle (1994) talks about religious quarrels, and Hoyle (1999) discusses the logical structure of the universe.

Gold (1982) describes working with Bondi and Hoyle, and Gold (1999) talks about the movie *Dead of Night*. Hoyle (1999) talks about the rejection of his steady state paper; see also page 53 of Hoyle (1982a).

Hoyle (1950a,b,c,d,e) delivered five BBC broadcasts, later collected as Hoyle (1950f). Pages 253 to 255 of Hoyle (1994) talk about these broadcasts. Beale (1950), Carey (1950), and Copleston (1950) all criticized Hoyle's final broadcast; Anonymous (1950) defended him. Hoyle (1949b) coined "big bang"; Croswell (1995), pages 113 to 114, gives Hoyle's motivation for choosing that name (Hoyle 1993); Ferris (1993) criticized it; Robertson (1993) criticized Ferris's statement that the big bang was dark; and Beatty and Fienberg (1994) reported the results of the big bang contest, which Hoyle (1999) called a farce.

Hoyle's statement about being trained as a mathematician is from Hoyle (1999). The last lines of Gamow (1970) describe the success of his books.

Cornwell (1999) called Pius XII "Hitler's Pope." Part of the Pope's statement about big bang cosmology appears in Pius XII (1952); the quote in *Time* is from Anonymous (1951). Gamow (1952b)'s paper starts with words from the Pope; Gold (1999) was less impressed; and McMullin (1981), page 53, reports Lemaître's reaction. Pages 126 and 128 of Gamow (1970) make fun of England. Anonymous (1959) reported the results of the cosmology poll.

Osterbrock (2001) describes Baade's life, including his discovery of the two populations. Chapter 6 of Croswell (1995) also describes Baade's discovery of the two populations and includes words from Osterbrock (1993). Baade (1944a,b) reported his discovery of the two populations.

Fernie (1969) describes how astronomers since Hertzsprung (1913) and Shapley (1918a) have attempted to calibrate the Cepheid period-luminosity relation. Joy (1940) reported peculiarities in a "Cepheid" in M3. Smith (1995) reviews RR Lyrae stars; Williamina Fleming discovered RR Lyrae's variability, as reported by Pickering (1901). Baade (1956), page 10, mentions his failure to find RR Lyrae stars in Andromeda; Hoyle (1954) recorded Baade's doubling of the distance scale, which Thackeray and Wesselink (1953) confirmed. Shapley (1953a,b,c) attempted to claim credit for Baade's work, just as Shapley (1929) had done earlier to Hubble. P. (1953) and Anonymous (1953) reported Shapley's "discovery"; Sandage (1999) gave Baade's reaction. Stebbins, Whitford, and Johnson (1950) recognized that faint magnitudes were wrong; Humason, Mayall, and Sandage (1956) and Sandage (1958) revised the Hubble constant downward.

Sandage's quotes are from Sandage (1999); Bondi (1952b, 1955) said that observations can be just as bad as theories.

Edge and Mulkay (1976), Scheuer (1990), Smith and Lovell (1983), and Sullivan (1990) describe aspects of the history of the radio counts. Page 269 of Hoyle (1994) talks about the principle of maximum trivialization in astronomy. Gold (1951) advocated an extragalactic location for most radio sources against Ryle (1951); see also Ryle and Hewish (1955). Baade and Minkowski (1954) photographed Cygnus A, showing it to be extragalactic; Gold's quotes about this are from Gold (1999). Page 271 of Hoyle (1994) compares Ryle's group to Bach's violinist visitor. Ryle (1955), Ryle and Scheuer (1955), Scheuer (1957), Ryle (1958), Edge, Scheuer, and Shakeshaft (1958), Edge et al. (1959), Ryle and Clarke (1961), and Scott and Ryle (1961) all reported that the radio counts argued against the steady state theory. Pages 409 to 410 of Hoyle (1994) recount the 1961 announcement; page 324 of Kragh (1996) gives the "Bible was right" quote; and Hoyle (1994), page 326, explains why he couldn't get along with Ryle. Schmidt (1963) discovered that quasars have large redshifts. Chiu (1964) coined the term "quasar."

Chapters 8 and 9 of Croswell (1995) describe how astronomers recognized that stars created most elements on Earth. Hoyle (1946) already advocated this idea, even before Chamberlain and Aller (1951) discovered that halo stars have much lower abundances of heavy elements and before Merrill (1952) discovered technetium on red giant stars. Burbidge, Burbidge, Fowler, and Hoyle (1957) was the culmination of the idea; Cameron (1958) independently developed much of the same. On page 114 of Croswell (1995), Hoyle (1993) discusses his motivation for the idea. Truran, Hansen, and Cameron (1965) discuss the helium abundance. Hoyle and Tayler (1964) showed how the big bang could produce helium; Wagoner, Fowler, and Hoyle (1967) followed up on this work.

Penzias and Wilson (1965) discovered the cosmic microwave background; Sullivan (1965) reported on it for *The New York Times*; Peebles and Wilkinson (1967) barely mentioned Alpher, Herman, and Gamow. Gamow's quote about the lost nickel is on page 41 of Penzias (1972); the quote from his letter is on page 351 of Kragh (1996). Alpher and Herman discuss their unhappiness at the end of Chapter 6 of Ferris (1977); this discussion is much abridged in the 1983 paperback edition of the same book. McKellar (1941), following up on McKellar (1940), interpreted the observations of Adams (1941) to derive a temperature of 2.3 Kelvin for interstellar CN molecules. Hoyle (1981) recounts discussing this work with Gamow, which Hoyle (1949a) had cited.

Hoyle (1965) conceded that the big bang was right, but he later reversed himself. His quote about psychology is from Hoyle (1981); his comparison of the big bang to fundamentalist religion is from Hoyle (1982b), page 23. Gold's closing quote is from Gold (1999); Bondi's is from Bondi (1999).

## 5 THE BIG BANG FOR BEGINNERS

Hubble (1926b) classified galaxies into ellipticals, spirals, and irregulars. Sandage (1961), Sandage and Bedke (1994), van den Bergh (1998), and Chapter 3 of Croswell (1999) also describe galaxies.

For an introductory textbook on cosmology, see Harrison (2000); Weinberg (1972), Peebles (1993), and Peacock (1999) are much more advanced.

Odenwald and Fienberg (1993) contrast the expansion of space with the Doppler shift. Van den Bergh (2000a,b) describes the Local Group. Zwicky (1929) proposed the tired-light explanation for the distance-redshift relation. Leibundgut et al. (1996) used a high-redshift supernova to demonstrate the universe's expansion. Rees (1969) describes the collapse of the universe.

## 6 DARK MATTERS

Most quotes are from Rubin (2000), Freeman (2000), Roberts (2000), and Ostriker (2000). Tucker and Tucker (1988) and van den Bergh (1999) describe many of the developments in this chapter.

Slipher (1914) and Wolf (1914) detected the rotation of the Sombrero Galaxy and M81, respectively. Rubin and Ford (1970) measured Andromeda's optical rotation curve; Rubin (1973) provides a more popular account. Freeman (1970) recognized the implications of flat rotation curves in M33 and NGC 300. Following Roberts (1966), Roberts and Whitehurst (1975) reported a flat radio rotation curve for Andromeda, which contradicted Emerson and Baldwin (1973). Roberts (1976) reviews rotation curves of galaxies.

Ostriker and Peebles (1973) suggested that halos could stabilize galactic disks. Einasto, Kaasik, and Saar (1974) and Ostriker, Peebles, and Yahil (1974) proposed that galaxies had dark halos; Tremaine (1999) describes this work. Page 3 of Gilmore, King, and van der Kruit (1990) distinguishes the stellar halo from the dark halo.

Rubin, Ford, and Thonnard (1978) found flat optical rotation curves for several spiral galaxies, as did Rubin, Ford, and Thonnard (1980), Rubin, Ford, Thonnard, and Burstein (1982), and Rubin, Burstein, Ford, and Thonnard (1985). Bosma (1978) found flat radio rotation curves for many spiral galaxies; his quote is from page 149. Many of his rotation curves appear in Figure 2 of Faber and Gallagher (1979).

Zwicky (1933, 1937c) and Smith (1936) found dark matter in galaxy clusters; Ostriker (1999) describes this work. Zwicky (1971) lambasted the astronomical profession. Babcock (1939) measured Andromeda's optical rotation curve. Kahn and Woltjer (1959) proposed the timing argument for measuring the masses of the Milky Way and Andromeda. Hartwick and Sargent (1978) and Zaritsky et al. (1989) used the Milky Way's satellites to deduce its mass. Rubin et al. (1962) deduced a flat rotation curve for the outer Milky Way.

## 7 ASHES OF CREATION

All quotes are from Steigman (2000), Sargent (2000), York (2000), and Spite (2000). Boesgaard and Steigman (1985) review big bang nucleosynthesis. Hoyle and Tayler (1964), Peebles (1966a,b), and Wagoner, Fowler, and Hoyle (1967) examined the nuclear reactions that occurred shortly after the big bang.

Eggen and Sandage (1962) suggested that halo stars might have little helium; later, though, Faulkner (1967) and Cayrel (1968) showed that the stars did have helium. Weak helium lines in blue halo stars (Searle and Rodgers 1966, Greenstein 1966, Sargent and Searle 1966) were reinterpreted to give normal helium abundances (Greenstein, Truran, and Cameron 1967; Sargent and Searle 1967). Christy (1966) used RR Lyrae stars to establish a large halo helium level; O'Dell, Peimbert, and Kinman (1964) found large quantities of helium in M15's planetary nebula; and Faulkner and Iben

(1966) said that high helium levels fit the color-magnitude diagram of another metal-poor globular, *M92*. Searle and Sargent (1972) discovered that I Zwicky 18 and II Zwicky 40 were oxygen- and neon-poor but helium-rich. Kunth and Östlin (2000) review metal-poor galaxies.

Black (1971, 1972) and Geiss and Reeves (1972) used helium-3 to infer a lower deuterium abundance in space than in terrestrial water. Rogerson and York (1973) detected interstellar atomic deuterium, and Trauger, Roesler, Carleton, and Traub (1973) detected HD on Jupiter.

M. Spite and F. Spite (1982) and F. Spite and M. Spite (1982) discovered lithium in halo stars.

## 8 UNIVERSE IN HYPERDRIVE

Most quotes are from Guth (2000) and Linde (2000), except for Guth's quote about the beauty of inflation, which is from page 250 of Guth (1997), and Linde's quote about simplifying the work of God, which is from page 486 of Lightman and Brawer (1990). Tye (2000) supplied additional information.

Popular-level overviews of inflation appear by Guth (1997), Guth and Steinhardt (1984), and Linde (1994). Preskill (1979) found that grand unified theories would produce too many magnetic monopoles, and Guth and Tye (1980) looked for a solution by delaying the universe's phase transition. Guth (1981) found that this led to inflation, which solved the flatness problem (Dicke 1970, page 62) and the horizon problem (Misner 1969); Dicke and Peebles (1979) discuss both problems. Linde (1982) and Albrecht and Steinhardt (1982) proposed new inflation. Linde (1983) proposed chaotic inflation, and Linde (1986) eternal inflation in the context of chaotic inflationary theory. Page 73 of Godart and Heller (1985) gives the statement that Lemaître deleted from his 1931 paper. Tryon (1973) suggested that the entire universe might be a quantum fluctuation. Ostriker's quote is from Ostriker (2000).

## 9 THE ARCHITECTURE OF THE COSMOS

Most quotes are from Gregory (2000), Thompson (2000), Olszewski (2000), and Dressler (2000).

Hubble (1934) concluded that the distribution of galaxies was fairly homogeneous; his words are from page 62 of his paper. Bernheimer (1932) and Tombaugh (1937) independently discovered the Pisces-Perseus supercluster; Tombaugh's quote is from page 113 of Levy (1991). Page 135 of Herschel (1847) and page 46 of Stratonoff (1900) briefly mention an abundance of nebulae in this part of the sky. Shapley (1930, 1933, 1934) and Zwicky (1938) also reported galactic irregularities.

De Vaucouleurs (1953) proposed the Local Supercluster; his quote is from page 92 of Lightman and Brawer (1990). De Vaucouleurs (1989) briefly discusses the history of the idea. Abell (1958) published his catalogue of galaxy clusters and found evidence for what he called second-order clusters—that is, superclusters; see also Abell (1961). Shane and Wirtanen (1967) counted over a million galaxies.

Popular-level reviews of superclusters and voids appear by Chincarini and Rood (1980), Gregory and Thompson (1982), and Gregory (1988); technical reviews by Oort (1983) and Rood (1988). Gregory and Thompson (1978) discovered the Coma supercluster and the void in front of it. Precursors to this work include Gregory (1975), Chincarini and Rood (1975, 1976), and Tifft and Gregory (1976). Seldner, Siebers, Groth, and Peebles (1977) turned the Shane-Wirtanen counts into a map, yet argued against the reality of the features it revealed. Tarenghi et al. (1979) confirmed the Hercules supercluster and discovered a large void in front of it; Gregory, Thompson, and Tifft (1981) and Haynes and Giovanelli (1986) did the same for the Pisces-Perseus supercluster. Jôeveer, Einasto, and Tago (1978) and Einasto, Jôeveer, and Saar (1980) argued for superclusters and voids, too. Kirshner, Oemler, Schechter, and Shectman (1981) discovered the void in Boötes; Davis, Huchra, Latham, and Tonry (1982) found a "frothy" galactic distribution; Giovanelli and Haynes (1982) discovered the Lynx-Ursa Major supercluster. De Lapparent, Geller, and Huchra (1986) found chains of galaxies and voids, but Geller, de Lapparent, and Kurtz (1984) had earlier questioned the reality of these features in the Shane-Wirtanen map. Geller (2000) denied failing to cite earlier work. Sandage's quote, and his comparison of Gregory and Thompson with Alpher and Herman, is on page 81 of Lightman and Brawer (1990).

The bottom-up model for galaxy formation (Peebles 1965) battled the top-down model (Zel'dovich 1970; Doroshkevich, Sunyaev, and Zel'dovich 1974). Peebles (1982) proposed the cold dark matter model, developed further by Blumenthal, Faber, Primack, and Rees (1984). Aaronson (1983) discovered dark matter in the Draco Galaxy, a discovery interpreted by Faber and Lin (1983) and Lin and Faber (1983).

Dressler (1994) provides a popular overview of the Great Attractor. Dressler et al. (1987) and Lynden-Bell et al. (1988) found the Great Attractor, and Dressler and Faber (1990) confirmed its existence. Later, Tonry, Blakeslee, Ajhar, and Dressler (2000) employed surface brightness fluctuations to measure more accurate distances to elliptical galaxies, again confirming the Great Attractor but cutting our velocity toward it. Kraan-Korteweg et al. (1996) found the Great Attractor's center. Geller and Huchra (1989) discovered the Great Wall.

## 10 RIPPLES IN SPACE

Most quotes are from Mather (2000), Silk (2000), Bennett (2000), and Smoot (2000); Wright (2000) supplied additional information.

Books about COBE include Chown (1993), Mather and Boslough (1996), and Smoot and Davidson (1993). Shortly after Penzias and Wilson (1965)'s discovery of the cosmic microwave background, Sachs and Wolfe (1967) and Silk (1968) explored possible imprints in that background. The infamous quote "My God, Thiokol, when do you want me to launch, next April?" is from Lawrence Mulloy and appears in many places, including page 196 of McConnell (1987) and page 111 of Lewis (1988).

Mather et al. (1990) reported that COBE had found a blackbody spectrum for the cosmic microwave background, contradicting Matsumoto et al. (1988). Burbidge's words were mentioned by both Mather (2000) and Smoot (2000).

Smoot's quote in *The New York Times* is from Leary (1990). Smoot et al. (1992) detected ripples in the cosmic microwave background. Smoot's words from the press conference appear in slightly different forms on page 288 of Smoot and Davidson (1993) and page 251 of Mather and Boslough (1996). His quote about seeing God appears in slightly different forms on page 289 of Smoot and Davidson (1993) and page 251 of Mather and Boslough (1996); many remember the quote as mentioning "the face of God"—see page 138 of Chown (1993)—but Smoot (2000) denied having used that phrase. The *New Scientist* quote is from Anonymous (1992). Ostriker (1993) evaluated the standard cold dark matter model in light of the COBE work. Horgan (1992) profiled Smoot for *Scientific American*. White (1992) and Konigsberg (1995) discuss Smoot's book deal; the latter is the source of the lament of Smoot's publisher.

## 11 MACHOs VERSUS WIMPs

Nearly all quotes are from Paczyński (1995, 2000) and Alcock (1995, 2000). Paczyński (1996) reviews microlensing.

Paczyński (1986b) proposed that gamma-ray bursts occur far outside the Galaxy. Paczyński (1986a) proposed observing gravitational microlensing toward the Magellanic Clouds in order to detect dark-halo MACHOs.

The MACHO (Alcock et al. 1993), EROS (Aubourg et al. 1993), and OGLE (Udalski et al. 1993) teams reported microlensing events, MACHO and EROS toward the Magellanic Clouds, OGLE toward the Galactic center. Schwarzschild (1994) described the acronyms as an assault on political correctness. Lasserre et al. (2000) and Alcock et al. (2000) give recent analyses. Ibata, Richer, Gilliland, and Scott (1999) said they had seen white dwarfs in the dark halo, and Oppenheimer et al. (2001) detected a large number of such stars. Sahu (1994) suggested that the microlensing events were caused by stars in the Magellanic Clouds, not MACHOs in the dark halo.

## 12 THE CONSTANT HUBBLE WAR

Most quotes are from Sandage (2000), Tammann (2000), Tully (2000), Jacoby (2000), Tonry (2000), Freedman (2000), Schild (2000), Kochanek (2000), Stanek (2000), and Guinan (2000), supplemented by Sandage (1993, 1994, 1998, 1999), Freedman (1998), and Bolte (1998). De Vaucouleurs' quotes are from pages 93 and 98 of Lightman and Brawer (1990) and from de Vaucouleurs (1983). The *New Scientist* quote appears in Seife (1999). VandenBerg (2000) provided information on globular cluster ages, Branch (2000) on type Ia supernovae, and Mason (2000) on the Sunyaev-Zel'dovich effect.

IC 4182 entered the Hubble war when Sandage and Tammann (1982) used red supergiants to estimate its distance and the luminosity of its type Ia supernova; Pierce, Ressler, and Shure (1992) found a much shorter distance, but Sandage et al. (1992), using Cepheids, confirmed the original distance. Then Pierce and Jacoby (1995) attacked the supernova's brightness; Schaefer (1996) attacked them; and Jacoby and Pierce (1996) attacked him back.

Sandage (1958) found the Hubble constant to be between 50 and 100. The Cepheid scene is on page 397 of Overbye (1991). Sandage (1968b, 1972) attempted to measure the universe's deceleration but was criticized by Tinsley (1972); by observing giant stars in elliptical galaxies, Frogel et al. (1975) showed that Tinsley was right.

Tammann and Sandage (1968) determined the distance of the spiral galaxy NGC 2403; Sandage and Tammann (1974a) determined the distance of M101; Sandage and Tammann (1974b, 1976) determined the distance of the Virgo cluster.

De Vaucouleurs (1977) attacked Sandage and Tammann's work, saying the Hubble constant was much higher. Earlier, Sandage (1968a) and de Vaucouleurs (1970) had briefly reversed roles, the former claiming a high Hubble constant and the latter a low one. Tully and Fisher (1977) used the relation that now bears their names to find a high Hubble constant, too. Sandage's comment about "Fishy-Tuller" appears on page 273 of Overbye (1991). Jacoby, Ciardullo, and Ford (1990) used planetary nebulae and Tonry, Ajhar, and Luppino (1990) surface brightness fluctuations to arrive at high values of the Hubble constant.

By detecting Cepheids, Cook, Aaronson, and Illingworth (1986) confirmed the Sandage-Tammann distance to M101; see also Kelson et al. (1996). Madore (1976) had attacked the Sandage-Tammann distance to NGC 2403, but Freedman and Madore (1988) proved it correct.

Pierce et al. (1994) and Freedman et al. (1994) reported Cepheids in the Virgo cluster. Sandage's team (Saha et al. 1997) discovered Cepheids in NGC 4639. Madore et al. (1998, 1999) discovered Cepheids in NGC 1365. Freedman et al. (2001) reported a Hubble constant of 72, while Sandage's team reported a Hubble constant of 58 (Parodi, Saha, Sandage, and Tammann 2000). Hillebrandt and Niemeyer (2000) and Leibundgut (2000) review type Ia supernovae, while Branch (1998) describes their application to the Hubble constant. Tully and Pierce (2000) and Freedman et al. (2001) use the Tully-Fisher relation to get Hubble constants in the 70s, but Sandage, Tammann, and Federspiel (1995) use it to get a Hubble constant around 50. Teerikorpi (1997) discusses Malmquist and other biases. Sakai et al. (2000) did the paper that Tammann blasted as "astrology."

Zwicky (1937a,b) suggested that galaxies could gravitationally lens more distant objects, and Refsdal (1964) proposed using such lensing to measure the Hubble constant. Walsh, Carswell, and Weymann (1979) discovered the double quasar and Young et al. (1980, 1981) the lensing galaxy. Schild and Cholfin (1986) discovered the time delay, but Press, Rybicki, and Hewitt (1992a,b) attacked it. In 1996 the definitive test convinced all that Schild was right (Kundić et al. 1997), but not before Turner, Cen, and Ostriker (1992) proposed a variable Hubble constant. Rephaeli (1995) reviews the Sunyaev-Zel'dovich effect, which Reese et al. (2000) and Mason, Myers, and Readhead (2001) use to obtain the Hubble constant.

Feast and Catchpole (1997) claimed that Hipparcos parallaxes of Cepheids implied a greater distance to the Large Magellanic Cloud, but few believed this (Croswell 1998). Stanek, Zaritsky, and Harris (1998) and Stanek, Kaluzny, Wysocka, and Thompson (2000) used red clump stars to find a short distance to the Large Magellanic Cloud. Using eclipsing binaries, Guinan et al. (1998) and Fitzpatrick et al. (2001) also found a short distance, but Groenewegen and Salaris (2001) used an eclipsing binary to find an intermediate distance.

## 13 OMEGA: THE WEIGHT OF THE UNIVERSE

Most quotes are from Jacoby (2000), Tully (2000), Bahcall (2000), Tyson (2000), and Daly (2000).

From observations of the baryon content in galaxy clusters, White, Navarro, Evrard, and Frenk (1993) concluded that omega is much less than 1. Tyson, Kochanski, and Dell'Antonio (1998) used gravitational lensing to deduce a mass-to-light ratio for CL 0024+1654 and a low omega. Bahcall and Fan (1998) used the distant galaxy cluster MS 1054–03 to argue against an omega of 1. Computer simulations of galaxies, groups, clusters, and superclusters also indicated a low omega (Bahcall et al. 2000), as did a large redshift survey (Peacock et al. 2001). Guerra and Daly (1998) used radio galaxies to do the same. The quote from Kolb appears in Anonymous (1998). Efstathiou, Sutherland, and Maddox (1990) advocated a flat, lambda-dominated universe.

## 14 LAMBDA: EINSTEIN'S CURSE

Most quotes are from Perlmutter (2000), Turner (1995), Kochanek (1995), Schmidt (2000), Branch (2000), and Tammann (2000); Filippenko (2000) supplied additional information.

Hillebrandt and Niemeyer (2000) and Leibundgut (2000) review type Ia supernovae. Carroll, Press, and Turner (1992) describe the cosmological constant. The *New Scientist* quote about the supernova at redshift 0.46 is from Croswell (1992); the supernova was discovered by Perlmutter et al. (1995).

Fukugita, Futamase, and Kasai (1990) and Turner (1990) independently realized that the number of gravitationally lensed quasars probed the numerical value of lambda. The *Sky and Telescope* quote is from Maran (1992). Kochanek (1996) concluded that if the universe was flat, lambda was less than 0.66.

Phillips (1993) found that more luminous type Ia supernovae take longer to fade. From observations of distant type Ia supernovae, Perlmutter et al. (1997) concluded that omega was around 1; however, Schmidt's team came to the opposite conclusion (Garnavich et al. 1998). Subsequent reports indicated low omega (Perlmutter et al. 1998) and high lambda (Riess et al. 1998; Schmidt et al. 1998; Perlmutter et al. 1999).

Riess, Filippenko, Li, and Schmidt (1999) claimed that distant type Ia supernovae rose to maximum brightness faster than nearby ones; however, Aldering, Knop, and Nugent (2000) refuted this. Aguirre (1999) proposed that gray dust could dim the distant supernovae and mimic lambda. BOOMERANG (de Bernardis et al. 2000) and MAXIMA observations of the cosmic microwave background indicated a flat universe. Page 349 of Peacock (1999) has the quote about nothing.

## 15 THE FUTURE OF THE UNIVERSE

Sackmann, Boothroyd, and Kraemer (1993) predict the future of the Sun. Davies (1994), Adams and Laughlin (1999), and Krauss and Starkman (2000) describe the future of the universe.

# BIBLIOGRAPHY

Aaronson, Marc, 1983. Accurate Radial Velocities for Carbon Stars in Draco and Ursa Minor: The First Hint of a Dwarf Spheroidal Mass-to-Light Ratio. *Astrophysical Journal Letters*, 266, L11.

Abbe, Cleveland, 1867. On the Distribution of the Nebulae in Space. *Monthly Notices of the Royal Astronomical Society*, 27, 257.

Abell, George O., 1958. The Distribution of Rich Clusters of Galaxies. *Astrophysical Journal Supplement Series*, 3, 211.

———, 1961. Evidence Regarding Second-Order Clustering of Galaxies and Interactions Between Clusters of Galaxies. *Astronomical Journal*, 66, 607.

Abell, George O., Corwin, Harold G., Jr., and Olowin, Ronald P., 1989. A Catalog of Rich Clusters of Galaxies. *Astrophysical Journal Supplement Series*, 70, 1.

Adams, Fred, and Laughlin, Greg, 1999. *The Five Ages of the Universe* (New York: The Free Press).

Adams, Walter S., 1941. Some Results with the Coudé Spectrograph of the Mount Wilson Observatory. *Astrophysical Journal*, 93, 11.

Aguirre, Anthony N., 1999. Dust Versus Cosmic Acceleration. *Astrophysical Journal Letters*, 512, L19.

Albrecht, Andreas, and Steinhardt, Paul J., 1982. Cosmology for Grand Unified Theories with Radiatively Induced Symmetry Breaking. *Physical Review Letters*, 48, 1220.

Alcock, Charles, 1995. Interview with Ken Croswell: November 13, 1995.

———, 2000. Interview with Ken Croswell: June 22, 2000.

Alcock, C., Akerlof, C. W., Allsman, R. A., Axelrod, T. S., Bennett, D. P., Chan, S., Cook, K. H., Freeman, K. C., Griest, K., Marshall, S. L., Park, H-S., Perlmutter, S., Peterson, B. A., Pratt, M. R., Quinn, P. J., Rodgers, A. W., Stubbs, C. W., and Sutherland, W., 1993. Possible Gravitational Microlensing of a Star in the Large Magellanic Cloud. *Nature*, 365, 621.

Alcock, C., Allsman, R. A., Alves, D. R., Axelrod, T. S., Becker, A. C., Bennett, D. P., Cook, K. H., Dalal, N., Drake, A. J., Freeman, K. C., Geha, M., Griest, K., Lehner, M. J., Marshall, S. L., Minniti, D., Nelson, C. A., Peterson, B. A., Popowski, P., Pratt, M. R., Quinn, P. J., Stubbs, C. W., Sutherland, W., Tomaney, A. B., Vandehei, T., and Welch, D., 2000. The MACHO Project:

Microlensing Results from 5.7 Years of Large Magellanic Cloud Observations. *Astrophysical Journal*, 542, 281.

Aldering, Greg, Knop, Robert, and Nugent, Peter, 2000. The Rise Times of High- and Low-Redshift Type Ia Supernovae Are Consistent. *Astronomical Journal*, 119, 2110.

Alpher, Ralph, 1999. Interview with Ken Croswell: August 16, 1999.

Alpher, R. A., Bethe, H., and Gamow, G., 1948. The Origin of Chemical Elements. *Physical Review*, 73, 803.

Alpher, Ralph A., and Herman, Robert, 1948. Evolution of the Universe. *Nature*, 162, 774.

———, 1972a. Reflections on "Big Bang" Cosmology. In *Cosmology, Fusion and Other Matters*, edited by Frederick Reines (Boulder, Colorado: Colorado Associated University Press), p. 1.

———, 1972b. Memories of Gamow. In *Cosmology, Fusion and Other Matters*, edited by Frederick Reines (Boulder, Colorado: Colorado Associated University Press), p. 304.

———, 1988. Reflections on Early Work on "Big Bang" Cosmology. *Physics Today*, 41, No. 8 (August 1988), p. 24.

———, 1990. Early Work on "Big Bang" Cosmology and the Cosmic Blackbody Radiation. In *Modern Cosmology in Retrospect*, edited by B. Bertotti, R. Balbinot, S. Bergia, and A. Messina (Cambridge: Cambridge University Press), p. 129.

———, 2001. *Genesis of the Big Bang* (New York: Oxford University Press).

Anonymous, 1864. Analysis of Light from the Nebulae. *Monthly Notices of the Royal Astronomical Society*, 25, 112.

———, 1931. Le Maitre Suggests One, Single, Great Atom, Embracing All Energy, Started the Universe. *New York Times*, May 19, 1931, p. 3.

———, 1950. Mortals and Morals. *The Listener*, 43, 416.

———, 1951. Behind Every Door: God. *Time*, 58, No. 23 (December 3, 1951), p. 75.

———, 1953. Double the Universe. *Time*, 61, No. 2 (January 12, 1953), p. 60.

———, 1959. Discuss Origin of Universe. *Science News Letter*, 76, No. 2 (July 11, 1959), p. 22.

———, 1992. Hold the Front Page! *New Scientist*, 134, No. 1819 (May 2, 1992), p. 3.

———, 1998. Neutrinos Have Mass. *Astronomy*, 26, No. 9 (September 1998), p. 24.

Aubourg, E., Bareyre, P., Bréhin, S., Gros, M., Lachièze-Rey, M., Laurent, B., Lesquoy, E., Magneville, C., Milsztajn, A., Moscoso, L., Queinnec, F., Rich, J., Spiro, M., Vigroux, L., Zylberajch, S., Ansari, R., Cavalier, F., Moniez, M., Beaulieu, J.-P., Ferlet, R., Grison, Ph., Vidal-Madjar, A., Guibert, J., Moreau, O., Tajahmady, F., Maurice, E., Prévôt, L., and Gry, C., 1993. Evidence for Gravitational Microlensing by Dark Objects in the Galactic Halo. *Nature*, 365, 623.

Baade, W., 1944a. The Resolution of Messier 32, NGC 205, and the Central Region of the Andromeda Nebula. *Astrophysical Journal*, 100, 137.

———, 1944b. NGC 147 and NGC 185, Two New Members of the Local Group of Galaxies. *Astrophysical Journal*, 100, 147.

———, 1956. The Period-Luminosity Relation of the Cepheids. *Publications of the Astronomical Society of the Pacific*, 68, 5.

Baade, W., and Minkowski, R., 1954. Identification of the Radio Sources in Cassiopeia, Cygnus A, and Puppis A. *Astrophysical Journal*, 119, 206.

Babcock, Horace W., 1939. The Rotation of the Andromeda Nebula. *Lick Observatory Bulletin*, 19, 41 (No. 498).

Bahcall, Neta, 2000. Interview with Ken Croswell: July 18, 2000.

Bahcall, N. A., Cen, R., Davé, R., Ostriker, J. P., and Yu, Q., 2000. The Mass-to-Light Function: Antibias and $\Omega_m$. *Astrophysical Journal*, 541, 1.

Bahcall, Neta A., and Fan, Xiaohui, 1998. The Most Massive Distant Clusters: Determining $\Omega$ and $\sigma_8$. *Astrophysical Journal*, 504, 1.

Beale, W. J., 1950. The Nature of the Universe. *The Listener*, 43, 433.

Beatty, Cheryl J., and Tresch Fienberg, Richard, 1994. Participatory Cosmology: The Big Bang Challenge. *Sky and Telescope*, 87, No. 3 (March 1994), p. 20.

Becker, Barbara J., 2001. Visionary Memories: William Huggins and the Origins of Astrophysics. *Journal for the History of Astronomy*, 32, 43.

Bennett, Charles, 2000. Interview with Ken Croswell: June 13, 2000.

Berendzen, R., and Hart, R., 1973. Adriaan van Maanen's Influence on the Island Universe Theory. *Journal for the History of Astronomy*, 4, 46; 4, 73.

Berendzen, Richard, Hart, Richard, and Seeley, Daniel, 1976. *Man Discovers the Galaxies* (New York: Science History Publications).

Berendzen, Richard, and Hoskin, Michael, 1971. Hubble's Announcement of Cepheids in Spiral Nebulae. *Astronomical Society of the Pacific Leaflet* No. 504.

Bernheimer, W. E., 1932. A Metagalactic Cloud Between Perseus and Pegasus. *Nature*, 130, 132.

Bessel, F. W., 1838. A Letter from Professor Bessel to Sir J. Herschel. *Monthly Notices of the Royal Astronomical Society*, 4, 152.

Black, David C., 1971. Some Implications of a New Value for the Primordial Solar Deuterium-Hydrogen Ratio. *Nature Physical Science*, 234, 148.

———, 1972. On the Origins of Trapped Helium, Neon and Argon Isotopic Variations in Meteorites—I. Gas-Rich Meteorites, Lunar Soil and Breccia. *Geochimica et Cosmochimica Acta*, 36, 347.

Blumenthal, George R., Faber, S. M., Primack, Joel R., and Rees, Martin J., 1984. Formation of Galaxies and Large-Scale Structure with Cold Dark Matter. *Nature*, 311, 517.

Boesgaard, Ann Merchant, and Steigman, Gary, 1985. Big Bang Nucleosynthesis: Theories and Observations. *Annual Review of Astronomy and Astrophysics*, 23, 319.

Bohlin, Karl, 1907. Versuch einer Bestimmung der Parallaxe des Andromedanebels. *Astronomische Nachrichten*, 176, 205.

Bolte, Michael, 1998. Interview with Ken Croswell: February 4, 1998; February 5, 1998.

Bondi, Hermann, 1952a. *Cosmology* (Cambridge: Cambridge University Press).

———, 1952b. Fact and Inference in Experiment and Observation. *Observatory*, 72, 89.

———, 1955. Fact and Inference in Theory and in Observation. *Vistas in Astronomy*, 1, 155.

———, 1982. Steady State Origins: Comments I. In *Cosmology and Astrophysics*, edited by Yervant Terzian and Elizabeth M. Bilson (Ithaca: Cornell University Press), p. 58.

———, 1988. Steady-State Cosmology. *Quarterly Journal of the Royal Astronomical Society*, 29, 65.

———, 1990a. *Science, Churchill and Me* (Oxford: Pergamon).

———, 1990b. The Cosmological Scene 1945–1952. In *Modern Cosmology in Retrospect*, edited by B. Bertotti, R. Balbinot, S. Bergia, and A. Messina (Cambridge: Cambridge University Press), p. 189.

———, 1999. Interview with Ken Croswell: August 24, 1999.

Bondi, H., and Gold, T., 1948. The Steady-State Theory of the Expanding Universe. *Monthly Notices of the Royal Astronomical Society*, 108, 252.

Bondi, Hermann, Gold, Thomas, and Hoyle, Fred, 1995. Origins of Steady-State Theory. *Nature*, 373, 10.

Bosma, Albert, 1978. *The Distribution and Kinematics of Neutral Hydrogen in Spiral Galaxies of Various Morphological Types*. Ph.D. thesis (Groningen: University of Groningen).

Branch, David, 1998. Type Ia Supernovae and the Hubble Constant. *Annual Review of Astronomy and Astrophysics*, 36, 17.

———, 2000. Interview with Ken Croswell: July 19, 2000.

Brashear, Ronald W., and Hetherington, Norriss S., 1991. The Hubble-van Maanen Conflict Over Internal Motions in Spiral Nebulae: Yet More New Information on an Already Old Topic. *Vistas in Astronomy*, 34, 415.

Burbidge, E. Margaret, Burbidge, G. R., Fowler, William A., and Hoyle, F., 1957. Synthesis of the Elements in Stars. *Reviews of Modern Physics*, 29, 547.

Buta, Ronald J., and McCall, Marshall L., 1999. The IC 342/Maffei Group Revealed. *Astrophysical Journal Supplement Series*, 124, 33.

C., J., 1869. William Parsons. *Monthly Notices of the Royal Astronomical Society*, 29, 123.

Calinger, Ronald, 1979. Kant and Newtonian Science: The Pre-Critical Period. *Isis*, 70, 349.

Cameron, A. G. W., 1958. *Stellar Evolution, Nuclear Astrophysics, and Nucleogenesis*, CRL-41, Second Edition (Chalk River, Ontario: Atomic Energy of Canada Limited).

Campbell, W. W., 1914. Historical Quotations. *Publications of the Astronomical Society of the Pacific*, 26, 87.

Carey, Walter, 1950. The Nature of the Universe. *The Listener*, 43, 477.

Carroll, Sean M., Press, William H., and Turner, Edwin L., 1992. The Cosmological Constant. *Annual Review of Astronomy and Astrophysics*, 30, 499.

Cayrel, Roger, 1968. The Location of a Few Subdwarfs in the Theoretical H-R Diagram and the Helium Content of Population II. *Astrophysical Journal*, 151, 997.

Chamberlain, Joseph W., and Aller, Lawrence H., 1951. The Atmospheres of A-Type Subdwarfs and 95 Leonis. *Astrophysical Journal*, 114, 52.

Charlier, C. V. L., 1925. On the Structure of the Universe. *Publications of the Astronomical Society of the Pacific*, 37, 53.

Chincarini, Guido, and Rood, Herbert J., 1975. Size of the Coma Cluster. *Nature*, 257, 294.

———, 1976. The Coma Supercluster: Analysis of Zwicky-Herzog Cluster 16 in Field 158. *Astrophysical Journal*, 206, 30.

———, 1980. The Cosmic Tapestry. *Sky and Telescope*, 59, 364 (May 1980).

Chiu, Hong-Yee, 1964. Gravitational Collapse. *Physics Today*, 17, No. 5 (May 1964), p. 21.

Chown, Marcus, 1993. *Afterglow of Creation* (London: Arrow).

Christianson, Gale E., 1984. *In the Presence of the Creator* (New York: The Free Press).

———, 1995. *Edwin Hubble* (New York: Farrar, Straus and Giroux).

Christy, Robert F., 1966. A Study of Pulsation in RR Lyrae Models. *Astrophysical Journal*, 144, 108.

Clark, Ronald W., 1971. *Einstein* (New York: World Publishing).

Clerke, Agnes M., 1890. *The System of the Stars* (London: Longmans, Green, and Company).

Cook, Kem H., Aaronson, Marc, and Illingworth, Garth, 1986. Discovery of Cepheids in M101. *Astrophysical Journal Letters*, 301, L45.

Copleston, Frederick C., 1950. The Nature of the Universe. *The Listener*, 43, 477.

Cornwell, John, 1999. *Hitler's Pope* (New York: Viking).

Croswell, Ken, 1992. The Supernova at the End of the Universe. *New Scientist*, 136, No. 1846 (November 7, 1992), p. 17.

———, 1995. *The Alchemy of the Heavens* (New York: Doubleday/Anchor).

———, 1997. *Planet Quest* (New York: The Free Press).

———, 1998. Uneasy Truce. *New Scientist*, 158, No. 2136 (May 30, 1998), p. 42.

———, 1999. *Magnificent Universe* (New York: Simon and Schuster).

Curtis, H. D., 1917a. New Stars in Spiral Nebulae. *Publications of the Astronomical Society of the Pacific*, 29, 180.

———, 1917b. Novae in Spiral Nebulae and the Island Universe Theory. *Publications of the Astronomical Society of the Pacific*, 29, 206.

———, 1917c. Three Novae in Spiral Nebulae. *Lick Observatory Bulletin*, 9, 108 (No. 300).

D'Agnese, Joseph, 1999. The Last Big Bang Man Left Standing. *Discover*, 20, No. 7 (July 1999), p. 60.

Daly, Ruth, 2000. Interview with Ken Croswell: July 17, 2000.

Davies, Paul, 1994. *The Last Three Minutes* (New York: Basic).

Davis, Marc, Huchra, John, Latham, David W., and Tonry, John, 1982. A Survey of Galaxy Redshifts. II. The Large Scale Space Distribution. *Astrophysical Journal*, 253, 423.

De Bernardis, P., Ade, P. A. R., Bock, J. J., Bond, J. R., Borrill, J., Boscaleri, A., Coble, K., Crill, B. P., De Gasperis, G., Farese, P. C., Ferreira, P. G., Ganga, K., Giacometti, M., Hivon, E., Hristov, V. V., Iacoangeli, A., Jaffe, A. H., Lange, A. E., Martinis, L., Masi, S., Mason, P. V., Mauskopf, P. D., Melchiorri, A., Miglio, L., Montroy, T., Netterfield, C. B., Pascale, E., Piacentini, F., Pogosyan, D., Prunet, S., Rao, S., Romeo, G., Ruhl, J. E., Scaramuzzi, F., Sforna, D., and Vittorio, N., 2000. A Flat Universe from High-Resolution Maps of the Cosmic Microwave Background Radiation. *Nature*, 404, 955.

De Lapparent, Valérie, Geller, Margaret J., and Huchra, John P., 1986. A Slice of the Universe. *Astrophysical Journal Letters*, 302, L1.

Deprit, Andre, 1984. Monsignor Georges Lemaitre. In *The Big Bang and Georges Lemaître*, edited by A. Berger (Dordrecht: D. Reidel), p. 363.

De Sitter, W., 1917. On Einstein's Theory of Gravitation, and Its Astronomical Consequences. Third Paper. *Monthly Notices of the Royal Astronomical Society*, 78, 3.

De Vaucouleurs, Gérard, 1953. Evidence for a Local Supergalaxy. *Astronomical Journal*, 58, 30.

———, 1970. The Brightest Star Clusters in Galaxies as Distance Indicators. *Astrophysical Journal*, 159, 435.

———, 1977. Distances of the Virgo, Fornax and Hydra Clusters of Galaxies and the Local Value of the Hubble Ratio. *Nature*, 266, 126.

———, 1983. The Distance Scale of the Universe. *Sky and Telescope*, 66, 511 (December 1983).

———, 1985. The Supernova of 1885 in Messier 31. *Sky and Telescope*, 70, 115 (August 1985).

———, 1987. Discovering M31's Spiral Shape. *Sky and Telescope*, 74, 595 (December 1987).

———, 1989. Who Discovered the Local Supercluster of Galaxies? *Observatory*, 109, 237.

De Vaucouleurs, G., and Corwin, H. G., Jr., 1985. S Andromedae 1885: A Centennial Review. *Astrophysical Journal*, 295, 287.

Dewhirst, David W., and Hoskin, Michael, 1991. The Rosse Spirals. *Journal for the History of Astronomy*, 22, 257.

Dicke, Robert H., 1970. *Gravitation and the Universe* (Philadelphia: American Philosophical Society).

Dicke, R. H., and Peebles, P. J. E., 1979. The Big Bang Cosmology—Enigmas and Nostrums. In *General Relativity*, edited by S. W. Hawking and W. Israel (Cambridge: Cambridge University Press), p. 504.

Disney, M. J., 2000. The Case Against Cosmology. *General Relativity and Gravitation*, 32, 1125.

Doroshkevich, A. G., Sunyaev, R. A., and Zel'dovich, Ya. B., 1974. The Formation of Galaxies in Friedmannian Universes. In *Confrontation of Cosmological*

*Theories with Observational Data*, IAU Symposium 63, edited by M. S. Longair (Dordrecht: D. Reidel), p. 213.

Dressler, Alan, 1994. *Voyage to the Great Attractor* (New York: Knopf).

———, 2000. Interview with Ken Croswell: May 23, 2000.

Dressler, Alan, and Faber, S. M., 1990. Confirmation of a Large-Scale, Large-Amplitude Flow in the Direction of the Great Attractor. *Astrophysical Journal*, 354, 13.

Dressler, Alan, Faber, S. M., Burstein, David, Davies, Roger L., Lynden-Bell, Donald, Terlevich, R. J., and Wegner, Gary, 1987. Spectroscopy and Photometry of Elliptical Galaxies: A Large-Scale Streaming Motion in the Local Universe. *Astrophysical Journal Letters*, 313, L37.

Eddington, Arthur, 1928. *The Nature of the Physical World* (New York: Macmillan).

———, 1933. *The Expanding Universe* (New York: Macmillan).

———, 1939. *The Philosophy of Physical Science* (New York: Macmillan).

Edge, David O., and Mulkay, Michael J., 1976. *Astronomy Transformed* (New York: Wiley).

Edge, D. O., Scheuer, P. A. G., and Shakeshaft, J. R., 1958. Evidence on the Spatial Distribution of Radio Sources Derived from a Survey at a Frequency of 159 Mc/s. *Monthly Notices of the Royal Astronomical Society*, 118, 183.

Edge, D. O., Shakeshaft, J. R., McAdam, W. B., Baldwin, J. E., and Archer, S., 1959. A Survey of Radio Sources at a Frequency of 159 Mc/s. *Memoirs of the Royal Astronomical Society*, 68, 37.

Efstathiou, G., Sutherland, W. J., and Maddox, S. J., 1990. The Cosmological Constant and Cold Dark Matter. *Nature*, 348, 705.

Eggen, O. J., and Sandage, A. R., 1962. On the Existence of Subdwarfs in the ($M_{bol}$, Log $T_e$)-Plane. II. *Astrophysical Journal*, 136, 735.

Einasto, Jaan, Jôeveer, Mihkel, and Saar, Enn, 1980. Structure of Superclusters and Supercluster Formation. *Monthly Notices of the Royal Astronomical Society*, 193, 353.

Einasto, Jaan, Kaasik, Ants, and Saar, Enn, 1974. Dynamic Evidence on Massive Coronas of Galaxies. *Nature*, 250, 309.

Einstein, Albert, 1917. Kosmologische Betrachtungen zur allgemeinen Relativitätstheorie. *Sitzungsberichte der Königlich Preussische Akademie der Wissenschaften zu Berlin*, 142.

Ellis, G. F. R., 1990. Innovation, Resistance and Change: The Transition to the Expanding Universe. In *Modern Cosmology in Retrospect*, edited by B. Bertotti, R. Balbinot, S. Bergia, and A. Messina (Cambridge: Cambridge University Press), p. 97.

Emerson, D. T., and Baldwin, J. E., 1973. The Rotation Curve and Mass Distribution in M31. *Monthly Notices of the Royal Astronomical Society*, 165, 9P.

Engel, Alfred (translator), and Schucking, Engelbert (consultant), 1997. *The Collected Papers of Albert Einstein, Volume 6. The Berlin Years: Writings, 1914–1917* (Princeton: Princeton University Press).

Evans, David S., 1980. Nicolas de la Caille and the Southern Sky. *Sky and Telescope*, 60, 4 (July 1980).

Faber, S. M., and Gallagher, J. S., 1979. Masses and Mass-to-Light Ratios of Galaxies. *Annual Review of Astronomy and Astrophysics*, 17, 135.

Faber, S. M., and Lin, D. N. C., 1983. Is There Nonluminous Matter in Dwarf Spheroidal Galaxies? *Astrophysical Journal Letters*, 266, L17.

Faulkner, John, 1967. Quasi-Homology Relations and the Eggen-Sandage Residue. *Astrophysical Journal*, 147, 617.

Faulkner, J., and Iben, Icko, Jr., 1966. The Evolution of Population II Stars. *Astrophysical Journal*, 144, 995.

Feast, M. W., and Catchpole, R. M., 1997. The Cepheid Period-Luminosity Zero-Point from Hipparcos Trigonometrical Parallaxes. *Monthly Notices of the Royal Astronomical Society*, 286, L1.

Fernie, J. D., 1969. The Period-Luminosity Relation: A Historical Review. *Publications of the Astronomical Society of the Pacific*, 81, 707.

———, 1970. The Historical Quest for the Nature of the Spiral Nebulae. *Publications of the Astronomical Society of the Pacific*, 82, 1189.

Ferris, Timothy, 1977. *The Red Limit* (New York: Morrow).

———, 1993. Needed: A Better Name for the Big Bang. *Sky and Telescope*, 86, No. 2 (August 1993), p. 4.

Filippenko, Alex, 2000. Interview with Ken Croswell: September 22, 2000.

Fitzpatrick, E. L., Ribas, I., Guinan, E. F., DeWarf, L. E., Maloney, F. P., and Massa, D., 2001. The Distance to the Large Magellanic Cloud from Eclipsing Binaries. II. HV 982. *Astrophysical Journal*, in press.

Freedman, Wendy, 1998. Interview with Ken Croswell: February 6, 1998; February 11, 1998.

———, 2000. Interview with Ken Croswell: July 19, 2000.

Freedman, Wendy L., and Madore, Barry F., 1988. Distances to the Galaxies M81 and NGC 2403 from CCD I Band Photometry of Cepheids. *Astrophysical Journal Letters*, 332, L63.

Freedman, W. L., Madore, B. F., Gibson, B. K., Ferrarese, L., Kelson, D. D., Sakai, S., Mould, J. R., Kennicutt, R. C., Jr., Ford, H. C., Graham, J. A., Huchra, J. P., Hughes, S. M. G., Illingworth, G. D., Macri, L. M., and Stetson, P. B., 2001. Final Results from the Hubble Space Telescope Key Project to Measure the Hubble Constant. *Astrophysical Journal*, 553, 47.

Freedman, Wendy L., Madore, Barry F., Mould, Jeremy R., Hill, Robert, Ferrarese, Laura, Kennicutt, Robert C., Jr., Saha, Abhijit, Stetson, Peter B., Graham, John A., Ford, Holland, Hoessel, John G., Huchra, John, Hughes, Shaun M., and Illingworth, Garth D., 1994. Distance to the Virgo Cluster Galaxy M100 from Hubble Space Telescope Observations of Cepheids. *Nature*, 371, 757.

Freeman, K. C., 1970. On the Disks of Spiral and S0 Galaxies. *Astrophysical Journal*, 160, 811.

———, 2000. Interview with Ken Croswell: February 6, 2000.

Friedmann, A., 1922. On the Curvature of Space. 1986 reprint in *Cosmological Constants*, edited by Jeremy Bernstein and Gerald Feinberg (New York: Columbia University Press), p. 49.

———, 1924. On the Possibility of a World with Constant Negative Curvature. 1986 reprint in *Cosmological Constants*, edited by Jeremy Bernstein and Gerald Feinberg (New York: Columbia University Press), p. 59.

Frogel, J. A., Persson, S. E., Aaronson, M., Becklin, E. E., Matthews, K., and Neugebauer, G., 1975. Stellar Content of the Nuclei of Elliptical Galaxies Determined from 2.3-Micron CO Band Strengths. *Astrophysical Journal Letters*, 195, L15.

Fukugita, M., Futamase, T., and Kasai, M., 1990. A Possible Test for the Cosmological Constant with Gravitational Lenses. *Monthly Notices of the Royal Astronomical Society*, 246, 24P.

Gamow, George, 1940. *The Birth and Death of the Sun* (New York: Viking).

———, 1952a. *The Creation of the Universe* (New York: Viking).

———, 1952b. The Role of Turbulence in the Evolution of the Universe. *Physical Review*, 86, 251.

———, 1970. *My World Line* (New York: Viking).

Garnavich, P. M., Kirshner, R. P., Challis, P., Tonry, J., Gilliland, R. L., Smith, R. C., Clocchiatti, A., Diercks, A., Filippenko, A. V., Hamuy, M., Hogan, C. J., Leibundgut, B., Phillips, M. M., Reiss, D., Riess, A. G., Schmidt, B. P., Schommer, R. A., Spyromilio, J., Stubbs, C., Suntzeff, N. B., and Wells, L., 1998. Constraints on Cosmological Models from Hubble Space Telescope Observations of High-z Supernovae. *Astrophysical Journal Letters*, 493, L53.

Geiss, J., and Reeves, H., 1972. Cosmic and Solar System Abundances of Deuterium and Helium-3. *Astronomy and Astrophysics*, 18, 126.

Geller, Margaret, 2000. Interview with Ken Croswell: June 14, 2000.

Geller, Margaret J., de Lapparent, Valerie, and Kurtz, Michael J., 1984. The Shane-Wirtanen Counts. *Astrophysical Journal Letters*, 287, L55.

Geller, Margaret J., and Huchra, John P., 1989. Mapping the Universe. *Science*, 246, 897.

Gilman, Carolyn, 1978. John Goodricke and His Variable Stars. *Sky and Telescope*, 56, 400 (November 1978).

Gilmore, Gerard, King, Ivan R., and van der Kruit, Pieter C., 1990. *The Milky Way as a Galaxy* (Mill Valley, California: University Science Books).

Gingerich, Owen, 1953. Messier and His Catalogue. *Sky and Telescope*, 12, 255 (August 1953); 12, 288 (September 1953).

———, 1960. The Missing Messier Objects. *Sky and Telescope*, 20, 196 (October 1960).

———, 1987. The Mysterious Nebulae, 1610–1924. *Journal of the Royal Astronomical Society of Canada*, 81, 113.

———, 1988. J. L. E. Dreyer and His NGC. *Sky and Telescope*, 76, 621 (December 1988).

Gingerich, Owen, and Welther, Barbara, 1985. Harlow Shapley and the Cepheids. *Sky and Telescope*, 70, 540 (December 1985).

Giovanelli, Riccardo, and Haynes, Martha P., 1982. The Lynx-Ursa Major Supercluster. *Astronomical Journal*, 87, 1355.

Glyn Jones, Kenneth, 1971. The Observational Basis for Kant's *Cosmogony*: A Critical Analysis. *Journal for the History of Astronomy*, 2, 29.

———, 1976. S Andromedae, 1885: An Analysis of Contemporary Reports and a Reconstruction. *Journal for the History of Astronomy*, 7, 27.

———, 1991. *Messier's Nebulae and Star Clusters*, Second Edition (Cambridge: Cambridge University Press).

Godart, O., and Heller, M., 1985. *Cosmology of Lemaître* (Tucson: Pachart).

Gold, Thomas, 1951. The Origin of Cosmic Radio Noise. In *A Source Book in Astronomy and Astrophysics, 1900–1975*, edited by Kenneth R. Lang and Owen Gingerich (Cambridge: Harvard University Press), p. 782.

———, 1968. Rotating Neutron Stars as the Origin of the Pulsating Radio Sources. *Nature*, 218, 731.

———, 1982. Steady State Origins: Comments II. In *Cosmology and Astrophysics*, edited by Yervant Terzian and Elizabeth M. Bilson (Ithaca: Cornell University Press), p. 62.

———, 1999. Interview with Ken Croswell: August 13, 1999.

Goodricke, C. A., 1912. Gift to the Society of a Portrait of John Goodricke. *Monthly Notices of the Royal Astronomical Society*, 73, 3.

Greenstein, G. S., Truran, J. W., and Cameron, A. G. W., 1967. Helium Deficiency in Old Halo B Stars. *Nature*, 213, 871.

Greenstein, Jesse L., 1966. The Nature of the Faint Blue Stars. *Astrophysical Journal*, 144, 496.

Gregory, Stephen A., 1975. Redshifts and Morphology of Galaxies in the Coma Cluster. *Astrophysical Journal*, 199, 1.

———, 1988. The Structure of the Visible Universe. *Astronomy*, 16, No. 4 (April 1988), p. 42.

———, 2000. Interview with Ken Croswell: May 22, 2000.

Gregory, Stephen A., and Thompson, Laird A., 1978. The Coma/A1367 Supercluster and Its Environs. *Astrophysical Journal*, 222, 784.

———, 1982. Superclusters and Voids in the Distribution of Galaxies. *Scientific American*, 246, No. 3 (March 1982), p. 106.

Gregory, Stephen A., Thompson, Laird A., and Tifft, William G., 1981. The Perseus Supercluster. *Astrophysical Journal*, 243, 411.

Groenewegen, M. A. T., and Salaris, M., 2001. The LMC Eclipsing Binary HV 2274 Revisited. *Astronomy and Astrophysics*, 366, 752.

Guerra, Erick J., and Daly, Ruth A., 1998. Central Engines of Active Galactic Nuclei: Properties of Collimated Outflows and Applications for Cosmology. *Astrophysical Journal*, 493, 536.

Guinan, Edward, 2000. Interview with Ken Croswell: August 3, 2000.

Guinan, E. F., Fitzpatrick, E. L., DeWarf, L. E., Maloney, F. P., Maurone, P. A., Ribas, I., Pritchard, J. D., Bradstreet, D. H., and Giménez, A., 1998. The Distance to the Large Magellanic Cloud from the Eclipsing Binary HV 2274. *Astrophysical Journal Letters*, 509, L21.

Guth, Alan H., 1981. Inflationary Universe: A Possible Solution to the Horizon and Flatness Problems. *Physical Review D*, 23, 347.

———, 1997. *The Inflationary Universe* (Reading, Massachusetts: Addison-Wesley).

———, 2000. Interview with Ken Croswell: April 29, 2000.

Guth, Alan H., and Steinhardt, Paul J., 1984. The Inflationary Universe. *Scientific American*, 250, No. 5 (May 1984), p. 116.

Guth, Alan H., and Tye, S.-H. H., 1980. Phase Transitions and Magnetic Monopole Production in the Very Early Universe. *Physical Review Letters*, 44, 631.

Hall, John S., 1970. V. M. Slipher's Trailblazing Career. *Sky and Telescope*, 39, 84 (February 1970).

Halley, Edmond, 1721a. Of the Infinity of the Sphere of Fix'd Stars. *Philosophical Transactions*, 31, 22.

———, 1721b. Of the Number, Order, and Light of the Fix'd Stars. *Philosophical Transactions*, 31, 24.

Harrison, E. R., 1964. Olbers' Paradox. *Nature*, 204, 271.

———, 1965. Olbers' Paradox and the Background Radiation Density in an Isotropic Homogeneous Universe. *Monthly Notices of the Royal Astronomical Society*, 131, 1.

———, 1974. Why the Sky Is Dark at Night. *Physics Today*, 27, 30.

———, 1981. *Cosmology* (Cambridge: Cambridge University Press).

———, 1986. Newton and the Infinite Universe. *Physics Today*, 39, No. 2 (February 1986), p. 24.

———, 1987. *Darkness at Night* (Cambridge: Harvard University Press).

———, 1990a. The Dark Night-Sky Riddle, "Olbers's Paradox." In *The Galactic and Extragalactic Background Radiation*, edited by Stuart Bowyer and Christoph Leinert (Dordrecht: Kluwer), p. 3.

———, 1990b. Olbers' Paradox in Recent Times. In *Modern Cosmology in Retrospect*, edited by B. Bertotti, R. Balbinot, S. Bergia, and A. Messina (Cambridge: Cambridge University Press), p. 33.

———, 1998. Cosmology Comes of Age. *Sky and Telescope*, 96, No. 2 (August 1998), p. 79.

———, 1999. Interview with Ken Croswell: February 15, 1999.

———, 2000. *Cosmology*, Second Edition (Cambridge: Cambridge University Press).

Hartwick, F. D. A., and Sargent, W. L. W., 1978. Radial Velocities for Outlying Satellites and Their Implications for the Mass of the Galaxy. *Astrophysical Journal*, 221, 512.

Haynes, Martha P., and Giovanelli, Riccardo, 1986. The Connection Between Pisces-Perseus and the Local Supercluster. *Astrophysical Journal Letters*, 306, L55.

Herschel, John F. W., 1847. *Results of Astronomical Observations Made During the Years 1834, 5, 6, 7, 8, at the Cape of Good Hope* (London: Smith, Elder and Company).

Hertzsprung, E., 1913. Über die räumliche Verteilung der Veränderlichen vom Delta Cephei-Typus. *Astronomische Nachrichten*, 196, 201.

Hetherington, Norriss S., 1971. The Measurement of Radial Velocities of Spiral Nebulae. *Isis*, 62, 309.

———, 1974a. Adriaan van Maanen on the Significance of Internal Motions in Spiral Nebulae. *Journal for the History of Astronomy*, 5, 52.

———, 1974b. Edwin Hubble on Adriaan van Maanen's Internal Motions in Spiral Nebulae. *Isis*, 65, 390.

———, 1975. The Simultaneous 'Discovery' of Internal Motions in Spiral Nebulae. *Journal for the History of Astronomy*, 6, 115.

———, 1990. *The Edwin Hubble Papers* (Tucson: Pachart).

———, 1996. *Hubble's Cosmology* (Tucson: Pachart).

Hetherington, Norriss S., and Brashear, Ronald S., 1992. Walter S. Adams and the Imposed Settlement Between Edwin Hubble and Adriaan van Maanen. *Journal for the History of Astronomy*, 23, 53.

Hillebrandt, Wolfgang, and Niemeyer, Jens C., 2000. Type Ia Supernova Explosion Models. *Annual Review of Astronomy and Astrophysics*, 38, 191.

Horgan, John, 1992. COBE's Cosmic Cartographer. *Scientific American*, 267, No. 1 (July 1992), p. 34.

Hoskin, Michael, 1970. The Cosmology of Thomas Wright of Durham. *Journal for the History of Astronomy*, 1, 44.

———, 1976. Ritchey, Curtis and the Discovery of Novae in Spiral Nebulae. *Journal for the History of Astronomy*, 7, 47.

———, 1977. Newton, Providence and the Universe of Stars. *Journal for the History of Astronomy*, 8, 77.

———, 1979. Goodricke, Pigott and the Quest for Variable Stars. *Journal for the History of Astronomy*, 10, 23.

———, 1982a. *Stellar Astronomy* (Bucks, England: Science History Publications).

———, 1982b. The First Drawing of a Spiral Nebula. *Journal for the History of Astronomy*, 13, 97.

———, 1985. Stukeley's Cosmology and the Newtonian Origins of Olbers's Paradox. *Journal for the History of Astronomy*, 16, 77.

———, 1990. Rosse, Robinson, and the Resolution of the Nebulae. *Journal for the History of Astronomy*, 21, 331.

Hoskin, Michael, and Rochester, George D., 1992. Thomas Wright and the Royal Society. *Journal for the History of Astronomy*, 23, 167.

Hoyle, F., 1946. The Synthesis of the Elements from Hydrogen. *Monthly Notices of the Royal Astronomical Society*, 106, 343.

———, 1948. A New Model for the Expanding Universe. *Monthly Notices of the Royal Astronomical Society*, 108, 372.

———, 1949a. Stellar Evolution and the Expanding Universe. *Nature*, 163, 196.

———, 1949b. Continuous Creation. *The Listener*, 41, 567.

———, 1950a. The Earth and Nearby Space. *The Listener*, 43, 227.

———, 1950b. The Sun and the Stars. *The Listener*, 43, 271.

———, 1950c. The Origin and Evolution of the Stars. *The Listener*, 43, 321.

———, 1950d. Origin of the Earth and the Planets. *The Listener*, 43, 375.

———, 1950e. Man's Place in the Expanding Universe. *The Listener*, 43, 419.

———, 1950f. *The Nature of the Universe* (New York: Harper).

———, 1954. Commission des Nebuleuses Extragalactiques. *Transactions of the International Astronomical Union*, 8, 397.

———, 1955. *Frontiers of Astronomy* (New York: Harper).

———, 1965. Recent Developments in Cosmology. *Nature*, 208, 111.

———, 1977. *Ten Faces of the Universe* (San Francisco: Freeman).

———, 1981. The Big Bang in Astronomy. *New Scientist*, 92, 521 (November 19, 1981).

———, 1982a. Steady State Cosmology Revisited. In *Cosmology and Astrophysics*, edited by Yervant Terzian and Elizabeth M. Bilson (Ithaca: Cornell University Press), p. 17.

———, 1982b. The Universe: Past and Present Reflections. *Annual Review of Astronomy and Astrophysics*, 20, 1.

———, 1993. Interview with Ken Croswell: August 5, 1993.

———, 1994. *Home Is Where the Wind Blows* (Mill Valley, California: University Science Books).

———, 1999. Interview with Ken Croswell: August 31, 1999.

Hoyle, F., and Tayler, R. J., 1964. The Mystery of the Cosmic Helium Abundance. *Nature*, 203, 1108.

Hoyt, William Graves, 1976. *Lowell and Mars* (Tucson: University of Arizona Press).

———, 1980. *Planets X and Pluto* (Tucson: University of Arizona Press).

Hubble, Edwin P., 1925. Cepheids in Spiral Nebulae. *Popular Astronomy*, 33, 252.

———, 1926a. A Spiral Nebula as a Stellar System, Messier 33. *Astrophysical Journal*, 63, 236.

———, 1926b. Extra-Galactic Nebulae. *Astrophysical Journal*, 64, 321.

———, 1929a. A Spiral Nebula as a Stellar System, Messier 31. *Astrophysical Journal*, 69, 103.

———, 1929b. A Relation Between Distance and Radial Velocity Among Extra-Galactic Nebulae. *Proceedings of the National Academy of Sciences*, 15, 168.

———, 1934. The Distribution of Extra-Galactic Nebulae. *Astrophysical Journal*, 79, 8.

———, 1936. *The Realm of the Nebulae* (New Haven: Yale University Press).

Hubble, Edwin, and Humason, Milton L., 1931. The Velocity-Distance Relation Among Extra-Galactic Nebulae. *Astrophysical Journal*, 74, 43.

Huggins, William, 1864. *Monthly Notices of the Royal Astronomical Society*, 25, 155.

Humason, M. L., Mayall, N. U., and Sandage, A. R., 1956. Redshifts and Magnitudes of Extragalactic Nebulae. *Astronomical Journal*, 61, 97.

Ibata, R. A., Gilmore, G., and Irwin, M. J., 1994. A Dwarf Satellite Galaxy in Sagittarius. *Nature*, 370, 194.

Ibata, Rodrigo A., Richer, Harvey B., Gilliland, Ronald L., and Scott, Douglas, 1999. Faint, Moving Objects in the Hubble Deep Field: Components of the Dark Halo? *Astrophysical Journal Letters*, 524, L95.

Jacoby, George, 2000. Interview with Ken Croswell: July 14, 2000.

Jacoby, George H., Ciardullo, Robin, and Ford, Holland C., 1990. Planetary Nebulae as Standard Candles. V. The Distance to the Virgo Cluster. *Astrophysical Journal*, 356, 332.

Jacoby, George H., and Pierce, Michael J., 1996. Response to Schaefer's Comments on Pierce & Jacoby (1995) Regarding the Type Ia Supernova 1937C. *Astronomical Journal*, 112, 723.

Jaki, Stanley L., 1969. *The Paradox of Olbers' Paradox* (New York: Herder and Herder).

———, 1991. *Olbers Studies* (Tucson: Pachart).

Jôeveer, Mihkel, Einasto, Jaan, and Tago, Erik, 1978. Spatial Distribution of Galaxies and of Clusters of Galaxies in the Southern Galactic Hemisphere. *Monthly Notices of the Royal Astronomical Society*, 185, 357.

Johnson, Francis R., 1937. *Astronomical Thought in Renaissance England* (Baltimore: The John Hopkins Press).

———, 1957. Thomas Digges and the Infinity of the Universe. In *Theories of the Universe*, edited by Milton K. Munitz (New York: The Free Press), p. 184.

Joy, Alfred H., 1937. Some Early Variable Star Observers. *Astronomical Society of the Pacific Leaflet* No. 99.

———, 1940. Spectroscopic Observations of Barnard's Variable in Messier 3. *Astrophysical Journal*, 92, 396.

Kahn, F. D., and Woltjer, L., 1959. Intergalactic Matter and the Galaxy. *Astrophysical Journal*, 130, 705.

Kant, Immanuel, 1755. *Kant's Cosmogony*, translated by W. Hastie and revised by Willy Ley (1968 reprint; New York: Greenwood).

Kelson, Daniel D., Illingworth, Garth D., Freedman, Wendy F., Graham, John A., Hill, Robert, Madore, Barry F., Saha, Abhijit, Stetson, Peter B., Kennicutt, Robert C., Jr., Mould, Jeremy R., Hughes, Shaun M., Ferrarese, Laura, Phelps, Randy, Turner, Anne, Cook, Kem H., Ford, Holland, Hoessel, John G., and Huchra, John, 1996. The Extragalactic Distance Scale Key Project. III. The Discovery of Cepheids and a New Distance to M101 Using the Hubble Space Telescope. *Astrophysical Journal*, 463, 26.

Kirshner, Robert P., Oemler, Augustus, Jr., Schechter, Paul L., and Shectman, Stephen A., 1981. A Million Cubic Megaparsec Void in Boötes? *Astrophysical Journal Letters*, 248, L57.

Kochanek, Christopher, 1995. Interview with Ken Croswell: August 8, 1995.

———, 1996. Is There a Cosmological Constant? *Astrophysical Journal*, 466, 638.

———, 2000. Interview with Ken Croswell: July 11, 2000.

Konigsberg, Eric, 1995. Science Made Easy. *New Republic*, 212, No. 11 (March 13, 1995), p. 16.

Kraan-Korteweg, Renée, 2000. Interview with Ken Croswell: September 15, 2000.

Kraan-Korteweg, Renée C., and Lahav, Ofer, 1998. Galaxies Behind the Milky Way. *Scientific American*, 279, No. 4 (October 1998), p. 50.

———, 2000. The Universe Behind the Milky Way. *Astronomy and Astrophysics Review*, 10, 211.

Kraan-Korteweg, R. C., Loan, A. J., Burton, W. B., Lahav, O., Ferguson, H. C., Henning, P. A., and Lynden-Bell, D., 1994. Discovery of a Nearby Spiral Galaxy Behind the Milky Way. *Nature*, 372, 77.

Kraan-Korteweg, R. C., Woudt, P. A., Cayatte, V., Fairall, A. P., Balkowski, C., and Henning, P. A., 1996. A Nearby Massive Galaxy Cluster Behind the Milky Way. *Nature*, 379, 519.

Kragh, Helge, 1987. The Beginning of the World: Georges Lemaître and the Expanding Universe. *Centaurus*, 32, 114.

———, 1996. *Cosmology and Controversy* (Princeton: Princeton University Press).

Krauss, Lawrence M., and Starkman, Glenn D., 2000. Life, the Universe, and Nothing: Life and Death in an Ever-Expanding Universe. *Astrophysical Journal*, 531, 22.

Kundić, Tomislav, Turner, Edwin L., Colley, Wesley N., Gott, J. Richard III, Rhoads, James E., Wang, Yun, Bergeron, Louis E., Gloria, Karen A., Long, Daniel C., Malhotra, Sangeeta, and Wambsganss, Joachim, 1997. A Robust Determination of the Time Delay in 0957+561A, B and a Measurement of the Global Value of Hubble's Constant. *Astrophysical Journal*, 482, 75.

Kunth, D., and Östlin, G., 2000. The Most Metal-Poor Galaxies. *Astronomy and Astrophysics Review*, 10, 1.

Lasserre, T., Afonso, C., Albert, J. N., Andersen, J., Ansari, R., Aubourg, É., Bareyre, P., Bauer, F., Beaulieu, J. P., Blanc, G., Bouquet, A., Char, S., Charlot, X., Couchot, F., Coutures, C., Derue, F., Ferlet, R., Glicenstein, J. F., Goldman, B., Gould, A., Graff, D., Gros, M., Haïssinski, J., Hamilton, J. C., Hardin, D., de Kat, J., Kim, A., Lesquoy, É., Loup, C., Magneville, C., Mansoux, B., Marquette, J. B., Maurice, É., Milsztajn, A., Moniez, M., Palanque-Delabrouille, N., Perdereau, O., Prévot, L., Regnault, N., Rich, J., Spiro, M., Vidal-Madjar, A., Vigroux, L., and Zylberajch, S., 2000. Not Enough Stellar Mass Machos in the Galactic Halo. *Astronomy and Astrophysics*, 355, L39.

Leary, Warren E., 1990. Spacecraft Sees Few Traces of a Tumultuous Creation. *New York Times*, January 14, 1990, p. 22.

Leavitt, Henrietta S., 1907. 1777 Variables in the Magellanic Clouds. *Annals of Harvard College Observatory*, 60, 87.

———, 1912. Periods of 25 Variable Stars in the Small Magellanic Cloud. *Harvard College Observatory Circular* No. 173.

Leibundgut, Bruno, 2000. Type Ia Supernovae. *Astronomy and Astrophysics Review*, 10, 179.

Leibundgut, B., Schommer, R., Phillips, M., Riess, A., Schmidt, B., Spyromilio, J., Walsh, J., Suntzeff, N., Hamuy, M., Maza, J., Kirshner, R. P., Challis, P., Garnavich, P., Smith, R. C., Dressler, A., and Ciardullo, R., 1996. Time Dilation in the Light Curve of the Distant Type Ia Supernova SN 1995K. *Astrophysical Journal Letters*, 466, L21.

Lemaître, G., 1931a. A Homogeneous Universe of Constant Mass and Increasing Radius Accounting for the Radial Velocity of Extra-Galactic Nebulae. *Monthly Notices of the Royal Astronomical Society*, 91, 483.

———, 1931b. The Beginning of the World from the Point of View of Quantum Theory. *Nature*, 127, 706.

———, 1931c. *Nature*, 128, 704.

———, 1931d. L'expansion de l'espace. *Revue des Questions Scientifiques*, 17, 391.

———, 1950. *The Primeval Atom* (New York: D. Van Nostrand).

———, 1958. Rencontres avec A. Einstein. *Revue des Questions Scientifiques*, 129, 129.

Levy, David H., 1991. *Clyde Tombaugh* (Tucson: University of Arizona Press).

Lewis, Richard S., 1988. *Challenger: The Final Voyage* (New York: Columbia University Press).

Lightman, Alan, and Brawer, Roberta, 1990. *Origins* (Cambridge: Harvard University Press).

Lin, D. N. C., and Faber, S. M., 1983. Some Implications of Nonluminous Matter in Dwarf Spheroidal Galaxies. *Astrophysical Journal Letters*, 266, L21.

Linde, A. D., 1982. A New Inflationary Universe Scenario: A Possible Solution of the Horizon, Flatness, Homogeneity, Isotropy and Primordial Monopole Problems. *Physics Letters B*, 108, 389.

———, 1983. Chaotic Inflation. *Physics Letters B*, 129, 177.

———, 1986. Eternally Existing Self-Reproducing Chaotic Inflationary Universe. *Physics Letters B*, 175, 395.

———, 1994. The Self-Reproducing Inflationary Universe. *Scientific American*, 271, No. 5 (November 1994), p. 48.

———, 2000. Interview with Ken Croswell: April 28, 2000.

Lowell, Percival, 1906. *Mars and Its Canals* (New York: Macmillan).

Lynden-Bell, D., Faber, S. M., Burstein, David, Davies, Roger L., Dressler, Alan, Terlevich, R. J., and Wegner, Gary, 1988. Spectroscopy and Photometry of Elliptical Galaxies. V. Galaxy Streaming Toward the New Supergalactic Center. *Astrophysical Journal*, 326, 19.

Madore, Barry F., 1976. The Distance to NGC 2403. *Monthly Notices of the Royal Astronomical Society*, 177, 157.

Madore, Barry F., Freedman, Wendy L., Silbermann, N., Harding, Paul, Huchra, John, Mould, Jeremy R., Graham, John A., Ferrarese, Laura, Gibson, Brad K., Han, Mingsheng, Hoessel, John G., Hughes, Shaun M., Illingworth, Garth D., Phelps, Randy, Sakai, Shoko, and Stetson, Peter, 1998. A Cepheid Distance to the Fornax Cluster and the Local Expansion Rate of the Universe. *Nature*, 395, 47.

———, 1999. The Hubble Space Telescope Key Project on the Extragalactic Distance Scale. XV. A Cepheid Distance to the Fornax Cluster and Its Implications. *Astrophysical Journal*, 515, 29.

Maran, Stephen P., 1992. Hubble Illuminates the Universe. *Sky and Telescope*, 83, 619 (June 1992).

Mason, Brian, 2000. Interview with Ken Croswell: July 21, 2000.

Mason, Brian S., Myers, Steven T., and Readhead, A. C. S., 2001. An Improved Measurement of the Hubble Constant from the Sunyaev-Zeldovich Effect. *Astrophysical Journal Letters*, in press.

Mather, John, 2000. Interview with Ken Croswell: June 17, 2000.

Mather, John C., and Boslough, John, 1996. *The Very First Light* (New York: Basic).

Mather, J. C., Cheng, E. S., Eplee, R. E., Jr., Isaacman, R. B., Meyer, S. S., Shafer, R. A., Weiss, R., Wright, E. L., Bennett, C. L., Boggess, N. W., Dwek, E., Gulkis, S., Hauser, M. G., Janssen, M., Kelsall, T., Lubin, P. M., Moseley, S. H., Jr., Murdock, T. L., Silverberg, R. F., Smoot, G. F., and Wilkinson, D. T., 1990. A Preliminary Measurement of the Cosmic Microwave Background Spectrum by the Cosmic Background Explorer (COBE) Satellite. *Astrophysical Journal Letters*, 354, L37.

Matsumoto, T., Hayakawa, S., Matsuo, H., Murakami, H., Sato, S., Lange, A. E., and Richards, P. L., 1988. The Submillimeter Spectrum of the Cosmic Background Radiation. *Astrophysical Journal*, 329, 567.

Mayall, N. U., 1954. Edwin Hubble: Observational Cosmologist. *Sky and Telescope*, 13, 78 (January 1954).

McConnell, Malcolm, 1987. *Challenger: A Major Malfunction* (Garden City, New York: Doubleday).

McKellar, Andrew, 1940. Evidence for the Molecular Origin of Some Hitherto Unidentified Interstellar Lines. *Publications of the Astronomical Society of the Pacific*, 52, 187.

———, 1941. Molecular Lines from the Lowest States of Diatomic Molecules Composed of Atoms Probably Present in Interstellar Space. *Publications of the Dominion Astrophysical Observatory*, 7, 251.

McMullin, Ernan, 1981. How Should Cosmology Relate to Theology? In *The Sciences and Theology in the Twentieth Century*, edited by A. R. Peacocke (London: Oriel Press), p. 17.

Merrill, Paul W., 1952. Spectroscopic Observations of Stars of Class S. *Astrophysical Journal*, 116, 21.

Misner, Charles W., 1969. Mixmaster Universe. *Physical Review Letters*, 22, 1071.

Munitz, Milton K. (editor), 1957. *Theories of the Universe* (New York: The Free Press).

O'Dell, C. R., Peimbert, M., and Kinman, T. D., 1964. The Planetary Nebula in the Globular Cluster M15. *Astrophysical Journal*, 140, 119.

Odenwald, Sten, and Fienberg, Richard Tresch, 1993. Galaxy Redshifts Reconsidered. *Sky and Telescope*, 85, No. 2 (February 1993), p. 31.

Olszewski, Edward, 2000. Interview with Ken Croswell: May 18, 2000.

Oort, J. H., 1983. Superclusters. *Annual Review of Astronomy and Astrophysics*, 21, 373.

Oppenheimer, B. R., Hambly, N. C., Digby, A. P., Hodgkin, S. T., and Saumon, D., 2001. Direct Detection of Galactic Halo Dark Matter. *Science*, 292, 698.

Osterbrock, Donald, 1993. Interview with Ken Croswell: May 17, 1993.

———, 2001. *Walter Baade* (Princeton: Princeton University Press).

Ostriker, Jeremiah P., 1993. Astronomical Tests of the Cold Dark Matter Scenario. *Annual Review of Astronomy and Astrophysics*, 31, 689.

———, 1999. Discovery of "Dark Matter" in Clusters of Galaxies. *Astrophysical Journal*, 525, 297.

———, 2000. Interview with Ken Croswell: February 29, 2000; March 5, 2000.

Ostriker, J. P., and Peebles, P. J. E., 1973. A Numerical Study of the Stability of Flattened Galaxies: Or, Can Cold Galaxies Survive? *Astrophysical Journal*, 186, 467.

Ostriker, J. P., Peebles, P. J. E., and Yahil, A., 1974. The Size and Mass of Galaxies, and the Mass of the Universe. *Astrophysical Journal Letters*, 193, L1.

Overbye, Dennis, 1991. *Lonely Hearts of the Cosmos* (New York: HarperCollins).

P., R. K., 1953. A Wider and an Older Universe. *New York Times*, January 4, 1953, sec. 4, p. 7.

Paczyński, Bohdan, 1986a. Gravitational Microlensing by the Galactic Halo. *Astrophysical Journal*, 304, 1.

———, 1986b. Gamma-Ray Bursters at Cosmological Distances. *Astrophysical Journal Letters*, 308, L43.

———, 1995. Interview with Ken Croswell: November 13, 1995.

———, 1996. Gravitational Microlensing in the Local Group. *Annual Review of Astronomy and Astrophysics*, 34, 419.

———, 2000. Interview with Ken Croswell: June 19, 2000.

Parker, Barry, 1986. Discovery of the Expanding Universe. *Sky and Telescope*, 72, 227 (September 1986).

Parodi, B. R., Saha, A., Sandage, A., and Tammann, G. A., 2000. Supernova Type Ia Luminosities, Their Dependence on Second Parameters, and the Value of $H_0$. *Astrophysical Journal*, 540, 634.

Peacock, John A., 1999. *Cosmological Physics* (Cambridge: Cambridge University Press).

Peacock, John A., Cole, Shaun, Norberg, Peder, Baugh, Carlton M., Bland-Hawthorn, Joss, Bridges, Terry, Cannon, Russell D., Colless, Matthew, Collins, Chris, Couch, Warrick, Dalton, Gavin, Deeley, Kathryn, De Propris, Roberto, Driver, Simon P., Efstathiou, George, Ellis, Richard S., Frenk, Carlos S., Glazebrook, Karl, Jackson, Carole, Lahav, Ofer, Lewis, Ian, Lumsden, Stuart, Maddox, Steve, Percival, Will J., Peterson, Bruce A., Price, Ian, Sutherland, Will, and Taylor, Keith, 2001. A Measurement of the Cosmological Mass Density from Clustering in the 2dF Galaxy Redshift Survey. *Nature*, 410, 169.

Pease, Francis G., 1917. A Suspected Nova in N. G. C. 2841. *Publications of the Astronomical Society of the Pacific*, 29, 213.

Peebles, P. J. E., 1965. The Black-Body Radiation Content of the Universe and the Formation of Galaxies. *Astrophysical Journal*, 142, 1317.

———, 1966a. Primeval Helium Abundance and the Primeval Fireball. *Physical Review Letters*, 16, 410.

———, 1966b. Primordial Helium Abundance and the Primordial Fireball. II. *Astrophysical Journal*, 146, 542.

———, 1982. Large-Scale Background Temperature and Mass Fluctuations Due to Scale-Invariant Primeval Perturbations. *Astrophysical Journal Letters*, 263, L1.

———, 1993. *Principles of Physical Cosmology* (Princeton: Princeton University Press).

Peebles, P. J. E., and Wilkinson, David T., 1967. The Primeval Fireball. *Scientific American*, 21, No. 6 (June 1967), p. 28.

Penzias, Arno A., 1972. Cosmology and Microwave Astronomy. In *Cosmology, Fusion and Other Matters*, edited by Frederick Reines (Boulder, Colorado: Colorado Associated University Press), p. 29.

Penzias, A. A., and Wilson, R. W., 1965. A Measurement of Excess Antenna Temperature at 4080 Mc/s. *Astrophysical Journal*, 142, 419.

Perlmutter, Saul, 2000. Interview with Ken Croswell: July 17, 2000.

Perlmutter, S., Aldering, G., Della Valle, M., Deustua, S., Ellis, R. S., Fabbro, S., Fruchter, A., Goldhaber, G., Groom, D. E., Hook, I. M., Kim, A. G., Kim, M. Y., Knop, R. A., Lidman, C., McMahon, R. G., Nugent, P., Pain, R., Panagia, N., Pennypacker, C. R., Ruiz-Lapuente, P., Schaefer, B., and Walton, N., 1998. Discovery of a Supernova Explosion at Half the Age of the Universe. *Nature*, 391, 51.

Perlmutter, S., Aldering, G., Goldhaber, G., Knop, R. A., Nugent, P., Castro, P. G., Deustua, S., Fabbro, S., Goobar, A., Groom, D. E., Hook, I. M., Kim, A. G., Kim, M. Y., Lee, J. C., Nunes, N. J., Pain, R., Pennypacker, C. R., Quimby, R., Lidman, C., Ellis, R. S., Irwin, M., McMahon, R. G., Ruiz-Lapuente, P., Walton, N., Schaefer, B., Boyle, B. J., Filippenko, A. V., Matheson, T., Fruchter, A. S., Panagia, N., Newberg, H. J. M., and Couch, W. J., 1999. Measurements of $\Omega$ and $\Lambda$ from 42 High-Redshift Supernovae. *Astrophysical Journal*, 517, 565.

Perlmutter, S., Gabi, S., Goldhaber, G., Goobar, A., Groom, D. E., Hook, I. M., Kim, A. G., Kim, M. Y., Lee, J. C., Pain, R., Pennypacker, C. R., Small, I. A., Ellis, R. S., McMahon, R. G., Boyle, B. J., Bunclark, P. S., Carter, D., Irwin, M. J., Glazebrook, K., Newberg, H. J. M., Filippenko, A. V., Matheson, T., Dopita, M., and Couch, W. J., 1997. Measurements of the Cosmological Parameters $\Omega$ and $\Lambda$ from the First Seven Supernovae at $z \geq 0.35$. *Astrophysical Journal*, 483, 565.

Perlmutter, S., Pennypacker, C. R., Goldhaber, G., Goobar, A., Muller, R. A., Newberg, H. J. M., Desai, J., Kim, A. G., Kim, M. Y., Small, I. A., Boyle, B. J., Crawford, C. S., McMahon, R. G., Bunclark, P. S., Carter, D., Irwin, M. J., Terlevich, R. J., Ellis, R. S., Glazebrook, K., Couch, W. J., Mould, J. R., Small, T. A., and Abraham, R. G., 1995. A Supernova at $z = 0.458$ and Implications for Measuring the Cosmological Deceleration. *Astrophysical Journal Letters*, 440, L41.

Phillips, M. M., 1993. The Absolute Magnitudes of Type Ia Supernovae. *Astrophysical Journal Letters*, 413, L105.

Pickering, Edward C., 1895. A New Star in Centaurus. *Harvard College Observatory Circular* No. 4.

———, 1901. Sixty-Four New Variable Stars. *Harvard College Observatory Circular* No. 54.

Pierce, Michael J., and Jacoby, George H., 1995. "New" B and V Photometry of the "Old" Type Ia Supernova SN 1937C: Implications for $H_0$. *Astronomical Journal*, 110, 2885.

Pierce, Michael J., Ressler, Michael E., and Shure, Mark S., 1992. An Absolute Calibration of Type Ia Supernovae and the Value of $H_0$. *Astrophysical Journal Letters*, 390, L45.

Pierce, Michael J., Welch, Douglas L., McClure, Robert D., van den Bergh, Sidney, Racine, René, and Stetson, Peter B., 1994. The Hubble Constant and Virgo Cluster Distance from Observations of Cepheid Variables. *Nature*, 371, 385.

Pius XII, 1952. Theology and Modern Science. *Bulletin of the Atomic Scientists*, 8, 143.

Poe, Edgar Allan, 1848. *Eureka* (New York: Putnam).

———, 1973. *Eureka*. Edited by Richard P. Benton (Hartford, Connecticut: Transcendental Books).

Pollin, Burton R., 1975. Contemporary Reviews of *Eureka*: A Checklist. In *Poe as Literary Cosmologer*, edited by Richard P. Benton (Hartford, Connecticut: Transcendental Books), p. 26.

Preskill, John P., 1979. Cosmological Production of Superheavy Magnetic Monopoles. *Physical Review Letters*, 43, 1365.

Press, William H., Rybicki, George B., and Hewitt, Jacqueline N., 1992a. The Time Delay of Gravitational Lens 0957+561. I. Methodology and Analysis of Optical Photometric Data. *Astrophysical Journal*, 385, 404.

———, 1992b. The Time Delay of Gravitational Lens 0957+561. II. Analysis of Radio Data and Combined Optical-Radio Analysis. *Astrophysical Journal*, 385, 416.

Proctor, R. A., 1869. Distribution of the Nebulae. *Monthly Notices of the Royal Astronomical Society*, 29, 337.

Rees, Martin J., 1969. The Collapse of the Universe: An Eschatological Study. *Observatory*, 89, 193.

Reese, Erik D., Mohr, Joseph J., Carlstrom, John E., Joy, Marshall, Grego, Laura, Holder, Gilbert P., Holzapfel, William L., Hughes, John P., Patel, Sandeep K., and Donahue, Megan, 2000. Sunyaev-Zeldovich Effect-Derived Distances to the High-Redshift Clusters MS 0451.6–0305 and Cl 0016+16. *Astrophysical Journal*, 533, 38.

Refsdal, Sjur, 1964. On the Possibility of Determining Hubble's Parameter and the Masses of Galaxies from the Gravitational Lens Effect. *Monthly Notices of the Royal Astronomical Society*, 128, 307.

Rephaeli, Yoel, 1995. Comptonization of the Cosmic Microwave Background: The Sunyaev-Zeldovich Effect. *Annual Review of Astronomy and Astrophysics*, 33, 541.

Riess, Adam G., Filippenko, Alexei V., Challis, Peter, Clocchiatti, Alejandro, Diercks, Alan, Garnavich, Peter M., Gilliland, Ron L., Hogan, Craig J., Jha, Saurabh, Kirshner, Robert P., Leibundgut, B., Phillips, M. M., Reiss, David, Schmidt, Brian P., Schommer, Robert A., Smith, R. Chris, Spyromilio, J., Stubbs, Christopher, Suntzeff, Nicholas B., and Tonry, John, 1998. Observational Evidence from Supernovae for an Accelerating Universe and a Cosmological Constant. *Astrophysical Journal*, 116, 1009.

Riess, Adam G., Filippenko, Alexei V., Li, Weidong, and Schmidt, Brian P., 1999. Is There an Indication of Evolution of Type Ia Supernovae from Their Rise Times? *Astronomical Journal*, 118, 2668.

Ritchey, G. W., 1917a. Novae in Spiral Nebulae. *Publications of the Astronomical Society of the Pacific*, 29, 210.

———, 1917b. Another Faint Nova in the Andromeda Nebula. *Publications of the Astronomical Society of the Pacific*, 29, 257.

Roberts, Isaac, 1888. Photographs of the Nebulae M 31, h 44, and h 51 Andromedae, and M 27 Vulpeculae. *Monthly Notices of the Royal Astronomical Society*, 49, 65.

Roberts, Morton S., 1966. A High-Resolution 21-cm Hydrogen-Line Survey of the Andromeda Nebula. *Astrophysical Journal*, 144, 639.

———, 1976. The Rotation Curves of Galaxies. *Comments on Astrophysics*, 6, 105.

———, 2000. Interview with Ken Croswell: February 7, 2000.

Roberts, Morton S., and Whitehurst, Robert N., 1975. The Rotation Curve and Geometry of M31 at Large Galactocentric Distances. *Astrophysical Journal*, 201, 327.

Robertson, Douglas S., 1993. *Sky and Telescope*, 86, No. 6 (December 1993), p. 7.

Rogerson, John B., Jr., and York, Donald G., 1973. Interstellar Deuterium Abundance in the Direction of Beta Centauri. *Astrophysical Journal Letters*, 186, L95.

Rood, H. J., 1988. Voids. *Annual Review of Astronomy and Astrophysics*, 26, 245.

Rubin, Vera C., 1973. The Dynamics of the Andromeda Nebula. *Scientific American*, 228, No. 6 (June 1973), p. 30.

———, 2000. Interview with Ken Croswell: March 8, 2000.

Rubin, Vera C., Burley, Jaylee, Kiasatpoor, Ahmad, Klock, Benny, Pease, Gerald, Rutscheidt, Erich, and Smith, Clayton, 1962. Kinematic Studies of Early-Type Stars. I. Photometric Survey, Space Motions, and Comparison with Radio Observations. *Astronomical Journal*, 67, 491.

Rubin, Vera C., Burstein, David, Ford, W. Kent, Jr., and Thonnard, Norbert, 1985. Rotation Velocities of 16 Sa Galaxies and a Comparison of Sa, Sb, and Sc Rotation Properties. *Astrophysical Journal*, 289, 81.

Rubin, Vera C., and Ford, W. Kent, Jr., 1970. Rotation of the Andromeda Nebula from a Spectroscopic Survey of Emission Regions. *Astrophysical Journal*, 159, 379.

Rubin, Vera C., Ford, W. Kent, Jr., and Thonnard, Norbert, 1978. Extended Rotation Curves of High-Luminosity Spiral Galaxies. IV. Systematic Dynamical Properties, Sa—>Sc. *Astrophysical Journal Letters*, 225, L107.

———, 1980. Rotational Properties of 21 Sc Galaxies with a Large Range of Luminosities and Radii, from NGC 4605 (R = 4 kpc) to UGC 2885 (R = 122 kpc). *Astrophysical Journal*, 238, 471.

Rubin, Vera C., Ford, W. Kent, Jr., Thonnard, Norbert, and Burstein, David, 1982. Rotational Properties of 23 Sb Galaxies. *Astrophysical Journal*, 261, 439.

Russell, David, 1999. Island Universes from Wright to Hubble. *Sky and Telescope*, 97, No. 1 (January 1999), p. 56.

Ryle, Martin, 1951. *Observatory*, 71, 213.

———, 1955. Radio Stars and Their Cosmological Significance. *Observatory*, 75, 137.

———, 1958. The Nature of the Cosmic Radio Sources. *Proceedings of the Royal Society of London Series A*, 248, 289.

Ryle, M., and Clarke, R. W., 1961. An Examination of the Steady-State Model in the Light of Some Recent Observations of Radio Sources. *Monthly Notices of the Royal Astronomical Society*, 122, 349; and *Observatory*, 81, 57.

Ryle, M., and Hewish, A., 1955. The Cambridge Radio Telescope. *Observatory*, 75, 104.

Ryle, M., and Scheuer, P. A. G., 1955. The Spatial Distribution and the Nature of Radio Stars. *Proceedings of the Royal Society of London Series A*, 230, 448.

Sachs, R. K., and Wolfe, A. M., 1967. Perturbations of a Cosmological Model and Angular Variations of the Microwave Background. *Astrophysical Journal*, 147, 73.

Sackmann, I.-Juliana, Boothroyd, Arnold I., and Kraemer, Kathleen E., 1993. Our Sun. III. Present and Future. *Astrophysical Journal*, 418, 457.

Saha, A., Sandage, Allan, Labhardt, Lukas, Tammann, G. A., Macchetto, F. D., and Panagia, N., 1997. Cepheid Calibration of the Peak Brightness of Type Ia Supernovae. VIII. SN 1990N in NGC 4639. *Astrophysical Journal*, 486, 1.

Sahu, Kailash C., 1994. Stars Within the Large Magellanic Cloud as Potential Lenses for Observed Microlensing Events. *Nature*, 370, 275.

Sakai, Shoko, Mould, Jeremy R., Hughes, Shaun M. G., Huchra, John P., Macri, Lucas M., Kennicutt, Robert C., Jr., Gibson, Brad K., Ferrarese, Laura, Freedman, Wendy L., Han, Mingsheng, Ford, Holland C., Graham, John A., Illingworth, Garth D., Kelson, Daniel D., Madore, Barry F., Sebo, Kim, Silbermann, N. A., and Stetson, Peter B., 2000. The Hubble Space Telescope Key Project on the Extragalactic Distance Scale. XXIV. The Calibration of Tully-Fisher Relations and the Value of the Hubble Constant. *Astrophysical Journal*, 529, 698.

Sandage, Allan, 1958. Current Problems in the Extragalactic Distance Scale. *Astrophysical Journal*, 127, 513.

———, 1961. *The Hubble Atlas of Galaxies* (Washington, D.C.: Carnegie Institution of Washington).

———, 1968a. A New Determination of the Hubble Constant from Globular Clusters in M87. *Astrophysical Journal*, 152, L149.

———, 1968b. Observational Cosmology. *Observatory*, 88, 91.

———, 1972. The Redshift-Distance Relation. II. The Hubble Diagram and Its Scatter for First-Ranked Cluster Galaxies: A Formal Value for $q_o$. *Astrophysical Journal*, 178, 1.

———, 1993. Interview with Ken Croswell: September 15, 1993.

———, 1994. Interview with Ken Croswell: October 3, 1994.

———, 1998. Interview with Ken Croswell: February 5, 1998.

———, 1999. Interview with Ken Croswell: August 12, 1999.

———, 2000. Interview with Ken Croswell: July 26, 2000.

Sandage, Allan, and Bedke, John, 1994. *The Carnegie Atlas of Galaxies* (Washington, D.C.: Carnegie Institution of Washington).

Sandage, Allan, Saha, A., Tammann, G. A., Panagia, Nino, and Macchetto, D., 1992. The Cepheid Distance to IC 4182: Calibration of $M_V$(Max) for SN Ia 1937C and the Value of $H_o$. *Astrophysical Journal Letters*, 401, L7.

Sandage, Allan, and Tammann, G. A., 1974a. Steps Toward the Hubble Constant. III. The Distance and Stellar Content of the M101 Group of Galaxies. *Astrophysical Journal*, 194, 223.

———, 1974b. Steps Toward the Hubble Constant. IV. Distances to 39 Galaxies in the General Field Leading to a Calibration of the Galaxy Luminosity Classes and a First Hint of the Value of $H_o$. *Astrophysical Journal*, 194, 559.

———, 1976. Steps Toward the Hubble Constant. VII. Distances to NGC 2403, M101, and the Virgo Cluster Using 21 Centimeter Line Widths Compared with Optical Methods: The Global Value of $H_o$. *Astrophysical Journal*, 210, 7.

———, 1982. Steps Toward the Hubble Constant. VIII. The Global Value. *Astrophysical Journal*, 256, 339.

Sandage, Allan, Tammann, G. A., and Federspiel, Martin, 1995. Bias Properties of Extragalactic Distance Indicators. IV. Demonstration of the Population Incompleteness Bias Inherent in the Tully-Fisher Method Applied to Clusters. *Astrophysical Journal*, 452, 1.

Sanford, R. F., 1917. On Some Relations of the Spiral Nebulae to the Milky Way. *Lick Observatory Bulletin*, 9, 80 (No. 297).

Sargent, Wallace, 2000. Interview with Ken Croswell: April 10, 2000.

Sargent, Wallace L. W., and Searle, Leonard, 1966. Spectroscopic Evidence on the Helium Abundance of Stars in the Galactic Halo. *Astrophysical Journal*, 145, 652.

———, 1967. The Interpretation of the Helium Weakness in Halo Stars. *Astrophysical Journal Letters*, 150, L33.

Schaefer, Bradley E., 1996. The Peak Brightness of SN 1937C in IC 4182 and the Hubble Constant: Comments on Pierce & Jacoby [AJ, 110, 2885 (1995)]. *Astronomical Journal*, 111, 1668.

Scheuer, P. A. G., 1957. A Statistical Method for Analysing Observations of Faint Radio Stars. *Proceedings of the Cambridge Philosophical Society*, 53, 764.

———, 1990. Radio Source Counts. In *Modern Cosmology in Retrospect*, edited by B. Bertotti, R. Balbinot, S. Bergia, and A. Messina (Cambridge: Cambridge University Press), p. 331.

Schild, Rudolph, 2000. Interview with Ken Croswell: July 11, 2000.

Schild, Rudolph E., and Cholfin, Bryan, 1986. CCD Camera Brightness Monitoring of Q0957+561A, B. *Astrophysical Journal*, 300, 209.

Schmidt, Brian, 2000. Interview with Ken Croswell: July 17, 2000.

Schmidt, Brian P., Suntzeff, Nicholas B., Phillips, M. M., Schommer, Robert A., Clocchiatti, Alejandro, Kirshner, Robert P., Garnavich, Peter, Challis, Peter, Leibundgut, B., Spyromilio, J., Riess, Adam G., Filippenko, Alexei V., Hamuy, Mario, Smith, R. Chris, Hogan, Craig, Stubbs, Christopher, Diercks, Alan, Reiss, David, Gilliland, Ron, Tonry, John, Maza, José, Dressler, A., Walsh, J., and Ciardullo, R., 1998. The High-z Supernova Search: Measuring Cosmic Deceleration and Global Curvature of the Universe Using Type Ia Supernovae. *Astrophysical Journal*, 507, 46.

Schmidt, M., 1963. 3C 273: A Star-like Object with Large Red-shift. *Nature*, 197, 1040.

Schwarzschild, Bertram, 1994. MACHO Searches Find Most Candidates in Unexpected Places. *Physics Today*, 47, No. 7 (July 1994), p. 17.

Scott, P. F., and Ryle, M., 1961. The Number-Flux Density Relation for Radio Sources Away from the Galactic Plane. *Monthly Notices of the Royal Astronomical Society*, 122, 389.

Searle, Leonard, and Rodgers, A. W., 1966. The Horizontal-Branch Stars of the Globular Cluster NGC 6397. *Astrophysical Journal*, 143, 809.

Searle, Leonard, and Sargent, Wallace L. W., 1972. Inferences from the Composition of Two Dwarf Blue Galaxies. *Astrophysical Journal*, 173, 25.

Seife, Charles, 1999. Hubble Knows. *New Scientist*, 162, No. 2189 (June 5, 1999), p. 11.

Seldner, M., Siebers, B., Groth, Edward J., and Peebles, P. J. E., 1977. New Reduction of the Lick Catalog of Galaxies. *Astronomical Journal*, 82, 249.

Shane, C. D., and Wirtanen, C. A., 1967. The Distribution of Galaxies. *Publications of the Lick Observatory*, 22, 1.

Shapley, Harlow, 1916. Outline and Summary of a Study of Magnitudes in the Globular Cluster Messier 13. *Publications of the Astronomical Society of the Pacific*, 28, 171.

———, 1917a. A Faint Nova in the Andromeda Nebula. *Publications of the Astronomical Society of the Pacific*, 29, 213.

———, 1917b. Note on the Magnitudes of Novae in Spiral Nebulae. *Publications of the Astronomical Society of the Pacific*, 29, 213.

———, 1918a. Studies Based on the Colors and Magnitudes in Stellar Clusters. Sixth Paper: On the Determination of the Distances of Globular Clusters. *Astrophysical Journal*, 48, 89.

———, 1918b. Studies Based on the Colors and Magnitudes in Stellar Clusters. Seventh Paper: The Distances, Distribution in Space, and Dimensions of 69 Globular Clusters. *Astrophysical Journal*, 48, 154.

———, 1918c. Globular Clusters and the Structure of the Galactic System. *Publications of the Astronomical Society of the Pacific*, 30, 42.

———, 1918d. Studies Based on the Colors and Magnitudes in Stellar

Clusters. Eighth Paper: The Luminosities and Distances of 139 Cepheid Variables. *Astrophysical Journal*, 48, 279.

———, 1919. Studies Based on the Colors and Magnitudes in Stellar Clusters. Twelfth Paper: Remarks on the Arrangement of the Sidereal Universe. *Astrophysical Journal*, 49, 311.

———, 1929. Note on the Velocities and Magnitudes of External Galaxies. *Proceedings of the National Academy of Sciences*, 15, 565.

———, 1930. Note on a Remote Cloud of Galaxies in Centaurus. *Harvard College Observatory Bulletin* No. 874, p. 9.

———, 1933. Luminosity Distribution and Average Density of Matter in Twenty-Five Groups of Galaxies. *Proceedings of the National Academy of Sciences*, 19, 591.

———, 1934. On Some Structural Features of the Metagalaxy. *Monthly Notices of the Royal Astronomical Society*, 94, 791.

———, 1953a. Magellanic Clouds, VI. Revised Distances and Luminosities. *Proceedings of the National Academy of Sciences*, 39, 349.

———, 1953b. The Distance of the Magellanic Clouds. *Astronomical Journal*, 58, 47.

———, 1953c. Note on the Extragalactic Distance Scale. *Astronomical Journal*, 58, 227.

———, 1969. *Through Rugged Ways to the Stars* (New York: Charles Scribner's Sons).

Silk, Joseph, 1968. Cosmic Black-Body Radiation and Galaxy Formation. *Astrophysical Journal*, 151, 459.

———, 2000. Interview with Ken Croswell: June 15, 2000.

Singer, Dorothea Waley, 1950. *Giordano Bruno: His Life and Thought* (New York: Henry Schuman).

Skiff, Brian A., 1993. M Is for Messier. *Sky and Telescope*, 85, No. 4 (April 1993), p. 38.

Slipher, V. M., 1913. The Radial Velocity of the Andromeda Nebula. *Lowell Observatory Bulletin* No. 58.

———, 1914. The Detection of Nebular Rotation. *Lowell Observatory Bulletin* No. 62.

———, 1915. Spectrographic Observations of Nebulae. *Popular Astronomy*, 23, 21.

———, 1917. The Spectrum and Velocity of the Nebula N. G. C. 1068 (M 77). *Lowell Observatory Bulletin* No. 80.

Smith, F. Graham, and Lovell, Bernard, 1983. On the Discovery of Extragalactic Radio Sources. *Journal for the History of Astronomy*, 14, 155.

Smith, Horace A., 1995. *RR Lyrae Stars* (Cambridge: Cambridge University Press).

Smith, R. W., 1979. The Origins of the Velocity-Distance Relation. *Journal for the History of Astronomy*, 10, 133.

———, 1982. *The Expanding Universe* (Cambridge: Cambridge University Press).

———, 1990. Edwin P. Hubble and the Transformation of Cosmology. *Physics Today*, 43, No. 4 (April 1990), p. 52.

Smith, Sinclair, 1936. The Mass of the Virgo Cluster. *Astrophysical Journal*, 83, 23.

Smoot, George, 2000. Interview with Ken Croswell: June 27, 2000.

Smoot, G. F., Bennett, C. L., Kogut, A., Wright, E. L., Aymon, J., Boggess, N. W., Cheng, E. S., De Amici, G., Gulkis, S., Hauser, M. G., Hinshaw, G., Jackson, P. D., Janssen, M., Kaita, E., Kelsall, T., Keegstra, P., Lineweaver, C., Loewenstein, K., Lubin, P., Mather, J., Meyer, S. S., Moseley, S. H., Murdock, T., Rokke, L., Silverberg, R. F., Tenorio, L., Weiss, R., and Wilkinson, D. T., 1992. Structure in the COBE Differential Microwave Radiometer First-Year Maps. *Astrophysical Journal Letters*, 396, L1.

Smoot, George, and Davidson, Keay, 1993. *Wrinkles in Time* (New York: Morrow).

Spite, François, 2000. Interview with Ken Croswell: April 6, 2000.

Spite, F., and Spite, M., 1982. Abundance of Lithium in Unevolved Halo Stars and Old Disk Stars: Interpretation and Consequences. *Astronomy and Astrophysics*, 115, 357.

Spite, M., and Spite, F., 1982. Lithium Abundance at the Formation of the Galaxy. *Nature*, 297, 483.

Stachel, John, 1986. Eddington and Einstein. In *The Prism of Science*, edited by Edna Ullmann-Margalit (Dordrecht: D. Reidel), p. 225.

Stanek, Krzysztof, 2000. Interview with Ken Croswell: July 11, 2000.

Stanek, K. Z., Kaluzny, J., Wysocka, A., and Thompson, I., 2000. UBVI Color-Magnitude Diagrams in Baade's Window: Metallicity Range, Implications for the Red Clump Method, Color "Anomaly" and the Distances to the Galactic Center and the Large Magellanic Cloud. *Acta Astronomica*, 50, 191.

Stanek, K. Z., Zaritsky, D., and Harris, J., 1998. A "Short" Distance to the Large Magellanic Cloud with the Hipparcos Calibrated Red Clump Stars. *Astrophysical Journal Letters*, 500, L141.

Stebbins, Joel, Whitford, A. E., and Johnson, H. L., 1950. Photoelectric Magnitudes and Colors of Stars in Selected Areas 57, 61, and 68. *Astrophysical Journal*, 112, 469.

Steigman, Gary, 2000. Interview with Ken Croswell: April 7, 2000; April 10, 2000.

Stratonoff, W., 1900. Études sur la Structure de L'Univers. *Publications de L'Observatoire Astronomique et Physique de Tachkent*, No. 2.

Sullivan, Walter, 1965. Signals Imply a 'Big Bang' Universe. *New York Times*, May 21, 1965, p. 1.

Sullivan, Woodruff T., III, 1990. The Entry of Radio Astronomy into Cosmology: Radio Stars and Martin Ryle's 2C Survey. In *Modern Cosmology in Retrospect*, edited by B. Bertotti, R. Balbinot, S. Bergia, and A. Messina (Cambridge: Cambridge University Press), p. 309.

Swedenborg, Emanuel, 1734. *The Principia*, Volume 2, translated by James R. Rendell and Isaiah Tansley (1912 reprint; London: Swedenborg Society).

Tammann, Gustav, 2000. Interview with Ken Croswell: July 12, 2000.

Tammann, G. A., and Sandage, Allan, 1968. The Stellar Content and Distance of the Galaxy NGC 2403 in the M81 Group. *Astrophysical Journal*, 151, 825.

Tarenghi, Massimo, Tifft, William G., Chincarini, Guido, Rood, Herbert J., and Thompson, Laird A., 1979. The Hercules Supercluster. I. Basic Data. *Astrophysical Journal*, 234, 793.

Teerikorpi, P., 1997. Observational Selection Bias Affecting the Determination of the Extragalactic Distance Scale. *Annual Review of Astronomy and Astrophysics*, 35, 101.

Thackeray, A. D., and Wesselink, A. J., 1953. Distances of the Magellanic Clouds. *Nature*, 171, 693.

Thompson, Laird, 2000. Interview with Ken Croswell: May 15, 2000.

Tifft, W. G., and Gregory, S. A., 1976. Direct Observations of the Large-Scale Distribution of Galaxies. *Astrophysical Journal*, 205, 696.

Tinsley, Beatrice M., 1972. A First Approximation to the Effect of Evolution on $q_0$. *Astrophysical Journal Letters*, 173, L93.

Tipler, Frank J., 1988a. Olbers's Paradox, the Beginning of Creation, and Johann Mädler. *Journal for the History of Astronomy*, 19, 45.

———, 1988b. Johann Mädler's Resolution of Olbers' Paradox. *Quarterly Journal of the Royal Astronomical Society*, 29, 313.

Tombaugh, Clyde W., 1937. The Great Perseus-Andromeda Stratum of Extra-Galactic Nebulae and Certain Clusters of Nebulae Therein as Observed at the Lowell Observatory. *Publications of the Astronomical Society of the Pacific*, 49, 259.

Tombaugh, Clyde W., and Moore, Patrick, 1980. *Out of the Darkness: The Planet Pluto* (Harrisburg, Pennsylvania: Stackpole).

Tonry, John, 2000. Interview with Ken Croswell: July 12, 2000.

Tonry, John L., Ajhar, Edward A., and Luppino, Gerard A., 1990. Observations of Surface-Brightness Fluctuations in Virgo. *Astronomical Journal*, 100, 1416.

Tonry, John L., Blakeslee, John P., Ajhar, Edward A., and Dressler, Alan, 2000. The Surface Brightness Fluctuation Survey of Galaxy Distances. II. Local and Large-Scale Flows. *Astrophysical Journal*, 530, 625.

Trauger, J. T., Roesler, F. L., Carleton, N. P., and Traub, W. A., 1973. Observation of HD on Jupiter and the D/H Ratio. *Astrophysical Journal Letters*, 184, L137.

Tremaine, Scott, 1999. Comments on Ostriker, Peebles, & Yahil (1974) "The Size and Mass of Galaxies, and the Mass of the Universe," *Astrophysical Journal*, 525, 1223.

Truran, J. W., Hansen, C. J., and Cameron, A. G. W., 1965. The Helium Content of the Galaxy. *Canadian Journal of Physics*, 43, 1616.

Tryon, Edward P., 1973. Is the Universe a Vacuum Fluctuation? *Nature*, 246, 396.

Tucker, Wallace, and Tucker, Karen, 1988. *The Dark Matter* (New York: Morrow).

Tully, R. Brent, 2000. Interview with Ken Croswell: July 13, 2000.

Tully, R. Brent, and Fisher, J. Richard, 1977. A New Method of Determining Distances to Galaxies. *Astronomy and Astrophysics*, 54, 661.

Tully, R. Brent, and Pierce, Michael J., 2000. Distances to Galaxies from the Correlation Between Luminosities and Line Widths. III. Cluster Template and Global Measurement of $H_0$. *Astrophysical Journal*, 533, 744.

Turner, Edwin L., 1990. Gravitational Lensing Limits on the Cosmological Constant in a Flat Universe. *Astrophysical Journal Letters*, 365, L43.

———, 1995. Interview with Ken Croswell: August 8, 1995.

Turner, Edwin L., Cen, Renyue, and Ostriker, Jeremiah P., 1992. The Relation of Local Measures of Hubble's Constant to Its Global Value. *Astronomical Journal*, 103, 1427.

Tye, Henry, 2000. Interview with Ken Croswell: April 26, 2000.

Tyson, J. Anthony, 2000. Interview with Ken Croswell: July 12, 2000.

Tyson, J. Anthony, Kochanski, Greg P., and Dell'Antonio, Ian P., 1998. Detailed Mass Map of CL 0024+1654 from Strong Lensing. *Astrophysical Journal Letters*, 498, L107.

Udalski, A., Szymański, M., Kałużny, J., Kubiak, M., Krzemiński, W., Mateo, M., Preston, G. W., and Paczyński, B., 1993. The Optical Gravitational Lensing Experiment. Discovery of the First Candidate Microlensing Event in the Direction of the Galactic Bulge. *Acta Astronomica*, 43, 289.

VandenBerg, Don, 2000. Interview with Ken Croswell: July 11, 2000.

Van den Bergh, Sidney, 1998. *Galaxy Morphology and Classification* (Cambridge: Cambridge University Press).

———, 1999. The Early History of Dark Matter. *Publications of the Astronomical Society of the Pacific*, 111, 657.

———, 2000a. *The Galaxies of the Local Group* (Cambridge: Cambridge University Press).

———, 2000b. Updated Information on the Local Group. *Publications of the Astronomical Society of the Pacific*, 112, 529.

Van Maanen, A., 1916. Preliminary Evidence of Internal Motion in the Spiral Nebula Messier 101. *Astrophysical Journal*, 44, 210.

———, 1921a. Internal Motion in the Spiral Nebula Messier 33. *Proceedings of the National Academy of Sciences*, 7, 1.

———, 1921b. Investigations on Proper Motion. Fourth Paper: Internal Motion in the Spiral Nebula Messier 51. *Astrophysical Journal*, 54, 237.

———, 1921c. Investigations on Proper Motion. Fifth Paper: The Internal Motion in the Spiral Nebula Messier 81. *Astrophysical Journal*, 54, 347.

———, 1922a. Investigations on Proper Motion. Seventh Paper: Internal Motion in the Spiral Nebula N.G.C. 2403. *Astrophysical Journal*, 56, 200.

———, 1922b. Investigations on Proper Motion. Eighth Paper: Internal Motion in the Spiral Nebula M 94=N.G.C. 4736. *Astrophysical Journal*, 56, 208.

———, 1923a. Investigations on Proper Motion. Ninth Paper: Internal Motion in the Spiral Nebula Messier 63, N.G.C. 5055. *Astrophysical Journal*, 57, 49.

———, 1923b. Investigations on Proper Motion. Tenth Paper: Internal

Motion in the Spiral Nebula Messier 33, N.G.C. 598. *Astrophysical Journal*, 57, 264.

Wagoner, Robert V., Fowler, William A., and Hoyle, F., 1967. On the Synthesis of Elements at Very High Temperatures. *Astrophysical Journal*, 148, 3.

Walsh, D., Carswell, R. F., and Weymann, R. J., 1979. 0957+561 A, B: Twin Quasistellar Objects or Gravitational Lens? *Nature*, 279, 381.

Weinberg, Steven, 1972. *Gravitation and Cosmology* (New York: Wiley).

Wesson, Paul, 1999. Interview with Ken Croswell: February 12, 1999; February 13, 1999.

Wesson, Paul S., Valle, K., and Stabell, R., 1987. The Extragalactic Background Light and a Definitive Resolution of Olbers's Paradox. *Astrophysical Journal*, 317, 601.

Westfall, Richard S., 1980. *Never at Rest* (Cambridge: Cambridge University Press).

White, Michael, 1992. Eureka! They Like Science. *Times* (London), December 13, 1992, sec. 6, p. 8.

White, Simon D. M., Navarro, Julio F., Evrard, August E., and Frenk, Carlos S., 1993. The Baryon Content of Galaxy Clusters: A Challenge to Cosmological Orthodoxy. *Nature*, 366, 429.

Whitney, Charles A., 1971. *The Discovery of Our Galaxy* (New York: Knopf).

Whitrow, G. J., 1967. Kant and the Extragalactic Nebulae. *Quarterly Journal of the Royal Astronomical Society*, 8, 48.

Williams, William C., 1969. *Sky and Telescope*, 31, 376 (December 1969).

Wolf, M., 1914. *Vierteljahrsschrift der Astronomischen Gesellschaft*, 49, 162.

Woolf, Harry, 1959. *The Transits of Venus* (Princeton: Princeton University Press).

Wren, Christopher, and Wren, Stephen, 1750. *Parentalia: or Memoirs of the Family of the Wrens* (1965 reprint; Farnborough, Hants, England: Gregg Press Limited).

Wright, Ned, 2000. Interview with Ken Croswell: June 13, 2000.

Wright, Thomas, 1750. *An Original Theory or New Hypothesis of the Universe*, introduced and transcribed by Michael A. Hoskin (1971 reprint; London: MacDonald).

York, Donald, 2000. Interview with Ken Croswell: April 8, 2000.

Young, Peter, Gunn, James E., Kristian, Jerome, Oke, J. B., and Westphal, James A., 1980. The Double Quasar Q0957+561 A, B: A Gravitational Lens Image Formed by a Galaxy at z = 0.39. *Astrophysical Journal*, 241, 507.

———, 1981. Q0957+561: Detailed Models of the Gravitational Lens Effect. *Astrophysical Journal*, 244, 736.

Zaritsky, Dennis, Olszewski, Edward W., Schommer, Robert A., Peterson, Ruth C., and Aaronson, Marc, 1989. Velocities of Stars in Remote Galactic Satellites and the Mass of the Galaxy. *Astrophysical Journal*, 345, 759.

Zel'dovich, Ya. B., 1970. Gravitational Instability: An Approximate Theory for Large Density Perturbations. *Astronomy and Astrophysics*, 5, 84.

Zwicky, F., 1929. On the Red Shift of Spectral Lines Through Interstellar Space. *Proceedings of the National Academy of Sciences*, 15, 773.

———, 1933. Die Rotverschiebung von Extragalaktischen Nebeln. *Helvetica Physica Acta*, 6, 110.

———, 1937a. Nebulae as Gravitational Lenses. *Physical Review*, 51, 290.

———, 1937b. On the Probability of Detecting Nebulae Which Act as Gravitational Lenses. *Physical Review*, 51, 679.

———, 1937c. On the Masses of Nebulae and of Clusters of Nebulae. *Astrophysical Journal*, 86, 217.

———, 1938. On the Clustering of Nebulae. *Publications of the Astronomical Society of the Pacific*, 50, 218.

———, 1971. *Catalogue of Selected Compact Galaxies and of Post-Eruptive Galaxies* (Guemligen, Switzerland: F. Zwicky).

# CREDITS

Page ix: "Easter" by Marillion. From the 1989 album *Seasons End*. Lyrics by Steve Hogarth.

Page 1: "The Book of Love" by White Willow. From the 1998 album *Ex Tenebris*. Lyrics by Jacob C. Holm-Lupo.

Page 4: "The Raven" by Edgar Allan Poe. Published in 1845.

Page 70: "Close to the Edge" by Yes. From the 1972 album *Close to the Edge*. Lyrics by Jon Anderson and Steve Howe.

Page 104: "The Calling" by Yes. From the 1994 album *Talk*. Lyrics by Jon Anderson.

Page 118: Alfonso the Wise (Alfonso X) lived from 1221 to 1284.

Page 172: Mark Twain never said many of the things he is said to have said. This appears to be one of them.

Page 264: Thomas Carlyle's words, spoken in 1868, appear in *William Allingham: A Diary*, near the end of Chapter 10.

# INDEX

# ABOUT THE AUTHOR

Ken Croswell earned his Ph.D. in astronomy from Harvard University for his study of the Milky Way's halo. His first book, *The Alchemy of the Heavens*, explored the Milky Way Galaxy and was a *Los Angeles Times* Book Prize finalist. His second book, *Planet Quest*, described the discovery of new planets, in our solar system and beyond, and was a *New York Times* Notable Book of the Year. Dr. Croswell's third book, *Magnificent Universe*, featured a hundred stunning full-page, full-color photographs of the cosmos. In his book *See the Stars*, he introduced beginning stargazers to the constellations. He lives in Berkeley, California.